CISM International Centre for Mechanical Sciences

Courses and Lectures

Volume 625

For more than 40 years the book series edited by CISM, “International Centre for Mechanical Sciences: Courses and Lectures”, has presented groundbreaking developments in mechanics and computational engineering methods. It covers such fields as solid and fluid mechanics, mechanics of materials, micro- and nanomechanics, biomechanics, and mechatronics. The papers are written by international authorities in the field. The books are at graduate level but may include some introductory material.

Charalampos Baniotopoulos · Enzo Marino
Editors

Advances in Wind Energy in the Era of Artificial Intelligence

Springer

Editors
Charalampos Baniotopoulos
School of Engineering
University of Birmingham
Birmingham, UK

Enzo Marino
Department of Civil and Environmental Engineering (DICEA)
University of Florence
Firenze, Italy

ISSN 0254-1971 ISSN 2309-3706 (electronic)
CISM International Centre for Mechanical Sciences
ISBN 978-3-032-28814-1 ISBN 978-3-032-28815-8 (eBook)
https://doi.org/10.1007/978-3-032-28815-8

This Springer imprint is published by the registered company Springer Nature Switzerland AG
The registered company address is: Gewerbestrasse 11, 6330 Cham, Switzerland

Preface

The book *Advances in Wind Energy in the Era of Artificial Intelligence* presents a comprehensive overview of recent developments at the intersection of wind energy engineering and artificial intelligence (AI). The volume is the outcome of the Advanced Course bearing the same title, delivered at the International Centre of Mechanical Sciences (CISM) during the period 14–19 September 2025. Bringing together contributions from leading researchers and practitioners, the book reflects both the scientific foundations and the emerging technological trends that are shaping the future of wind energy systems.

The rapid growth of wind energy has introduced new challenges related to system complexity, scale, reliability and efficiency. At the same time, advances in AI, machine learning (ML) and data-driven methodologies have created unprecedented opportunities to address these challenges. This book aims to provide a structured and state-of-the-art account of how AI-based techniques can be leveraged to enhance the analysis, design, operation and maintenance of wind energy infrastructure.

The volume is composed of seven chapters, each addressing a key aspect of the application of AI in wind energy. Collectively, the chapters cover topics ranging from AI-enhanced computational modelling and offshore dynamics to digital twins, intelligent vibration control, data quality considerations and knowledge-enhanced machine learning. Together, they offer a coherent perspective on how artificial intelligence is transforming wind engineering across multiple scales and disciplines.

Chapter 1 introduces the overarching trends in the use of artificial intelligence for the advancement of wind energy infrastructure. It establishes the broader context in which AI-driven approaches are emerging, highlighting the increasing availability of data, advances in sensing technologies and the growing demand for intelligent decision-support systems. The chapter outlines the key motivations for adopting AI in wind energy applications and frames the subsequent chapters within a unified conceptual and technological landscape.

Chapter 2 provides an overview of the integration of machine learning and computational fluid dynamics (CFD) in the analysis and optimization of wind energy systems. By combining fundamental principles of fluid dynamics with emerging data-driven intelligence, the chapter illustrates how AI—particularly ML, deep learning and reinforcement learning—can accelerate simulations, enhance predictive fidelity and enable adaptive control strategies. Emphasis is placed on hybrid frameworks that merge physics-based modelling with data-driven methods, paving the way for next-generation AI-enhanced CFD tools capable of revolutionizing wind energy design, optimization and real-time operational efficiency.

Chapter 3 focuses on the dynamics of offshore wind turbines, with particular attention to floating systems. It reviews key aspects of wave–body interaction and frequency-domain modelling, which are essential for understanding the complex dynamic behaviour of offshore structures. The chapter further highlights optimization techniques and data-driven methods aimed at improving design efficiency and reliability assessment, thereby addressing the unique challenges associated with offshore wind energy deployment.

Chapter 4 examines the role of Digital Twins (DTs) and data-driven methodologies in establishing intelligent operation and maintenance (O&M) frameworks for wind energy systems. The evolution of DT technology is discussed in detail, including the integration of Internet of Things (IoT) devices, advanced simulation capabilities, artificial intelligence applications and the importance of model calibration and uncertainty quantification. Applications such as predictive maintenance, virtual sensing, lifecycle management and integrated aero-structural control are addressed, alongside challenges related to environmental effects and cybersecurity. Emerging innovations, including generative AI, are also considered. A detailed case study on corrosion–fatigue monitoring in offshore towers demonstrates the effectiveness of DT-informed maintenance, showing a nine-fold return on investment and an extension of the remaining useful life by several years.

Chapter 5 presents a comprehensive overview of state-of the-art vibration control technologies for wind turbines. Unlike conventional civil and mechanical structures, wind turbines operate under highly coupled dynamics involving aerodynamic loading, blade rotation and soil–structure interaction, while offering limited space for damper installation. These characteristics make vibration control particularly challenging. Following standard classification, the chapter discusses passive, semi-active and active control strategies. While passive methods are mature and widely implemented, semi-active and active approaches—classified as intelligent control—offer superior adaptability and performance, albeit mostly demonstrated at the laboratory scale. The chapter also introduces an innovative integrated control concept that combines structural vibration mitigation with blade pitch control to enhance overall system performance.

Chapter 6 addresses the quality of measurement data required for AI-based analyses of wind energy structures. It distinguishes between different objectives and data sources, including SCADA data for condition monitoring, additional sensors for structural health monitoring and experimental measurements for targeted investigations. AI methods, particularly machine learning and neural network training, require large volumes of high-quality data drawn from all these measurement classes, with full-scale testing identified as an optimal data source. The chapter introduces a data-driven diagnostic framework for simulating and tracking the temporal variability of wind turbine dynamics and discusses the fusion of surrogate models with simulation-based design procedures. Special attention is given to physics-informed neural networks (PINNs), which integrate physical laws with measurement data and applications based on wind tunnel measurements are presented.

The final Chap. 7 focuses on knowledge-enhanced machine learning (KEML) in wind engineering. It begins by introducing different forms of knowledge representation, including physical laws, empirical models and domain-specific expertise, as well as effective strategies for embedding such knowledge into ML pipelines through modifications of loss functions, model architectures and training processes. A systematic review of KEML applications in wind engineering is then presented, covering topics such as windstorms under climate change, bluff-body aerodynamics, aeroelasticity, structural dynamics, response mitigation and post-hazard recovery. The chapter concludes with a discussion of current challenges and future prospects, emphasizing the potential of KEML to improve data efficiency, extrapolation capability and model trustworthiness in wind engineering applications.

The Editors gratefully acknowledge the International Centre for Mechanical Sciences (CISM) for providing them and the contributing authors with the opportunity to disseminate this new knowledge in an effective and internationally visible manner. The Editors also wish to express their sincere appreciation to the International Association of Wind Engineering (IAWE), in particular to its President, Prof. Dr.-Ing. C. Borri, for the valuable support provided. Without the support of IAWE, the CISM Advanced Course, on which this book is based, could not have been organized. In addition, Prof. Charalampos Baniotopoulos gratefully acknowledges the support of the Swiss National Science Foundation (SNSF) through the project 'Data-Driven Framework for Predicting the Remaining Useful Life of Onshore Wind Turbines', as well as the support of the Steel Structures Laboratory (RESSLab) and the Civil Engineering Institute of EPFL. Special thanks are extended to the Director of the Institute and Director of the RESSLab, Prof. Dr. D. Lignos. This support was instrumental in the preparation of the present book.

Birmingham, UK — Charalampos Baniotopoulos
Firenze, Italy — Enzo Marino

Contents

Chapter 1
Current Trends on the Use of AI to Advance Wind Energy Infrastructures

Fengying Wu, Claudio Borri, and Lin Zhao

Abstract AI methods have been developed alongside innovations in computer technology during this century, providing new perspectives for solving engineering problems related to wind, particularly in response to challenges posed by extreme climate change. Global wind-induced disaster statistics are reviewed, and disaster-causing mechanism of wind effects on structures were introduced as well as utilization of wind energy have been highlighted worldwide. The advancements and challenges in Floating Offshore Wind Turbine technology are discussed from the perspective of potential wind power capacity and long-term planning, including the application of AI methods in onshore and offshore wind turbine design, wind-turbine interactions, and nonlinear dynamics of wind turbines. Furthermore, from the standpoint of wind resistance of existing infrastructure, attempts to develop new solutions for wind-induced structural dynamics using AI methods are introduced. Subsequently, wind field simulation and reproduction methods are summarized to clarify wind characteristics under changing climatic conditions. Moreover, in recent decades, intelligent experimental technologies have made progress in simulating wind environment characteristics and in addressing wind-induced vibrations of infrastructure as well as onshore and offshore wind turbines. Finally, long-term prospects for the potential application of AI methods in the development of infrastructure subject to wind impacts are presented.

F. Wu
Highway School, Chang'an University, Xi'an, China

C. Borri (✉)
Department of Civil and Environmental Engineering, University of Florence, Florence, Italy
e-mail: claudio.borri@unifi.it

L. Zhao
State Key Lab of Disaster Reduction in Civil Engineering, Tongji University, Shanghai, China

Guangxi Laboratory of Whole Life Safety for Land-Sea Corridor Engineering, Guangxi University, Nanning, China

C. Baniotopoulos and E. Marino (eds.), *Advances in Wind Energy in the Era of Artificial Intelligence*, CISM International Centre for Mechanical Sciences 625,
https://doi.org/10.1007/978-3-032-28815-8_1

1.1 Introduction of AI Development

With the advancement of computing, knowledge acquisition is increasingly relying on data-driven and machine learning methods rather than traditional rule-based or manual systems. The revolution of the scientific discovery is displayed in Fig. 1.1, developed from the physical observation in first paradigm, then move on to the second paradigm: mathematical derivations, and the CFD simulations at the third paradigm. Currently, a key focus of advanced AI research-aligning with the concept of the "fourth paradigm" is to develop adaptive and autonomous learning systems with high interpretability and strong generalization abilities, moving beyond current theoretical constraints.

Accordingly, AI-based methods have been intensively developed over the past decades. AI-based methods have been intensively developed over the past decades. The most frequently used ones include Neural Networks (NN), Artificial Neural Networks (ANN), Long Short-Term Memory (LSTM), Machine Learning (ML), Deep Learning (DL), and data-driven techniques. These AI-based methods are utilized widely, especially in solving engineering issues, and the machine learning owns strong capacity across engineering and wind engineering subdomains, for instance, the prediction of wind speed, wind power forecasting, wind farm design, wind pressures, and wind field reconstruction, etc. From the perspective of wind engineering and infrastructures, a word cloud of AI-based methods application in wind engineering and subdomains is generated in Fig. 1.2. The preliminary exploration and practical applications of the AI-based methods in wind engineering and infrastructures could be divided into the following aspects (Fig. 1.3): (1) wind field measurement to collect climate data; (2) wind force modeling based on aerodynamic parameter set in wind tunnel; (3) wind-induced responses prediction related to wind-induced vibrations and wind pressure characteristics; (4) wind-induced vibration control by utilizing intelligent active technology; (5) advanced intelligent experimental technologies, including automated wind tunnel testing technology and intelligent simulation of complex landforms, etc.

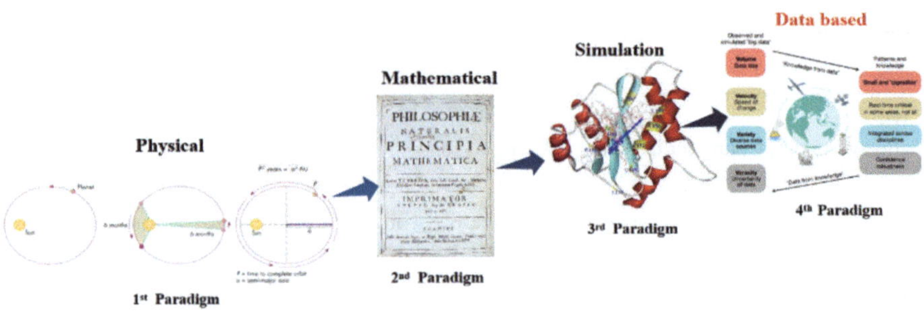

Fig. 1.1 Revolution map of the acknowledge acquisition

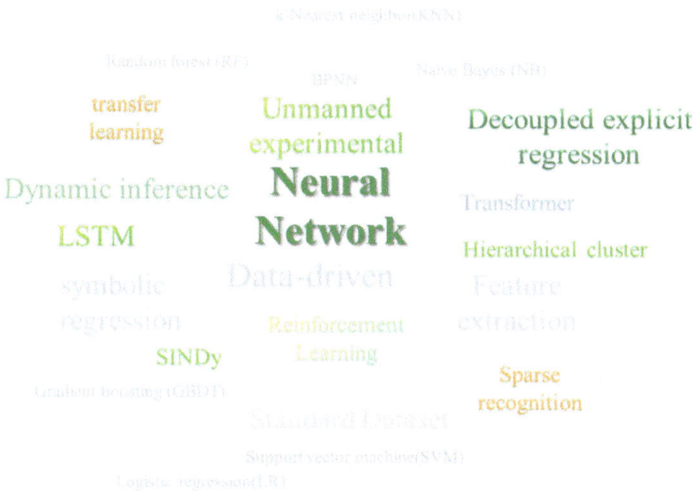

Fig. 1.2 Word cloud of AI-based methods applied in wind engineering

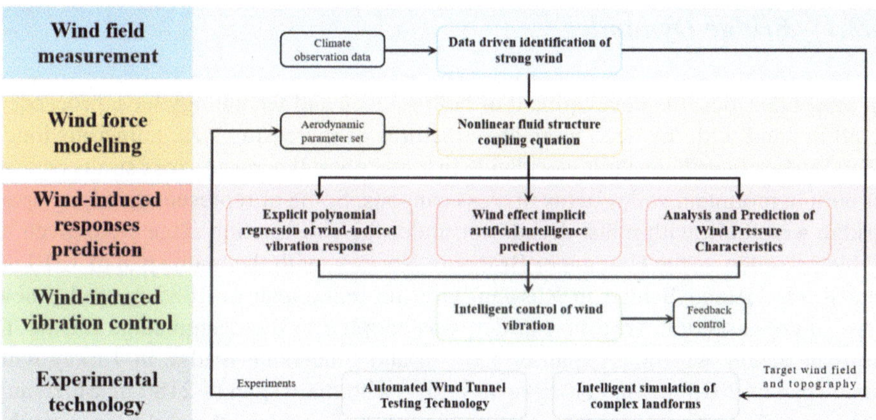

Fig. 1.3 AI-based methods preliminary exploration and practical applications in wind engineering

1.2 The Origin of AI in Structural Dynamics

According to statistical results from UNESCO (2022), extreme weather events are now widely observed across the globe, leading to substantial disaster-related losses. Among these, wind-induced disasters-notably typhoons and mountain windstorms-have become predominant. These disasters pose a serious threat to infrastructure due to the intense wind loads associated with increasingly severe climate conditions. Consequently, there is an urgent need to adopt AI methods to investigate the key causal factors of disasters under strong wind conditions (Fig. 1.4).

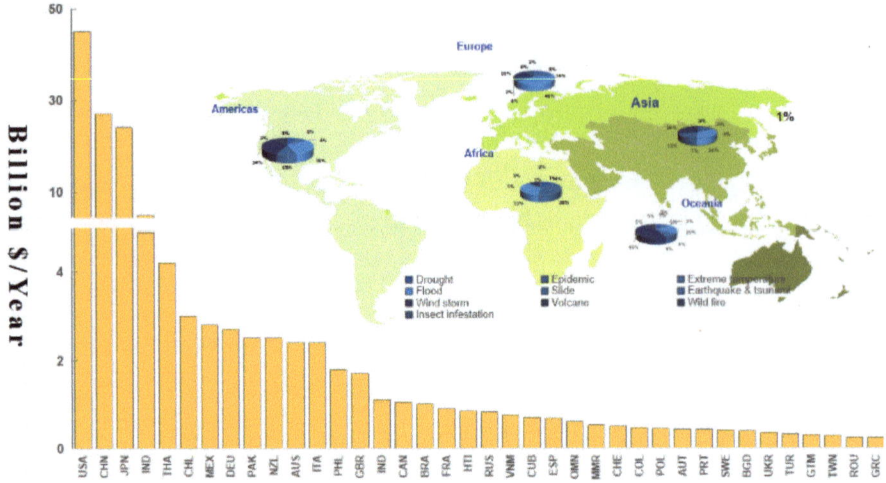

Fig. 1.4 Global disaster classification statistics (UNESCO, 2022)

1.2.1 Bridge Dynamics

In recent decades, the development of both science and technology have proceeded hand in hand with the need to create structures of increasing size, numerous long-span bridges have been built worldwide in many coastal aeras to cross straits or seas or built in mountain valley aeras to cross canyons. Some of representative long-span bridge were built within last decades or under the construction stage were given in Table 1.1, such as the Hardanger Bridge in Norway, with the main span of 1310 m, and Russki Island Bridge in Russian with the main span of 1885 m, and a new type of cross-section, twin-box girder, was adopted in the design stage for the Yi Sun-sin Bridge with main span of 1514 m and Canakkale Bridge in Turkey with main span of 2023 m, and a long-span bridge with main span of 2180 m, Shiziyang Bridge in China was under the construction stage. To achieve the ability to cross the canyons and straits, the frequently used structural types are cable-stayed bridge and suspension bridge, these long-span bridges are highly flexible structures with lower damping, which in turn makes them more vulnerable to vibrations due to the ambient environmental excitations, especially under the wind excitations (Fig. 1.5).

1.2.1.1 Vortex-Induced Vibration

Vortex-induced vibration (VIV) is a phenomenon triggered by the resonance between the Kármán vortex shedding forces and one of the natural frequencies of the structure, which indeed presents strong nonlinearities and where a major role is played by self-excitation. Large amplitude but usually non-catastrophic oscillations can occur in a limited range of wind speed, in which the spanwise correlation of the fluctuating

Table 1.1 Main performance parameters of the novel laboratory in Tongji University

Items	Performance parameters
main tank	Length of 80 m, Width of 4 m, Height of 3 m
wave generation system	Equipped with active absorption capability, maximum wave height of 1.0 m; Wave period range of 0.5 s-5.0 s; Capable of generating both regular and irregular waves, compatible with domestic and international wave standards (including JONSWAP, Pierson-Moskowitz, TMA, etc.) as well as user-defined irregular wave patterns
current generation system	Capable of bidirectional flow, maximum system flow velocity of 3.0 m/s; Working water depth of 1.5 m/s, flow speed range in tank of 0–3.0 m/s
wind generation system	Equipped with 12 fans (arranged 4 × 3), single fan power of 30 kW, total power of 360 kW; Maximum airflow per fan of 80,000 m^3/h; Maximum wind speed of 20 m/s; effective area: height of 2 m; width of 4 m

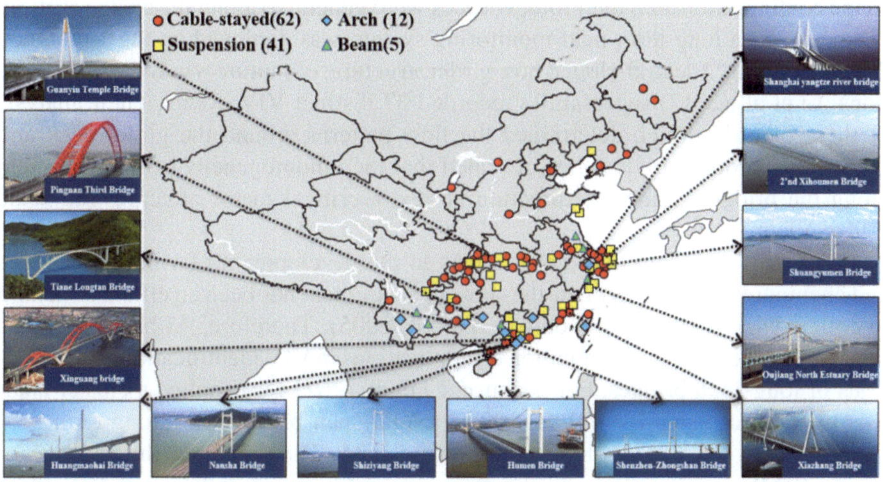

Fig. 1.5 Development of bridge engineering in China

pressures on the body surface is much higher. Also, the frequency of vortex shedding is not governed by the Strouhal law but is synchronized or "locked" to the vibration frequency of the structure ("lock-in"). Both the maximum amplitude of oscillation and the extension of the range of synchronization are dependent on a mass-damping parameter, the so-called Scruton number. For bridges this phenomenon can concern either an issue of resistance (fatigue) or serviceability (discomfort), given that it may occur at relatively low wind speed. Moreover, for bridge decks vortex-induced vibrations in a torsional mode are often possible (Baniotopoulos et al., 2011).

In recent decades, many cases of vortex-induced vibration of bridge structures have been registered. Regarding deck oscillations, after the first Tacoma Narrows Bridge, the most famous case of vortex- induced vibration was that of the Great Belt

East Bridge, Denmark, which during the final phases of deck erection showed low-frequency vertical oscillations of the main girder, that were considered un- acceptable with respect to the comfort of the bridge users (Larsen et al., 1999). VIV vibrations were observed not only for cable-supported bridges, as demonstrated by the case of the Trans-Tokyo Bay Crossing Bridge, Japan (Fujino & Yoshida, 2002). This is a straight ten-span continuous steel box-girder bridge with a main span of 240 m and variable cross-section height. The structure exhibited vibrations exceeding 50 cm, occurring for the first vertical mode in the wind-speed range 13–18 m/s. Another very recent example is Humen Bridge, which crosses the Pearl River to link Guangzhou and Dongguan, occurred the large amplitude vortex-induced vibration for the first time after 23 years of service life On May 5, 2020. Based on the fast results of field measurements by research team of Tongji University, Zhao et al. (2022) pointed out that the direct reason was that the water-filled barriers were placed along the two edges of the bridge deck upper surface, which destroyed the streamlined aerodynamic shape configuration and, therefore, vortices were generated along the bridge section. Furthermore, a long-term field monitoring system was deployed on the Xihoumen Bridge in China, a central-slotted box-girder structure, to capture wind-induced vibrations. Li et al. (2017) successfully recorded 37 distinct VIV events. Their analysis of these records, which determined the flow patterns around the girder deck and the vortex-shedding frequencies, revealed that the inhomogeneity of the wind field along the bridge's spanwise direction is also a critical factor affecting full-scale bridge VIVs.

It is worth declaring that many thanks to the development of vibration sensors and anemometers, long-term on-the spots measurement has been an efficient method to collect wind and vibration data (Ko & Ni, 2005). Therefore, AI-based methods nowadays arise much more attention on applications in VIV identification and predictions of long-span bridges. For instance, Li et al. (2021a, b) applied a cluster analysis method to recognize VIV events in a long-span bridge using long-term field measurements of girder vibration and wind data. They adopted the root-mean-square (RMS) acceleration and the frequency ratio as key indices for VIV identification. Besides, Liu et al. (2022) identified the vortex-excited force parameters by utilizing the Bayesian FFT inference method based on the on-site monitoring data (sensors installed on Humen Bridge), and a frequency domain formulation is proposed focusing on the structural vibration mode excited during VIV process, the flowchart is given in Fig. 1.6. Later on, Hou et al. (2023) proposed a migration learning method based on TCA (transfer component analysis) to extract vibration features of the long-span bridge based on the structural health monitoring data. More advanced research works by utilizing ML methods in VIV identification also based on the SHM data would be worth mentioning here but cannot be discussed for sake of brevity.

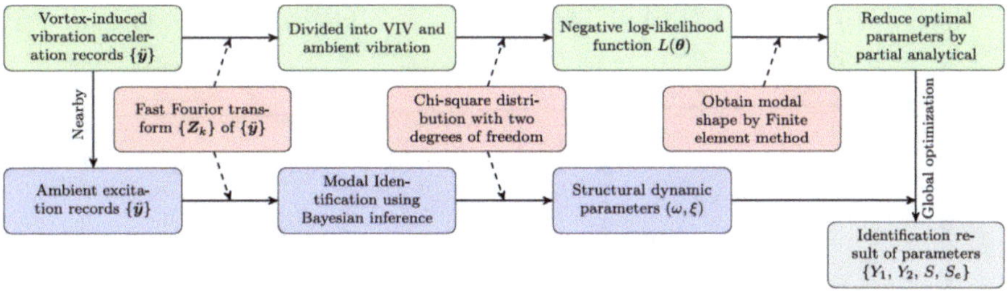

Fig. 1.6 Flowchart of VEF parameter identification of Humen Bridge (Liu et al., 2022)

1.2.1.2 Buffeting

In aeronautics buffeting denotes the dynamic excitation of the aircraft tail due to the interference with the wake of another part of the airplane, such as a stalled wing. By contrast, in wind engineering with buffeting one refers both to the excitation mechanism of a structure due to the wake of another structure (Scruton, 1965) and to the effect of turbulence naturally present in the atmospheric boundary layer. The response of bridges to turbulent wind is usually calculated by assuming the linear superposition of buffeting and self-excited forces and possibly using the flutter derivatives measured in turbulent flow (Scanlan, 1971). In fact, small-scale turbulence (say scales smaller than the deck dimensions) has mainly the effect of changing the aerodynamic behavior of the deck cross section, while large-scale turbulence represents the most energetic source of random excitation.

The analysis of wind-induced random vibration has become increasingly refined. Recent studies now account not only for normal seasonal winds but also for non-stationary and non-Gaussian phenomena associated with non-synoptic winds (Zhao et al., 2019). To predict buffeting acceleration responses under future potential typhoons, Liu et al. (2021) proposed a decoupling and regression modeling approach. Similarly, Li et al. (2021a, b) introduced a data-driven method to model the buffeting responses of a long-span bridge based on Structural Health Monitoring (SHM) data, employing a Long Short-Term Memory (LSTM) network to represent the bridge aerodynamic system. Their results demonstrate that this data-driven method achieves higher accuracy in modeling bridge aerodynamics than conventional methods. Liu (2025) adopted the Long Short-Term Memory (LSTM) sequence-to-sequence (seq2seq) method (Fig. 1.7) to predict the buffeting responses of the Xihoumen Bridge under both normal wind and typhoon conditions. The prediction efficacy was validated by comparing its results with those obtained from the Finite Element Method (FEM) and the Dynamic Relaxation Method (DRM). The results indicate that the conventional buffeting calculation method based on the Scanlan aerodynamic model exhibits deficiencies, particularly when the specific wind center of a typhoon passes the bridge. In contrast, predictions from the LSTM method align more closely with the actual measured values, especially in the large-amplitude region. A comparative visualization of the buffeting responses predicted by different

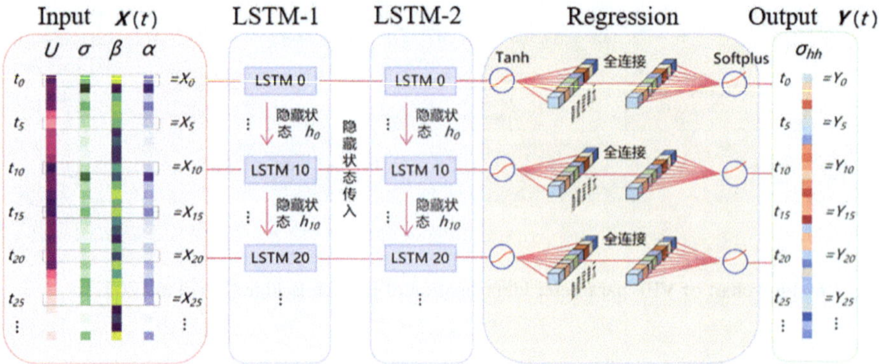

Fig. 1.7 Predicted buffeting responses under typhoon process (Liu, 2025)

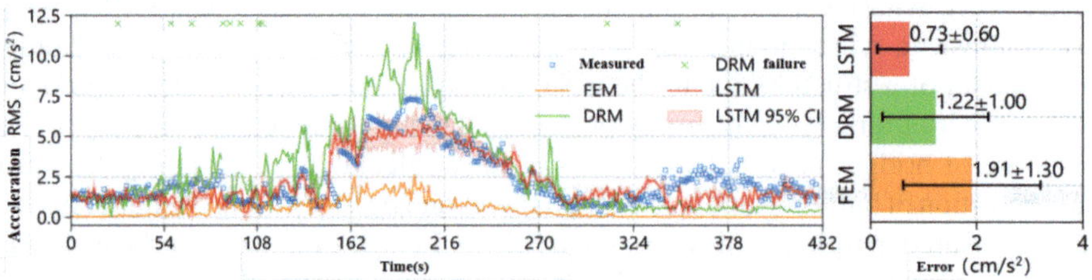

Fig. 1.8 Predicted buffeting responses under typhoon process (Liu, 2025)

methods under Typhoon Lekima is presented in Fig. 1.8. Similarly, Castellon et al. (2021) evaluated the efficacy of different models—including Multi-Layer Perceptron (MLP), FEM, and Support Vector Regression (SVR)—for estimating buffeting responses. Their results demonstrate that the SVR algorithm achieves superior accuracy for this task. In a related advancement, Laima et al. (2023) proposed a deep learning framework termed the "Buffeting-Net" method for predicting the buffeting responses of long-span bridges.

1.2.1.3 Flutter

Flutter is a dynamic instability basically involving a torsional and a vertical bending vibration mode of the deck with similar shapes. The mechanism through which energy is extracted from the flow and feeds the structural oscillations is related to the variation of modal frequencies, that tend to approach due to fluid–structure interaction, and to the phase between the two degrees of freedom. Flutter gives rise to limit-cycle oscillations of large amplitude, increasing with the wind speed, that can easily lead to the collapse of the structure. The importance of flutter for long-span bridges was already in the minds of the pioneers, from the evident work by Rocard

(1965), Scanlan (1971), Costa and Borri (2007) to the latest research work by Gao et al. (2020), Wu et al. (2020), and Lin et al. (2022), etc.

AI-based methods have emerged as effective solutions for predicting the flutter performance of bridges. In early research, the Artificial Neural Network (ANN) method was adopted to identify aeroelastic parameters from wind tunnel test results (Li, 2000). However, prediction errors exceeding 20% rendered this approach unsuitable for practical engineering applications. Subsequent advances in computational capacity enabled the use of ANN to analyze flutter derivatives for a range of bridge girder decks with varying heights and widths, and to predict corresponding flutter wind speeds (Chen et al., 2008). Similar research has proliferated in recent decades, with methods such as ANN, Neural Networks (NNs), and Support Vector Machines (SVMs) being widely employed to predict flutter wind speeds for long-span bridges (e.g., Chen & Ge, 2019; Jung et al., 2004; Wilms et al., 2019; Wu & Kareem, 2011). Beyond these, broader Machine Learning (ML) techniques have also become prominent tools for flutter prediction. For instance, Liao et al. (2021) developed a flutter prediction model using an ML strategy trained on a large dataset of flutter wind speeds. Figure 1.9 presents the flowchart for predicting critical flutter wind speed under various angles of attack. Using the Fuma Yangtze River Bridge as a case study, the authors validated their proposed ML method against other AI models. The results demonstrated that their method achieved superior prediction efficiency and accuracy compared to NN and Gradient Boosted Regression Tree (GBRT) models.

1.2.2 Tall Buildings

With the development of urbanization, numerous tall buildings are built with lightweight, flexible, and facing strong wind and especially under extreme wind conditions, such as typhoons. As a result of wind impacts, the vortex-induced vibration and the potentially galloping are the most severe wind-induced vibrations for tall buildings. A milestone research work on evaluating the wind acting on long-span bridges by Ohkuma and Kanya (1978), authors proposed theoretical wind forces models of tall buildings. Then Kareem (1982) developed a theoretical formular through the spatio-temporal measurement for wind forces of tall buildings based on the wind pressure. Later on, Solari (1985) also established a model for evaluating wind force acting on a tall building with a rectangular cross-section. Many researchers dedicated their time to wind force and wind-induced vibrations of the long-span buildings within the past decades. Kwok and Melbourne (1980), Zhou et al. (2003), Gu and Quan (2004), Liang et al. (2011), and Venanzi and Materazzi (2012) are focused on the vortex-induced vibrations and aeroelastic responses of the tall buildings and provided references for the tall building design.

The wealth of prior research findings and the extensive database from wind tunnel tests on tall buildings have created significant opportunities for applying AI-based methods to analyze wind-induced vibrations and wind forces on such structures. For instance, Hu et al. (2020) analyzed the interference effects of high-rise buildings

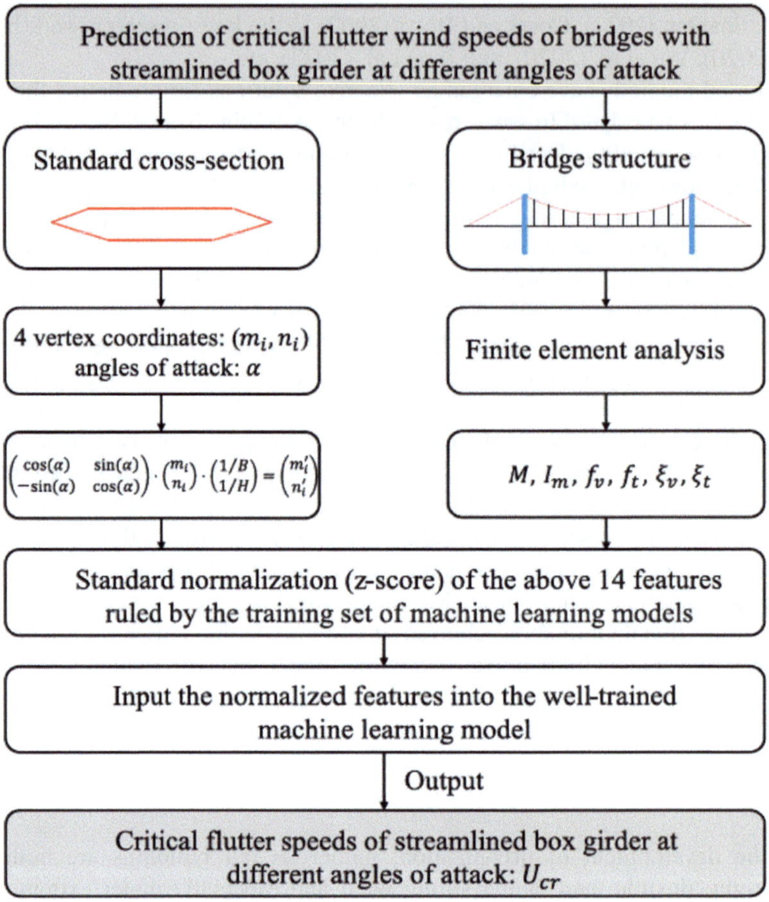

Fig. 1.9 Flowchart of VEF parameter identification of Humen Bridge (Liao et al., 2021)

by using generative adversarial neural networks. Recently, the deep reinforcement learning method is widely utilized in aeroelastic shape optimization and fluid mechanism. Therefore, Luo and Kareem (2020) predicted the mean and RMS value of the wind pressure on the tall buildings utilizing the probabilistic model developed based on the Bayesian deep learning. Concurrently, numerous studies have widely reported the application of machine learning in wind engineering related to tall buildings, including research by Wang et al. (2020), Lin et al. (2021), Liao et al. (2021), and Zhang et al. (2021).

1.2.3 Aeroelastic-Shape Optimization

Many researchers are dedicated to investigating the way to delay flutter instability beyond the design wind speed for the bridge site by passive or active devices able to improve the aerodynamic performance of the deck cross section. In this context the pioneering experimental work of Scruton (1952) for the Severn Bridge is an awe-inspiring example. Matsumoto et al. (2001) studied the effect on flutter stability of grating on plate-like bridge decks for various angles of attack, opening ratios and locations of the grids, also in the presence of fairings (Fig. 1.10).

Historically, passive aerodynamic control methods were widely used to mitigate flutters. However, it is important to note that these passive countermeasures face significant limitations. With the increasing span length and flexibility of modern bridges, they may no longer meet design requirements. Their effectiveness is typically confined to specific wind attack angles and speed ranges, posing a challenge for the survivability of long-span bridges in extreme wind environments, such as strong typhoons characterized by high wind speeds, large attack angles, and high turbulence intensities. Consequently, recent research has shifted focus toward active control mechanisms for flutter suppression. For example, Li et al. (2015a, b) analyzed the effect of a wind flap installed above the girder deck on critical flutter speed. Cross-validation of wind tunnel tests and theoretical analysis indicated that this configuration could improve the critical flutter speed by up to 80%. In a more recent study, Zhao et al. (2020) and Wang et al. (2023, 2024) proposed a highly effective active control device, investigating the improvement in critical wind speed achieved by a pair of wind flaps installed on the wind fairing. Critically, their work considered the interactions between the flaps and the surrounding flow field and implemented a real-time feedback control strategy during wind tunnel testing, ultimately obtaining a 25% improvement in critical flutter speed. Subsequently, Wang et al. (2025) analyzed the efficacy of active control flaps on the flutter performance of a streamlined bridge girder. They employed a Backpropagation Artificial Neural

Fig. 1.10 Active deck-flaps system in wind tunnel (Wang et al., 2023)

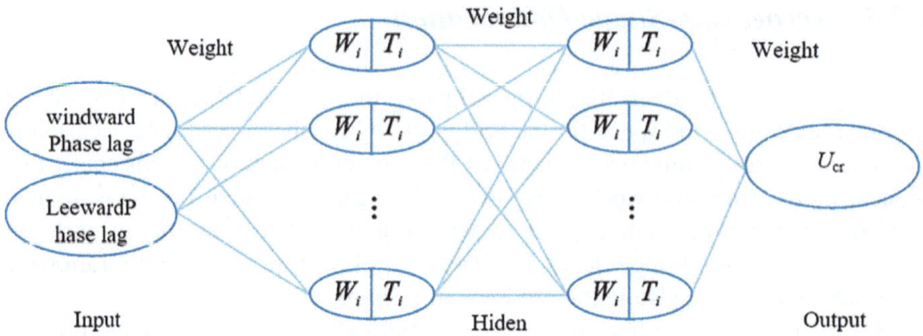

Fig. 1.11 The flowchart of the BP ANN method (Wang et al., 2025)

Network (BP ANN) method (see flowchart in Fig. 1.8) to optimize the control parameters for the active flaps. The results indicated that applying an optimized phase difference to the flaps, determined via the neural network, can effectively enhance flutter performance (Fig. 1.11).

1.3 AI Methodology for Wind Engineering and Wind Energy

1.3.1 Great Potential of Wind Power Application

Europe has launched a wind power initiative with the goal of establishing wind energy as the most competitive energy source on the market during the 2020–2030 period. A key interim target was to reduce wind energy costs by at least 20% by 2020. Consequently, strategies have been proposed to boost the development of both onshore and offshore wind turbines. The objectives include achieving a wind energy penetration rate exceeding 20% of European electricity consumption in the early 2020s, securing Europe's technological leadership in onshore and offshore wind sectors, and advancing large offshore wind turbine technologies, including concepts with capacities up to 20 MW. The progress toward achieving high competitiveness and penetration levels for both onshore and offshore wind power is illustrated in Fig. 1.12.

Consequently, there is a substantial demand for the installation and strategic planning of both offshore and onshore wind turbines. This is driven by the need to meet short-term targets—such as reducing greenhouse gas emissions by 20% and achieving a 20% share of renewable energy in Europe by 2020—as well as the long-term goal of deep decarbonization, which entails a 60–80% reduction in greenhouse gas emissions. To accelerate progress, priority should be given to offshore wind applications, including the development of turbines with double the power generation capacity of the largest current models. The cumulative installed capacity of onshore

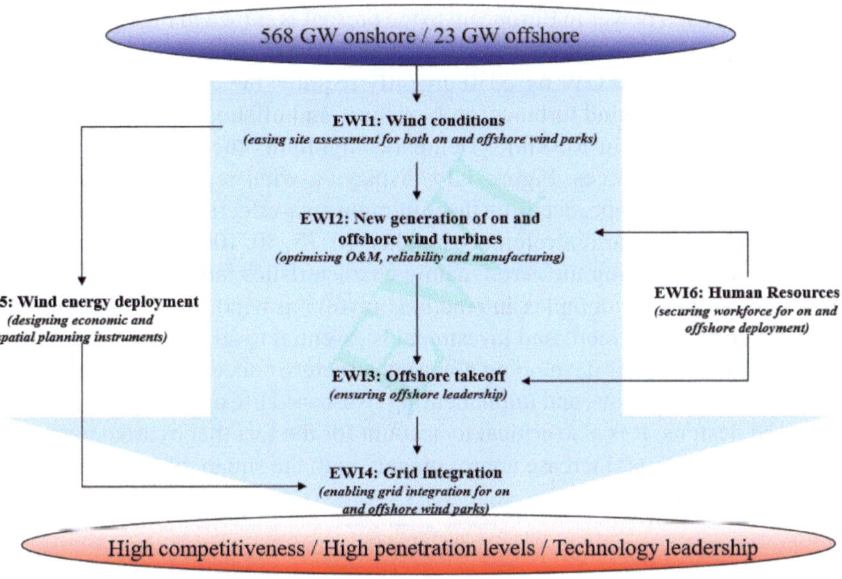

Fig. 1.12 Progress of high competitiveness and penetration levels onshore and offshore wind power

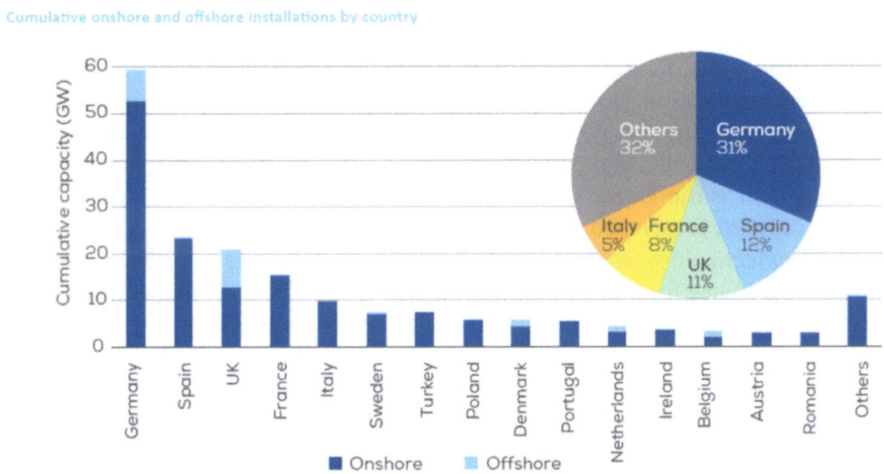

	Germany	Spain	UK	France	Italy	Sweden	Turkey	Poland	Denmark	Portugal	Netherlands	Ireland	Belgium	Austria	Romania	Others
Offshore	6.4	-	8.2	-	-	0.2	-	-	1.3	-	1.1	-	1.2	-	-	0.1
Onshore	52.9	23.5	12.8	15.3	10.0	7.2	7.4	5.9	4.4	5.4	3.4	3.5	2.2	3.0	3.0	10.8
Total	**59.3**	**23.5**	**21.0**	**15.3**	**10.0**	**7.4**	**7.4**	**5.9**	**5.7**	**5.4**	**4.5**	**3.5**	**3.4**	**3.0**	**3.0**	**10.9**

Source: WindEurope

Fig. 1.13 State by state capacity in Europe of installed and planned wind power

and offshore wind power in Europe up to the present is shown in Fig. 1.13. According to data from Chap. 3 of the Strategic Research Agenda, meeting the projected green power demand of 120 GW by 2030 urgently requires the accelerated development and construction of wind turbines, both onshore and offshore. Research by the Risø National Laboratory in Roskilde, Denmark, highlights the significant potential of European wind resources. Figure 1.14 displays a wind resource distribution map over open seas in Europe, detailing the minimum cost-effective wind speeds (ranging from 4 to 11 m/s) at various reference heights: 10, 25, 50, 100, and 200 m. Given this potential, understanding the aerodynamic characteristics induced by wind loads on blades, as well as the complex interactions involving wind, waves, and currents, is crucial. Furthermore, increased investment is essential to advance substructure technologies. This includes developing novel substructure concepts, extending structural lifespans, reducing costs, and implementing risk-based life cycle approaches through improved designs. It is also critical to account for the fact that hydrodynamic forces (and associated costs) increase approximately with the square of the combined wind and water flow velocity.

1.3.2 Dynamics Analysis of Wind Energy Structures

The deployment of onshore and, more demandingly, offshore wind turbines is a strategic priority, particularly given the vast potential for offshore wind energy development and application. However, constructing offshore wind turbines in deep water presents significant technological challenges. For instance, monopile substructure technology is typically limited to water depths of up to 30 m, while fixed-bottom solutions such as tripods are generally viable only up to about 80 m. Therefore, the floating supports is regarded as the solution while the water deep exceeds 40 m, and the floating technology seems to be the leading solution as deep waters represent a promising resource for many countries, e.g., Italy, US. The most frequently adopted floating systems are (1) Spar-buoy with two tier catenary mooring drag embedment anchors; (2) Tension-leg platform (TLP) with suction pile anchors; (3) Concrete with gravity base anchors. Given this extremely complex multi-physical environment, the loads acting on offshore wind turbines are a combination of aerodynamic, gravitational, inertial, and hydrodynamic forces. The latter are associated with waves, currents, and tides, as illustrated in Fig. 1.15. Crucially, the principal distinction between the loads on onshore and offshore wind turbines lies in the presence of these significant hydrodynamic loads.

The design loads on a wind turbine can be categorized into four primary components. The first is aerodynamic loads, as illustrated in Fig. 1.16a, which encompass steady uniform wind loads, steady wind shear, unsteady wind velocity at hub height, and unsteady three-dimensional wind fields. The second component comprises inertial loads, including gravitational, centrifugal, and gyroscopic forces. The third category is controlling loads, which involve braking and yawing forces, blade-pitching

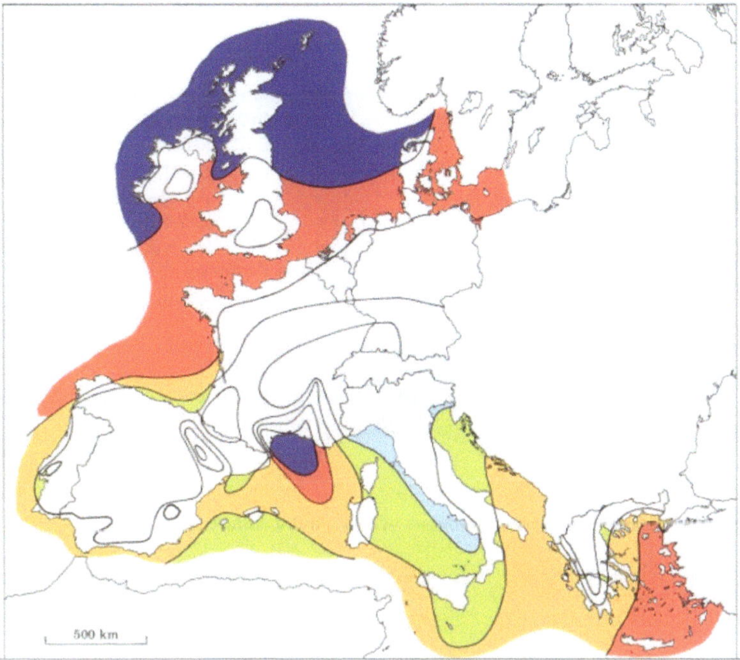

Wind resources over open sea (more than 10 km offshore) for five standard heights										
	10 m		25 m		50 m		100 m		200 m	
	$m\,s^{-1}$	Wm^{-2}	$m\,s^{-1}$	Wm^{-2}	$m\,s^{-1}$	Wm^{-2}	$m\,s^{-1}$	Wm^{-2}	$m\,s^{-1}$	Wm^{-2}
	> 8.0	> 600	> 8.5	> 700	> 9.0	> 800	> 10.0	> 1100	> 11.0	> 1500
	7.0-8.0	350-600	7.5-8.5	450-700	8.0-9.0	600-800	8.5-10.0	650-1100	9.5-11.0	900-1500
	6.0-7.0	250-300	6.5-7.5	300-450	7.0-8.0	400-600	7.5- 8.5	450- 650	8.0- 9.5	600- 900
	4.5-6.0	100-250	5.0-6.5	150-300	5.5-7.0	200-400	6.0- 7.5	250- 450	6.5- 8.0	300- 600
	< 4.5	< 100	< 5.0	< 150	< 5.5	< 200	< 6.0	< 250	< 6.5	< 300

Fig. 1.14 Wind resources map over open sea within the Europe distribution

forces, and forces due to grid loss. The fourth and final component consists of hydrodynamic loads, such as wave forces (both regular and irregular), current forces, and tidal forces. The overall modeling procedure for wind turbines is depicted in Fig. 1.17. It is important to note that while floating platforms have garnered significant attention in recent decades, current simulation methods often only account for linear loads. Given the prevalence of nonlinear waves in intermediate to shallow waters, the nonlinearities and complex load interactions warrant further investigation.

The nonlinear waves acting on a bottom-supported wind turbine were investigated by Marino et al. (2014) while the wind turbine in power production progress, the aeroelastic effects couldn't be neglected because of the aeroelastic interactions of the rotor. Furthermore, the nonlinear wave effects on the wind turbines and resulting resonant vibrations of the offshore wind turbines (OWT) are crucial for fatigue damage assessment. Marino et al. (2017) analyzed the impacts of nonlinear hydrodynamic force and aerodynamic loads with different wind speed and turbulence intensities on a

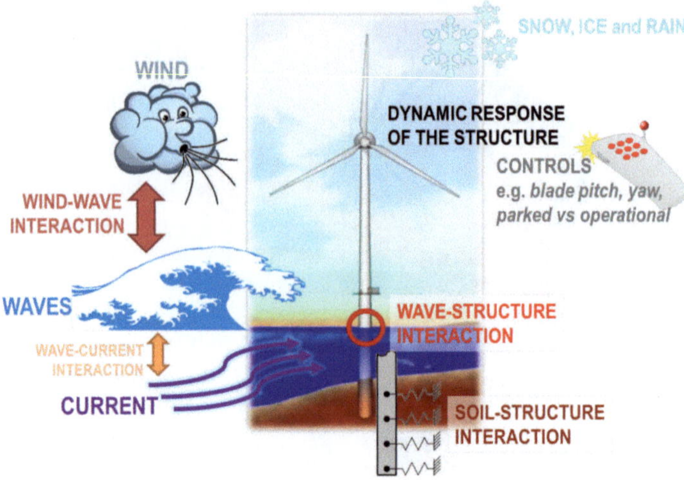

Fig. 1.15 The design loads and interactions act on wind turbines

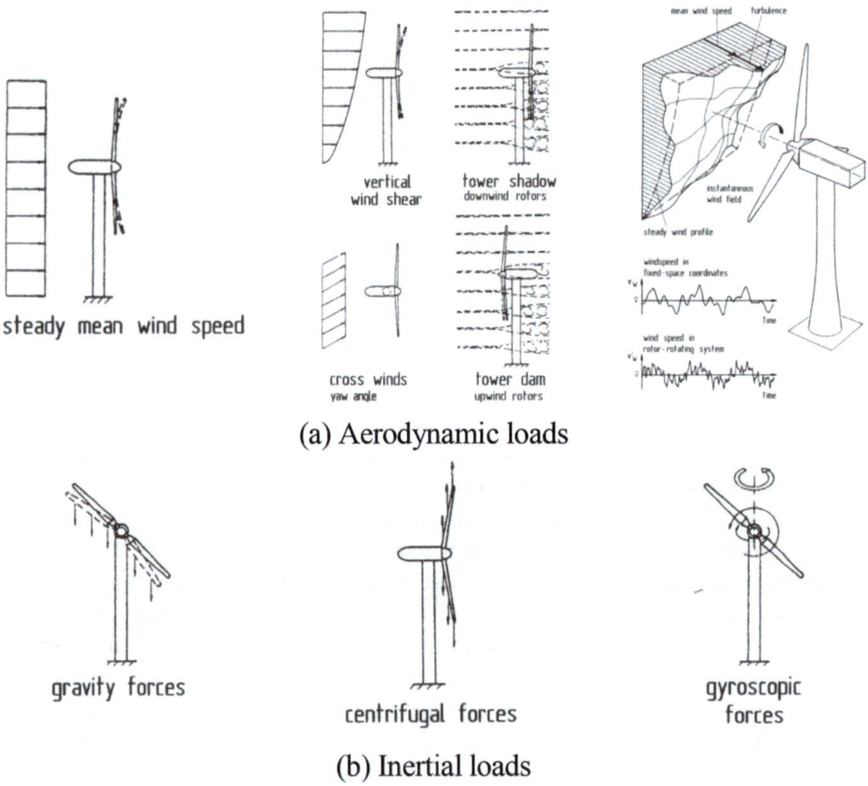

Fig. 1.16 Parts of the design loads of the wind turbine

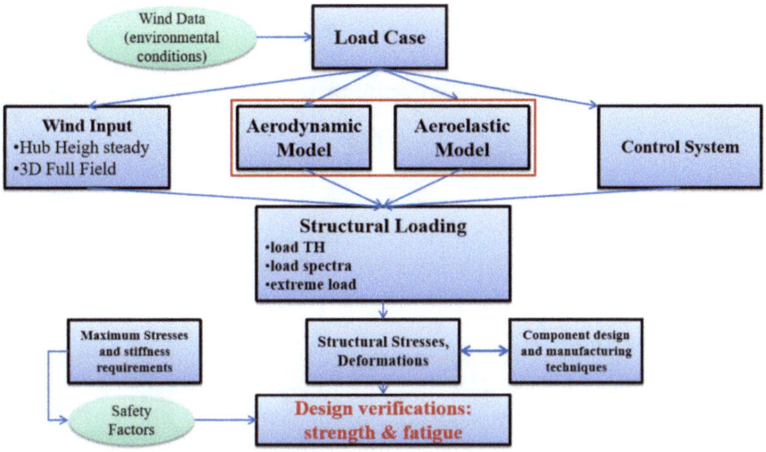

Fig. 1.17 The modeling procedure of the wind turbines

bottom-supported OWT under the parked state, based on a fully coupled hydro-aero-elastic solver. Results revealed that the widely used linear wave modeling approach significantly underestimates fatigue loads. However, the aerodynamic loads become dominant when the wind turbines are in the power production state.

Bruschi et al. (2020) analyzed the influence of position and weight on platform response and mooring line tension. Their results show that the mass of the clump weight is less influential than its position, which was found to be optimal for minimizing response-particularly in the surge degree of freedom-when placed closest to the platform. Ferri et al. (2020a,b) optimized the platform design by considering its shape and size, mooring line geometry, and positional parameters to enhance the overall structural performance of the offshore wind turbine (OWT) system. Ferri et al. (2020b) identified the diameter of the outer cylinders and their radial distance from the platform center as two crucial parameters for platform optimization. Their goal was to minimize both the amplitude of the peaks and the integral area of the Response Amplitude Operators (RAOs). They determined an optimal configuration with an outer column diameter (d) of 16 m and a platform radius of 35 m. Subsequently, Marino et al. (2024) presented an advanced, multi-objective, site-specific optimization procedure aimed at identifying the optimal substructure configuration for a 10 MW floating offshore wind turbine (FOWT). This procedure utilized a 20-year metocean database of wind and wave records to calculate the joint probability distributions of wind speed, significant wave height, and peak spectral period. Based on this analysis, they proposed an optimized substructure concept (Fig. 1.18) founded on the dual objectives of minimizing structural response under extreme conditions and minimizing substructure cost.

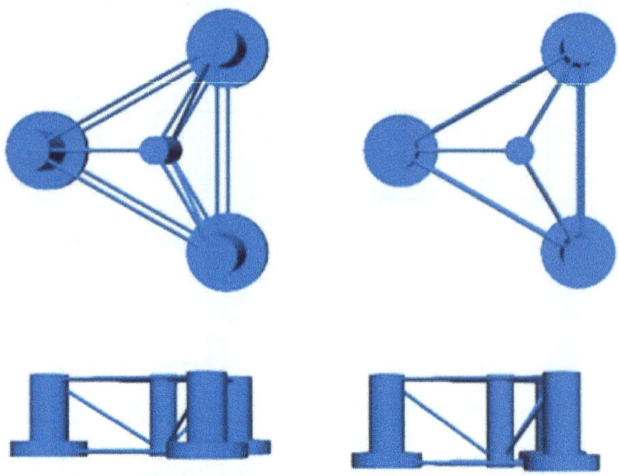

Fig. 1.18 The optimized substructure concept proposed by Ferri et al. (2024)

1.4 Wind Field Simulation and Reproduction

1.4.1 Introduction to Survey Methods

Field measurement has been the most efficient method to capture wind characteristics not only at the engineering structure site but also for the large-scale wind such as typhoon and downburst, the micro-wind data and the wind-induced responses measurements become reality thanks to the development of vibration sensors, anemometers, and the LiDAR, etc. The field measurements equipment in Tongji University, Southwest Jiaotong University, as well as Norwegian University of Science and Technology (Fenerci & Øiseth, 2018), etc., are reported within the past decades, but the most fully equipped is the Wind Runner Team in Tongji University. Zhao (2018) developed a complete set of field measurement equipment in Tongji University, named Movable Laboratory on Wind Effects Investigation and Measurement (the Wind Runner as well). The active and maneuverable are two advantages of the movable laboratory compared to other equipment. A medium size truck (Fig. 1.19) is adopted as the base equipped with multiple measurement devices, including a Doppler LiDAR, an aperture radar, two sets of mechanical anemometers, four ultrasonic anemometers, multiple acceleration sensors, multiple displacement testing systems, IBIS-FS (remote high-precision monitoring system for buildings), etc.

1.4.2 Wind Field Measurement

Utilizing the fully equipped mobile field measurement laboratory, Yi et al. (2022) captured the landing process along the southeast coastline of China of strong typhoons

Fig. 1.19 Movable laboratory on wind effects investigation and measurement

Haitang, Matsa, Khanun successfully (as depicted in Fig. 1.20), and the influence of typhoon intensity ΔP and the normalized distance D/Router on extreme winds under typhoon conditions, field measurements of mean wind speeds U and turbulence intensities I_u at different heights below 70 m during the strong typhoons are investigated. Furthermore, most of their exponents about mean wind α_U and ones about turbulence intensity α_{Iu} are significantly larger than the code value of 0.12 in terrain A. The α_U and α_{Iu} during the three strong typhoons fluctuated roughly before landing and linear stationary variation after landfall. After landing, α_U decreased with U and ΔP for the three typhoons. α_U increased with D during the typhoons Matsa and Khanun, whereas it decreased with D/Router during Haitang. After landfall, α_{Iu} decreased with ΔP and increased with D/Router during Matsa and Khanun, whereas it varied in a converse manner during Haitang. Subsequently, Zhao et al. (2024) developed a unified parametric model for typhoons based on prior measurements of Typhoon Hagupit (0814), incorporating key parameters such as the angle of attack (AoA), turbulence intensity, integral length scale, and mean wind speed. Furthermore, they systematically studied the structural performance, specifically the buffeting responses, of long-span bridges with main spans ranging from 1500 to 2500 m. The results revealed that the maximum buffeting response was not strongly correlated with wind speed alone; other wind characteristics, particularly turbulence intensity and AoA, also significantly influenced the outcomes. The overall framework for modeling the typhoon field and bridge buffeting response is presented in Fig. 1.21.

In addition to field measurements of large-scale weather phenomena such as typhoons and mountain winds, researchers have recently paid increasing attention to measuring wind characteristics at specific bridge sites and above bridge decks. For example, Zhao et al. (2025a, b) conducted long-term field measurements at the Murong Bridge site, recording 49 gust segments. These gusts were classified into

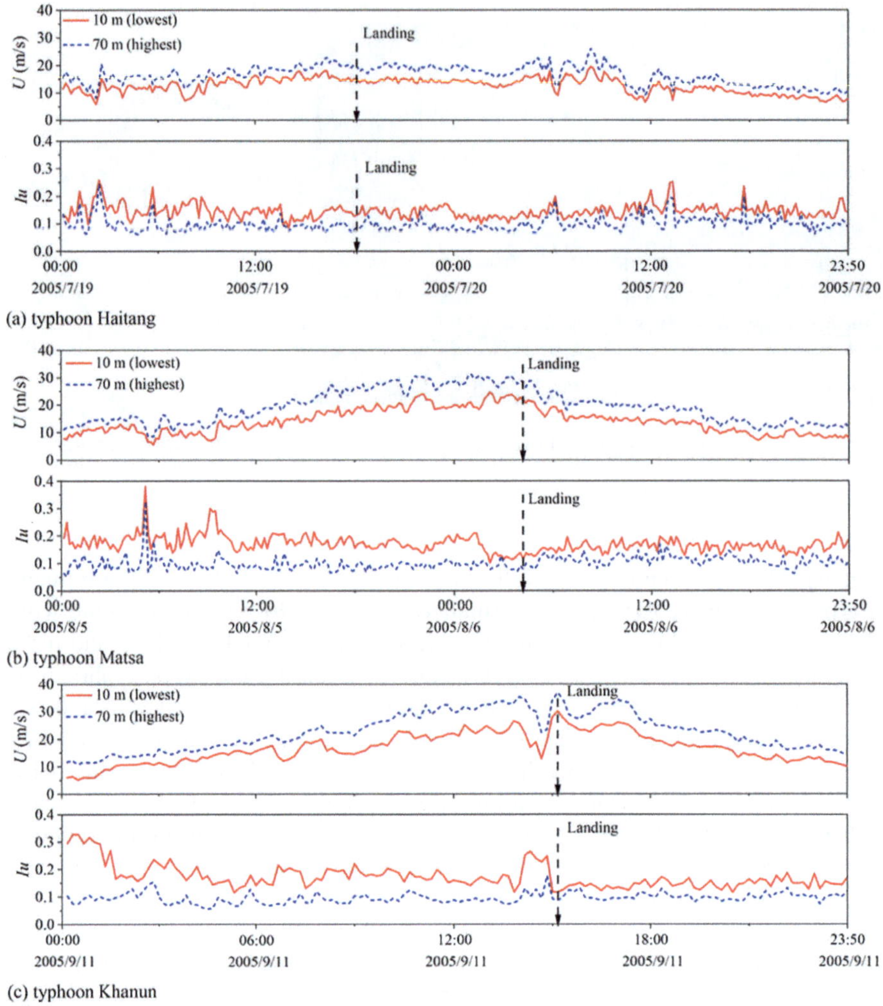

Fig. 1.20 Variations in U and Iu with time at five heights during typhoons Haitang, Matsa, and Khanun (Yi et al., 2022)

three types based on the longitudinal wind-speed time history: Rapid Rise, Slow Fall (RRSF); Slow Change (SC); and Slow Rise, Rapid Fall (SRRF), as shown in Fig. 1.22. The authors proposed a simplified non-stationary modeling method for the rapid reconstruction of measured short-rise-time gusts in high-altitude, deep-cut canyons. A case study demonstrated that this method effectively reproduces non-stationary characteristics. Separately, Wu et al. (2023) reported on microclimate characteristics at the vehicle driving safety level above a bridge deck. Based on on-site measurements, they discussed mean wind speed profiles, turbulence intensity, gust winds, and wind speed spectra. Furthermore, numerous studies conducted both before and

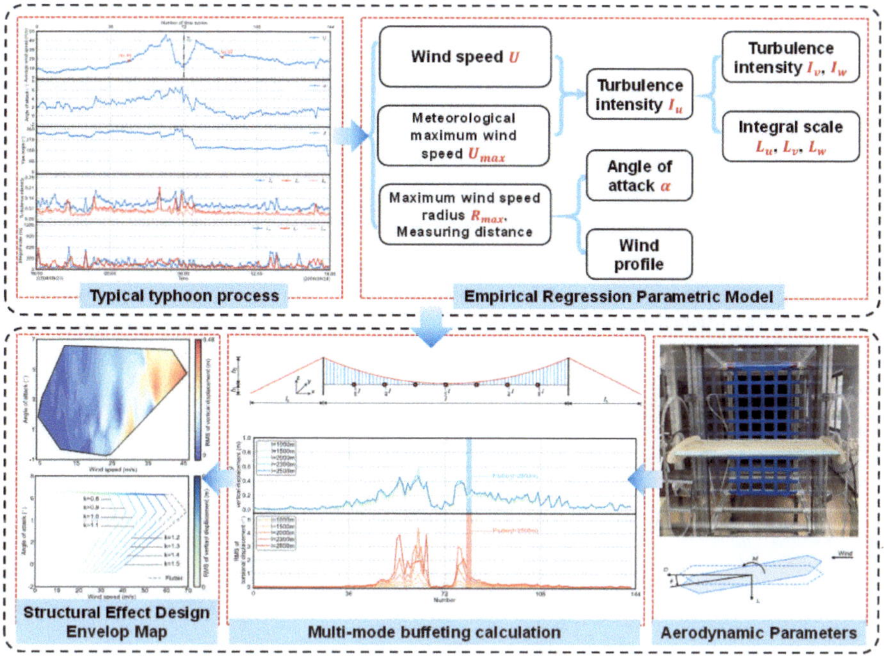

Fig. 1.21 Modeling of typhoon field and bridge buffeting response evolution (Zhao et al., 2024)

after these works have focused on typhoon field measurements, contributing to the modeling and analysis of typhoon structures and their induced buffeting responses on long-span bridges (e.g., Huang et al., 2021; Ishizaki, 1983; Li et al., 2015a, b; Tamura et al., 1993; Tao et al., 2020; Wang et al., 2009; Zhao et al., 2019).

1.4.3 Wind Field Reproduction

Currently, there is a scarcity of measured data for specific wind conditions, a lack of clarity regarding wind environment characteristics, and limited observation of structural wind effects. Information is also lacking in the wind climate categories to which extreme wind speeds belong in mixed climates. Previously, Wang et al. (2020) employed ensemble learning techniques and Gaussian process regression model to predict the distribution of future wind speed. Xie et al. (2021) proposed an adaptive adjustment strategy of constructing artificial prediction intervals to train a probabilistic model, where the original wind speed was decomposed by variational mode decomposition and the final prediction interval is a weighted sum of prediction intervals of different sub-sequences. Cui et al. (2021) proposed a windstorm type identification algorithm based on a numerical pattern recognition method that utilizes feature extraction and generalization, as shown in Fig. 1.23. Experiments include:

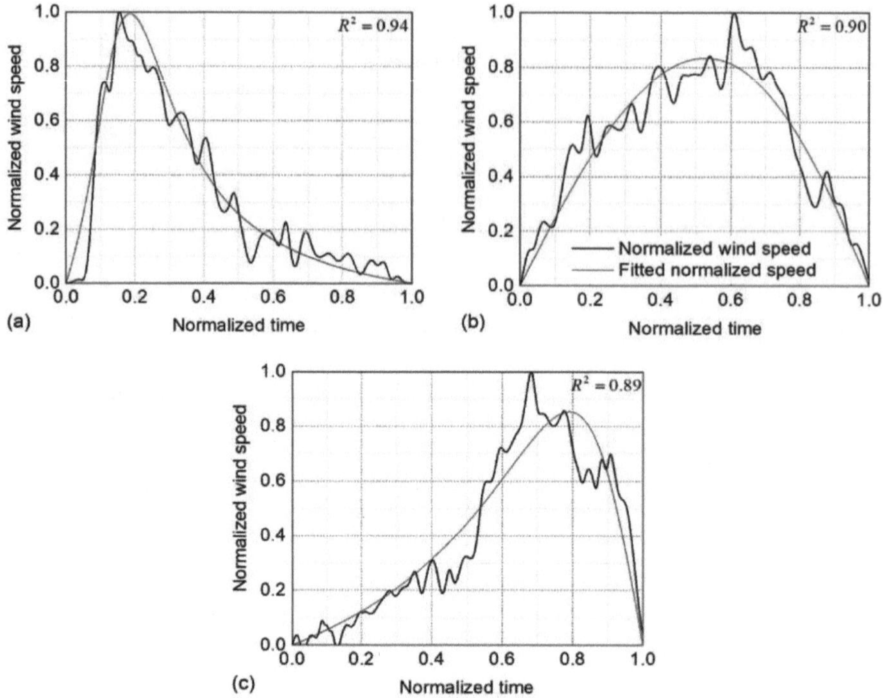

Fig. 1.22 The typical three kinds of short-rise-time gusts (Zhao et al., 2025a, b)

Experiment 1: A single station model predicts unknown data from itself; Experiment 2: The regional model predicts unknown data from known stations; Experiment 3: The regional model predicts unknown data from unknown stations. In each experiment, the performance of six machine learning algorithms is discussed based on the quantified index, and a representative algorithm is used as an example to give its confusion matrix on the test set. Based on identified windstorm data, extreme wind speed calculation results for different return periods of mixed distributions are compared with those obtained from traditional methods to verify the proposed method for extreme wind speed prediction in a mixed climate area. Later on, Ma et al. (2022) proposed a novel method for calculating extreme wind speeds in mountainous areas by using short-term field measurement data and long-term nearby meteorological observatory data (Fig. 1.24). The authors first provided a new method to extract wind speed time series of windstorms. Based on the different windstorm features, the local and synoptic winds are separated. Next, the synoptic wind speeds from nearby meteorological stations are converted and combined with local winds to derive the extreme wind speeds probability distribution function. The calculation results show that the extreme wind speed in the short return period is controlled by the local wind system, and the long-period extreme wind speed is determined by the synoptic wind system in the mountain area. The descending slope of the synoptic wind speed exceeding probability exceeds the local wind by about 20%. If the influence of mixed climate is not considered, and the wind speed samples are not divided by category, the decline slope will be 7% lower. Furthermore, wind climate can fluctuate seasonally or even monthly during the bridge erection process. To evaluate the flutter risk of suspension bridges under such varying conditions, Ma et al. (2025) proposed a framework to optimize the deck erection timeline under complex wind climates, summarizing the flutter risk assessment flow. In addition, data-driven methods—including Generative Adversarial Network (GAN)-based unsupervised learning, Graph Neural Network-based Adaptive Predictive Generative Adversarial Network (GAPGAN) models, and other machine learning techniques—are being utilized to reproduce and reconstruct wind fields at both large scales and specific bridge sites (e.g., Kim & Lee, 2020; Liu et al., 2024; Wold et al., 2024).

1.5 Intelligent Wind Tunnel Testing Methods for WE Structure

Long-span bridges exhibit complex stochastic and time-variant wind-induced vibration responses throughout their life cycles. These responses are characterized by large amplitudes and coupling between different degrees of freedom (DOFs), particularly when considering factors such as increasing span lengths, large angles of attack, and non-synoptic wind fields. Such complexities challenge the capability of traditional wind tunnel testing equipment, including free-vibration systems and forced-motion apparatuses. To address these challenges, intelligent wind tunnel testing methods have emerged as an effective tool for gaining deeper insights into aerodynamic mechanisms. As a specific solution, a novel Forced Motion Apparatus (FMA) has been proposed and designed. It incorporates Proportion Integration Differentiation (PID) technology to enable a wider variety of motion types, thereby allowing it to accurately replicate even stochastic structural movements with point-to-point coincidence between prescribed and simulated data.

1.5.1 Flow Field Vision Recognition

Over the past decades, the combination of forced motion apparatuses (FMAs) with wind tunnel testing has been established as a key method for reproducing wind-structure interactions. In addition to simulating large-amplitude, unsteady structural vibrations under conditions of high wind attack angles and multi-modal coupling, the forced vibration method offers several distinct advantages: (1) it enables testing across a wider range of reduced wind speeds; (2) it yields a higher signal-to-noise ratio in test signals; and (3) it facilitates the direct identification of self-excited aerodynamic forces acting on the model. The Norwegian University of Science and

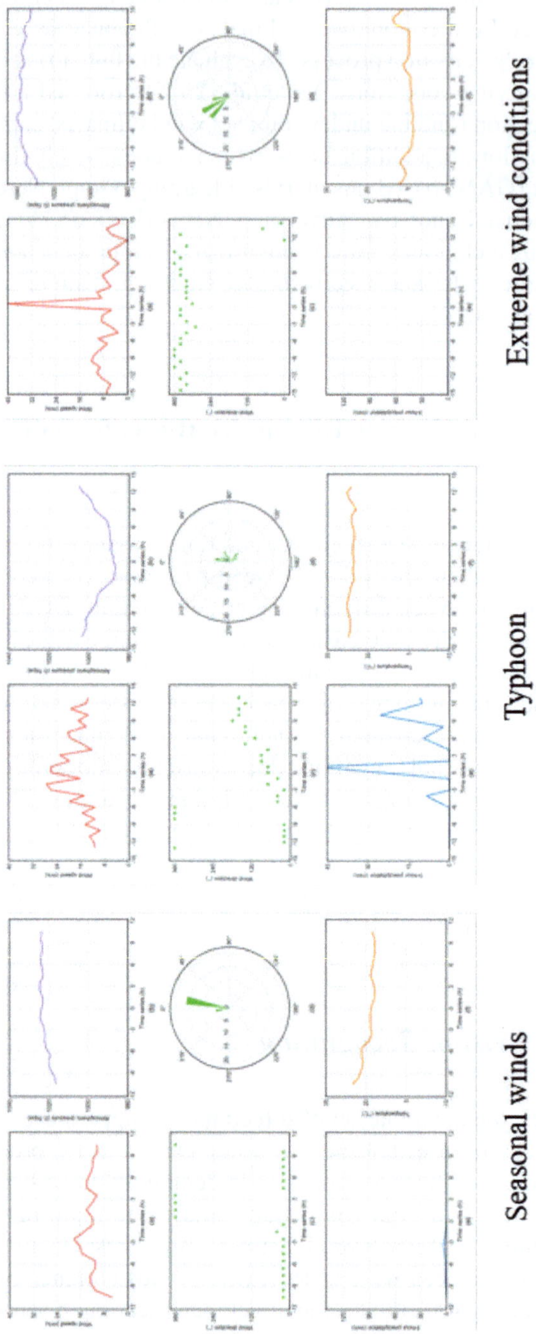

Fig. 1.23 Different types of winds recognition based on condensed multi-dimensional features (Ma et al., 2022)

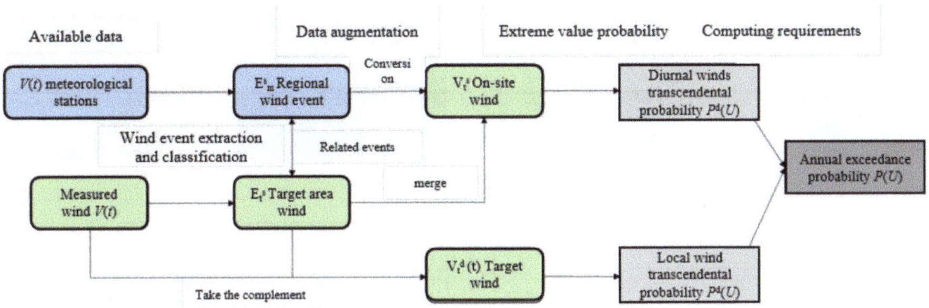

Fig. 1.24 Flowchart of extreme wind speed prediction in mountainous areas (Ma et al., 2022)

Technology invented an FMA which has a lateral and vertical amplitude of 100 mm, and a torsional amplitude that can reach 90°. The novel FMA of Tongji University, China, has larger amplitude limits with X = ± 200 mm, Y = ± 100 mm, and RZ = ± 360° (X is lateral, Y is vertical, and RZ is torsional amplitude). The higher amplitudes facilitate the study of large amplitude wind-induced vibration and nonlinear aerodynamics for bridge sections under extreme wind conditions. The representative FMA within worldwide are listed in Figure by Zhao et al. (2020). Then a novel FMA apparatus was developed in Tongji University with 3 DOF, namely horizontal (X), vertical (Y), and torsional (R_Z), are driven separately by two linear and one rotational motion actuators, as displayed in Fig. 1.25. Later on, a set of unmanned FMA with wind tunnel test system was developed to realize automation of the wind tunnel test by the developed FMA based on the user defined profile. The unmanned system has accumulated over 1,100 h of operation and has completed more than 33,000 test conditions across various bridge sections (Fig. 1.26).

The wind-bridge girder interactions have been successfully simulated, and a data set of wind-induced vibrations related to rectangular girder section, single-box girder deck, multi-girder deck has been built. As a result, a 5:1 rectangular section was chosen as the study object, and the nonlinear aerodynamic forces in the bending-torsional DOF independent and coupled motions were obtained using forced vibration wind tunnel tests, Zhao et al. (2025a, b) constructed a four-dimensional energy spectrum map of bending torsion coupling, as displayed in Fig. 1.27. Utilizing the nonlinear aerodynamic energy maps, the impacts of various motion parameters, especially the phase lag between motions, on the energy transformation of aerodynamic forces are clearly manifested with the aerodynamic energy evolutionary characteristics summarized. The mutual energy input effect of the bending and torsion 2 DOFs on the occurrence and development of coupled flutter is finally clarified by adopting the concept of energy maps.

Moreover, an Integrated Multi-Functional Wind, Wave and Current Flume Laboratory was developed in Tongji University recently to generate wind-wave interactions combined with the wind tunnel, wave making system, energy dissipation device, etc. This device aims to address challenges such as the safety and stability

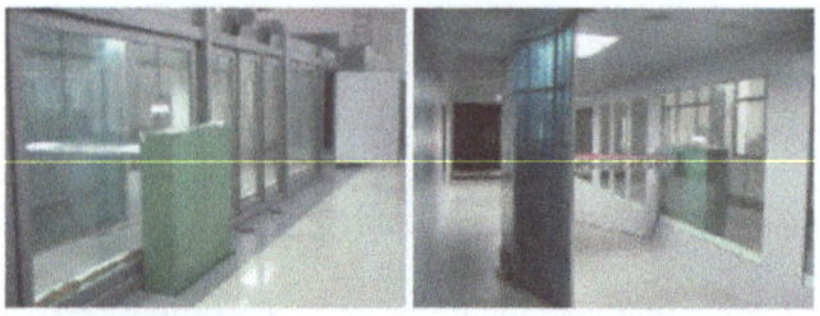

Number of DOFs	3
Frequency	0.1–3 Hz
Translational amplitude	0–±20 mm@4 mm
Rotational amplitude	0°–±5°@1°
Motion type	Sinusoidal

(a)

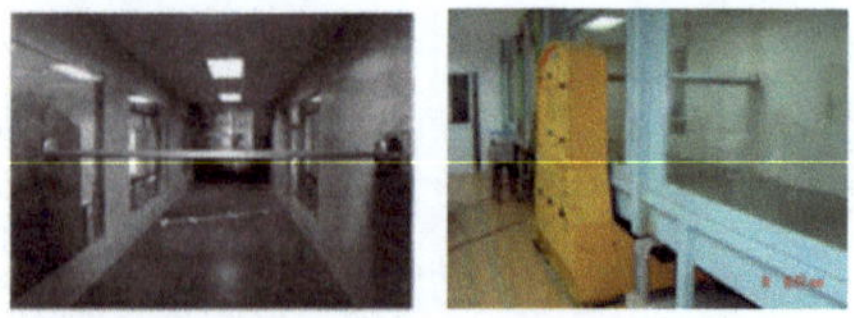

Number of DOFs	3
Frequency	0.1–3 Hz
Translational amplitude	0–±20 mm@4 mm
Rotational amplitude	0°–±5°@1°
Motion type	Sinusoidal

(b)

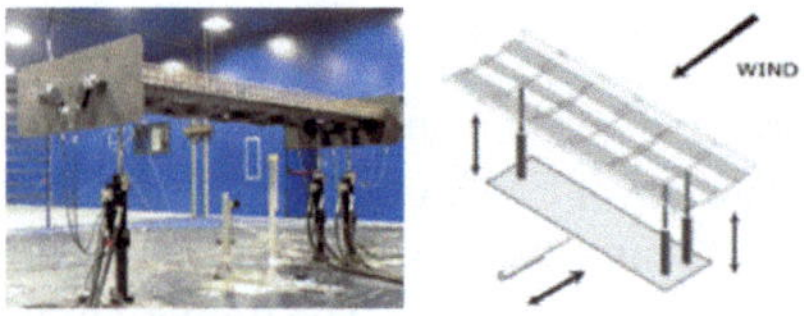

Number of DOFs	2
Frequency	0.025–4 Hz
Motion type	Sinusoidal

(c)

Number of DOFs	3
Lateral and vertical amplitudes	±100 mm
Torsional amplitude	±90°
Amplitude resolution	Continuous
Motion type	Stochastic

(d)

Fig. 1.25 Schematic of forced motion apparatuses: **a** FMA in Hunan University, China; **b** early FMA in Tongji University, China; **c** FMA in Milan Polytechnic University, Italy; **d** FMA in Norwegian University of Science and Technology, Norway

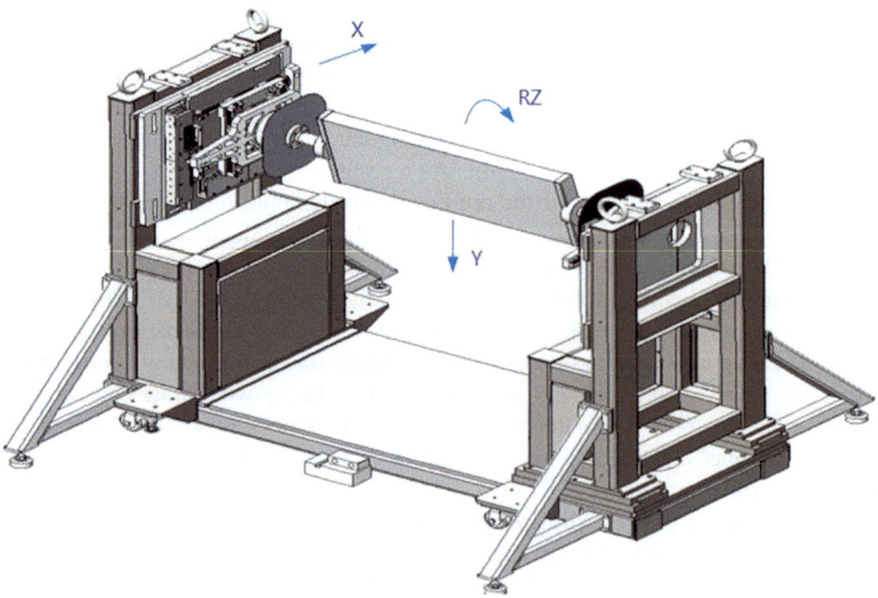

Fig. 1.26 Images of the novel FMA

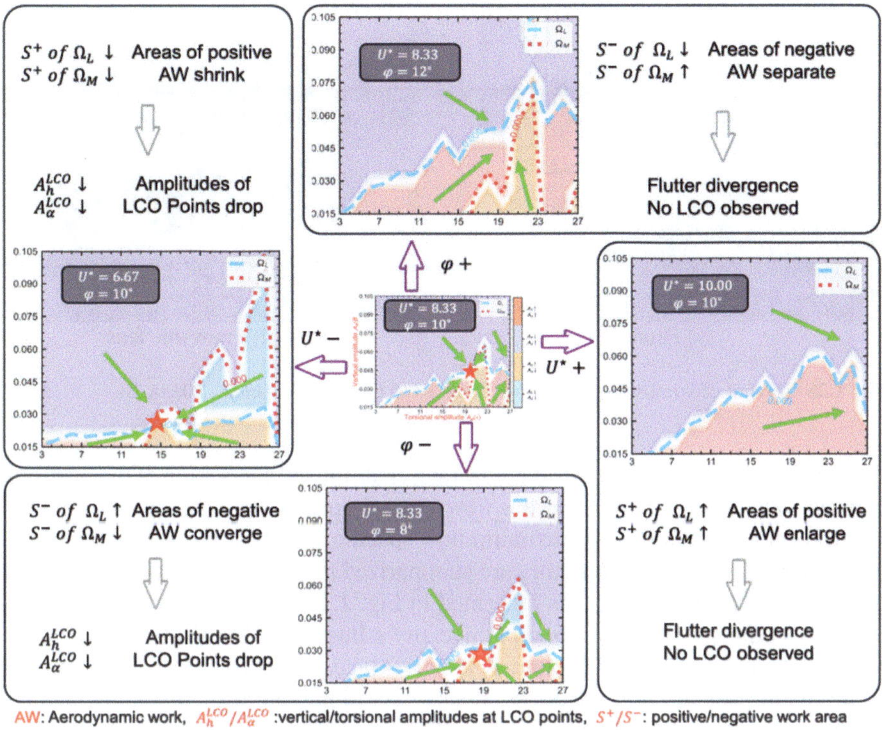

(b) Influences of the U^* and φ on the potential motion paths and LCO points in coupled flutter

Fig. 1.27 Schematic diagram of potential LCO points and motion development path under 4-D motion parameters (Zhao et al., 2025a, b)

of marine structures under strong nonlinear waves, storm conditions, and non-uniform currents; pollutant transport in coastal areas; bridge scouring under wind and wave actions; geological slope stability; and water pressure damage in underground tunnels. Furthermore, the system was equipped with lifting fans, which can simulate different water depths, thus, the sudden change in wind speed and direction generation becomes true. The diagram of the wind-wave multi-functional wind tunnel and movable fans are displayed in Fig. 1.28. Similar laboratories were built worldwide, for instance, FHWA HYDRAULIC LAB in America, Wind Tunnel and Water Flume in Harbin Institute of technology, and others in Zhejiang University and Dalian University of Technology, etc.

The multi-functional wind-wave tunnel can generate wind speeds ranging from 5 to 20 m/s with a low turbulence intensity of 2%. A push-plate wave maker is employed to generate both regular waves (e.g., sine and elliptic-cosine waves) and irregular waves based on various spectra, including the P-M, MPM, B, J spectra, as well as user-defined spectra. In experimental conditions, a maximum wave height of 1 m and a maximum water depth of 2.2 m can be achieved, with wave periods ranging from 0.5 to 5 s. The current flow generation system utilizes front and rear

(a) Flume (b) active movable fans

Fig. 1.28 Layout of wind-wave multi-functional wind tunnel (Zhao et al., 2025a, b)

flow-sharing devices to stabilize the flow rate, enabling the generation of bidirectional currents. The system can achieve a flow rate of up to 3 m^3/s at a water depth of 1.5 m, with a corresponding maximum flow speed of 0.5 m/s. The key performance parameters of this novel laboratory are summarized in Table 1.1. A comparison with other wind-wave laboratories is presented in Fig. 1.29, which clearly demonstrates the advanced capabilities of Tongji University's Integrated Multi-Functional Wind, Wave and Current Flume Laboratory. Its advantages include superior performance in simulating sudden wind changes, flexible positional adjustments, and enhanced specifications for tunnel length and wave height.

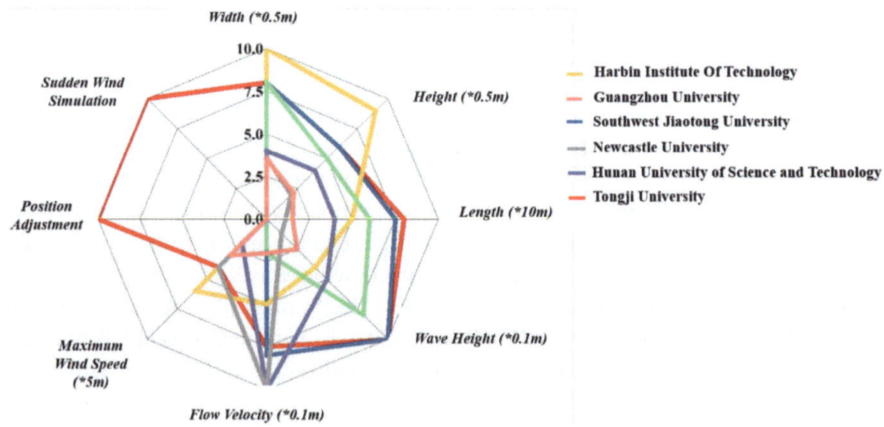

Fig. 1.29 Comparison of different wind-wave laboratories

1.5.2 Intelligent Control Technology

The complex terrain at engineering structure sites, especially for the long-span bridges, coupled with the strong local winds threatening the bridge. The traditional wind tunnel laboratory difficult to reproduce local wind characteristics at large-scale landform, accordingly, the intelligent control technology is applied to build new facilities or device in modern wind tunnels. The pioneer is the NHERI wind tunnel in University of Florida. The machine learning based method was adopted to achieve active control of the rough elements to build the different terrains, and the active control fans were adopted to generate target wind fields. Moreover, the automated measurement system was developed thus the wind velocity probes could be controlled by computer. More details and references of the NHERI wind tunnel can be found in the official web site: Natural Hazards Engineering Research Infrastructure (NHERI) Boundary Layer Wind Tunnel (Fig. 1.30).

Recently, an array-based, continuously adjustable three-dimensional (3D) dynamic boundary layer intelligent simulation device for complex terrain was developed at Tongji University; its layout is shown in Fig. 1.31a. This system is designed to simulate various terrain features at bridge sites, such as canyons of different shapes. It can also simulate wind-wave interactions when integrated with a wind tunnel system.

Through the integration of a remote-control platform, a custom operating system, and intelligent control elements, the simulation of random landforms has been achieved, as illustrated in Fig. 1.31. A prototype of this landform simulation device is currently operational at Tongji University. It comprises a 5×5 array of electric cylinders, over which a highly elastic skin is mounted. These cylinders are interconnected via data cables. Multi-axis synchronization and electronic cam linkage functions are implemented using a motion controller, enabling rapid switching between multiple terrain configurations.

Under low wind speeds, wind-induced bridge vibration may not directly compromise structural safety; however, it threatens driving safety and passenger comfort for road vehicles traveling on long-span bridges subjected to simultaneous wind and girder vibrations. This concern is particularly acute under extreme conditions, where sideslip and overturning accidents have been reported globally. To ensure driving safety, speed limits are commonly imposed. However, the wind speed thresholds for these restrictions often rely on empirical judgment, lacking a robust scientific basis. Consequently, standards can be overly conservative, leading to unnecessary traffic congestion and economic losses. Conversely, if the thresholds are too high, the combined effects of wind speed and vehicle speed can threaten driving safety. To address this challenge, a 6 DOF vibration platform integrated with a 7 DOF movable vehicle model was developed at Tongji University to physically reproduce wind-vehicle-bridge interactions. This system aims to simulate vehicle dynamics induced by wind and girder vibrations in a physical wind tunnel, moving beyond purely numerical simulations. The layout of the vehicle model and vibration platform is shown in Fig. 1.32.

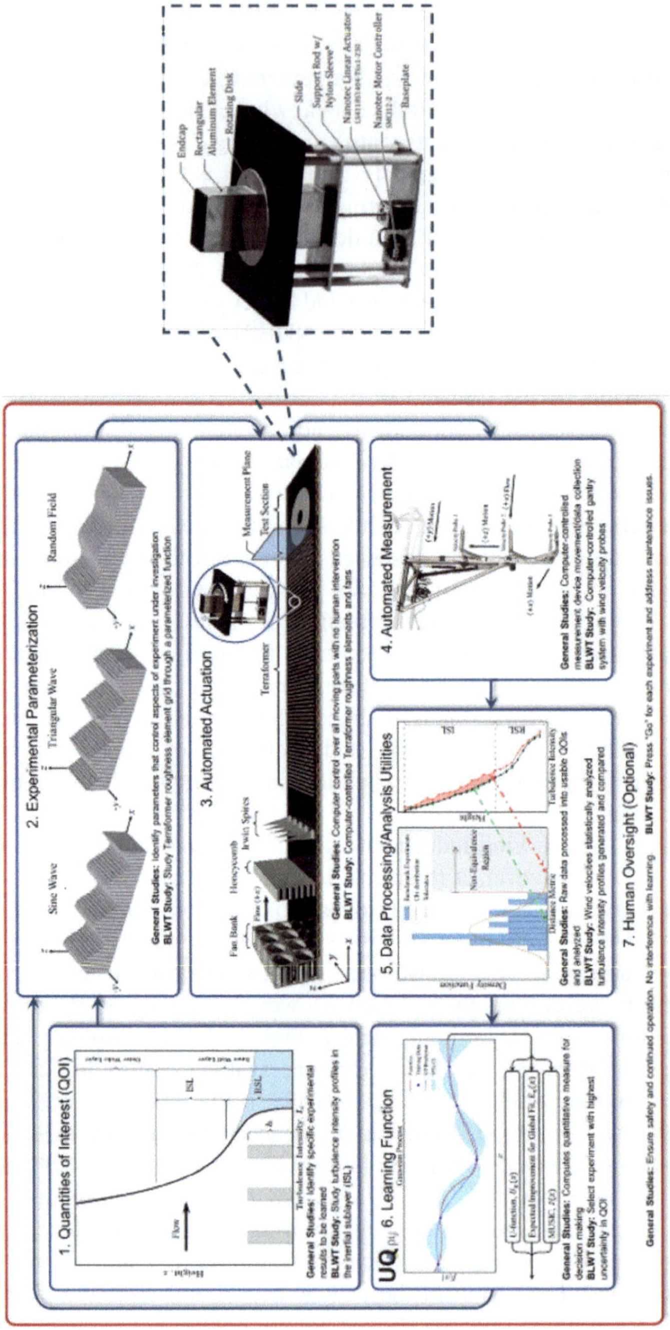

Fig. 1.30 NHERI wind tunnel in University of Florida

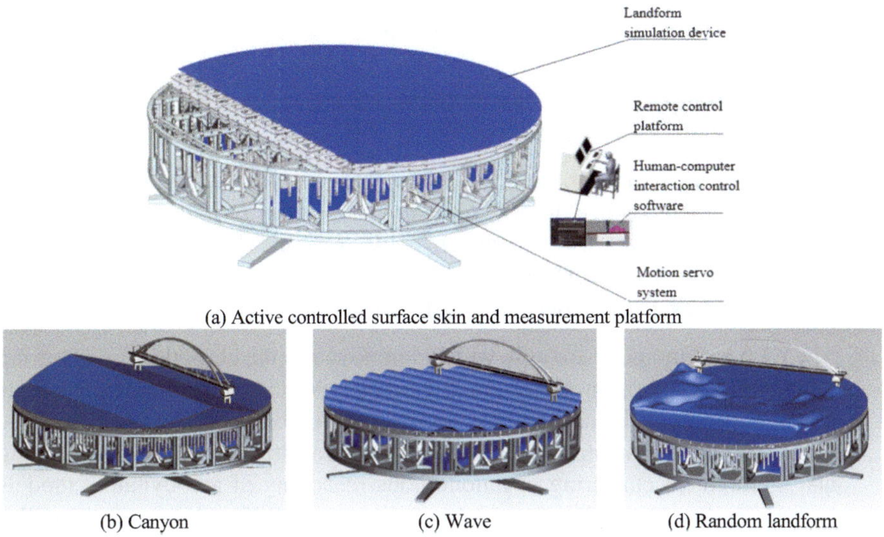

Fig. 1.31 Intelligent simulation device for 3-D dynamic boundary layer topography in TJ University

Fig. 1.32 Layout of multi-freedom vehicle model and vibration platform

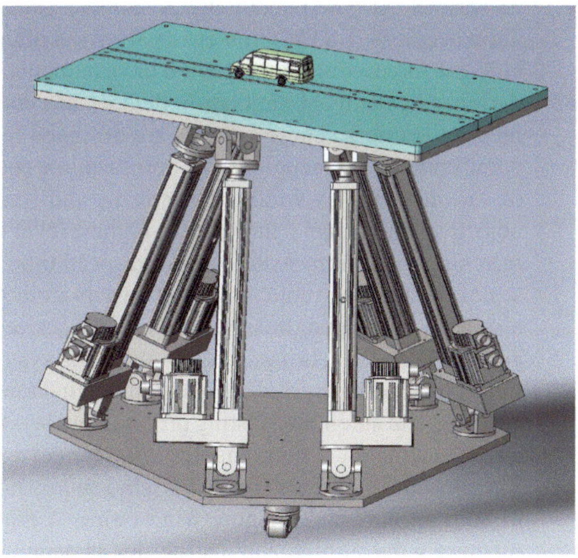

1.6 Concluding Remarks and Perspectives

1.6.1 Conclusions

Nowadays, AI and AI-based methods are no longer novel or unfamiliar topics. It is important to recognize that beyond any apprehensions, the vast potential of these methods across various research fields warrants significant focus. After reviewing the application of AI-based methods and algorithms in wind engineering and infrastructure over recent decades, the following key conclusions can be drawn:

(1) The AI-based methods are now widely employed in the classification of strong wind types based on meteorological datasets, especially for the ML methods that are successfully used in the recognition of different weather, such as monsoons, typhoons, and other storm types. Besides, the nonlinear mapping and hierarchical clustering of autonomous reasoning of aerodynamic models became reality, the reduced model complexity by 90% while retaining 99% accuracy, autonomously identified dominant aerodynamic modes and their governing formulas, and learned subtle aerodynamic and structural nonlinearities directly from vibration response data. Furthermore, the prediction of wind-induced structural responses has been advanced through the application of various AI-based methods. A notable example is the dual Long Short-Term Memory (LSTM) framework for wind effect prediction, which facilitates efficient forecasting of local wind environments.

(2) Advanced intelligent simulation devices are currently being developed based on deep learning. These systems are designed to perform real-time spatiotemporal continuous mapping and optimization control, enabling the simulation of various complex wind environments and static/dynamic boundaries. Their objective is to achieve the world's highest level of precision in local flow field matching (at a 5 mm scale) and wind speed time history matching (at a 100-ms scale). In addition, unmanned forced-vibration wind tunnel test systems have also been developed. In such systems, the active wind tunnel, forced-vibration apparatus, and synchronous data acquisition system are integrated. This integration enhances the level of automation in wind tunnel operations, realizes a standardized and automated testing process, and lays the foundation for establishing a reinforcement learning-enabled wind tunnel test environment.

(3) The substantial global demand for wind energy has accelerated the development of both onshore and offshore wind farms. A major technological challenge in offshore wind turbine application lies in accurately determining design loads under the coupled effects of wind, waves, and currents. While traditional analytical methods are effective for linear force analysis, AI-based methods offer a promising solution for handling complex nonlinear forces. Furthermore, both the spatial arrangement of turbines within a wind farm and the design parameters

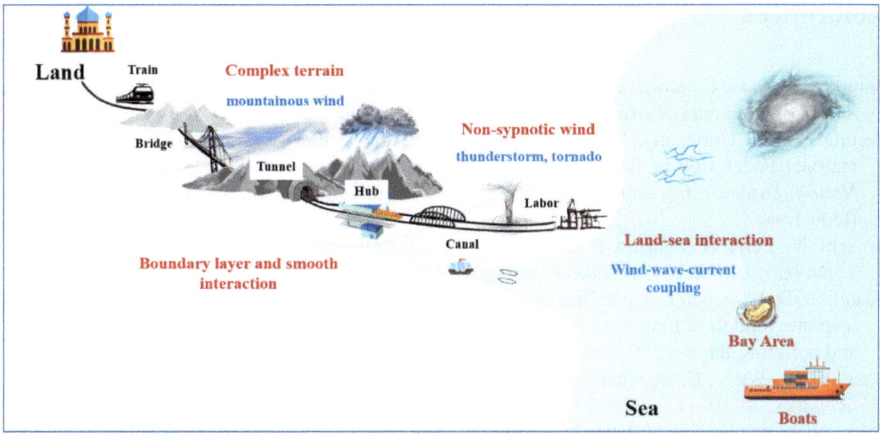

Fig. 1.33 Potential application fields of AI-based methods

of floating platforms significantly influence power production efficiency. Consequently, a growing body of research is dedicated to optimizing these layouts and parameters to maximize energy output.

1.6.2 Perspectives

The expansion of infrastructure from land to the deep ocean presents severe challenges. These arise not only from normal wind conditions but also from non-synoptic winds driven by intensive weather changes and land-sea interactions, as well as from coupled wind-wave-current interferences in open sea areas, as illustrated in Fig. 1.33. In this context, AI-based methods hold significant potential for diverse applications. These include reconstructing boundary layer characteristics; recognizing and reproducing non-synoptic wind fields; exploring multi-load interaction mechanisms; predicting the dynamic responses of infrastructure under combined loads; broadcasting safety status for road vehicle and train operability; and developing digital twin technologies to simulate the safety evolution of floating structures and vessels under wind-wave-current impacts in open sea areas.

Acknowledgements The authors gratefully acknowledge the support of the National Natural Science Foundation of China (52378527) and gratefully acknowledge the support of the Fundamental Research Funds for the Central Universities, CHD (FRFCU300102215111), and gratefully acknowledge the support of the Young Talent Fund of Association for Science and Technology in Shaanxi, China (20250457), and authors gratefully acknowledge the support of the International Association for Wind Engineering (IAWE) grant for the CISM scholarship. Furthermore, the authors thank Doctor Cui W., Doctor Liu P., Doctor Ding Y.J., Tan J.F., Wang Z.L. from the Tongji University, China, for their effort in the related research works.

References

Baniotopoulos, C. C., Borri, C., & Stathopoulos, T. (2011). *Environmental wind engineering and design of wind energy structures*. CISM, Springer Vienna

Beijing Normal University, National Disaster Reduction Center of China (NDRCC), and International Federation of Red Cross and Red Crescent Societies (IFRC). (2023). *2022 Global Natural Disaster Assessment Report*. PreventionWeb, United Nations Office for Disaster Risk Reduction.

Bruschi, N., Ferri, G., Marino, E., & Borri, C. (2020). Influence of clumps-weighted moorings on a spar buoy offshore wind turbine. *Energies, 13*(23), 6407.

Castellon, D. F., Fenerci, A., & Øiseth, O. (2021). A comparative study of wind-induced dynamic response models of long-span bridges using artificial neural networks, support vector regression and buffeting theory. *209*, Article 104484.

Chen, C. H., Wu, J. C., & Chen, J. H. (2008). Prediction of flutter derivatives by artificial neural networks. *96*(10–11), 1925–1937.

Chen, N. Y., & Ge, Y. J. (2019). Aerodynamic parameter identification of typical bridge sections based on artificial neural network. *China, 52*(8), 91–97.

Costa, C., & Borri, C. (2007). Full-scale identification of aeroelastic parameters for bridges. Pages 799–806 of: *Proceedings of the 12th International Conference on Wind Engineering,* Jul. 1–6, Cairns, Australia.

Cui, W., Ma, T., Zhao, L., & Ge Y. J. (2021). Data-based windstorm type identification algorithm and extreme. *Wind Speed Prediction Journal of Structural Engineering, 147*(5), 04021053.

Cui, W., Zhao, L., & Ge, Y. (2022). Non-Gaussian turbulence induced buffeting responses of long-span bridges based on state augmentation method. *Engineering Structures, 254*, Article 113774.

Fenerci, A., & Øiseth, O. (2018). Strong wind characteristics and dynamic response of a long-span suspension bridge during a storm. *172*, 116–138.

Ferri, G., Marino, E., & Borri, C. (2020a). Optimal dimensions of a semisubmersible floating platform for a 10 MW wind turbine. *Energies, 13*(12), 3092.

Ferri, G., Marino, E., & Borri, C. (2020b). Platform optimization of large floating offshore wind energy structures. In M. Papadrakakis, M. Fragiadakis, & C. Papadimitriou (Eds.), *EURODYN* (vol. 1, pp. 2175–2180). European Association for Structural Dynamics.

Ferri, G., Borri, C., & Marino, E. (2024). Substructure optimization of a 10MW floating wind turbine for installations in the Mediterranean Sea. *Journal of Physics: Conference Series, 2647*, Article 112011.

Fujino, Y., & Yoshida, Y. (2002). Wind-induced vibration and control of Trans-Tokyo Bay crossing bridge. *Journal of the Structural Engineering. American Society of Civil Engineers, 128*(8), 1012–1025.

Gao, G., Zhu, L., Li, J., Han, W., & Yao, B. (2020). A novel two-degree-of-freedom model of nonlinear self-excited force for coupled flutter instability of bridge decks. *Journal of Sound and Vibration, 480*, Article 115406.

Gu, M., & Quan, Y. (2004). Across-wind loads of typical tall buildings. *92*(13), 1147–1165.

Hou, J., Cao, S., Hu, H., Zhou, Z., Wan, C., Noori, M., Li, P., & Luo, Y. (2023). Vortex-induced vibration recognition for long-span bridges based on transfer component analysis. *Buildings, 13*(8), 2012.

Hu, G., Liu, L., Tao, D., Song, J., Tse, K. T., & Kwok, K. C. (2020). Deep learning-based investigation of wind pressures on tall building under interference effects. *201*, Article 104138.

Huang, Z., Xu, Y. L., & Zhan, S. (2021). Conditionally simulating nonstationary typhoon winds with time-varying coherences for long-span bridges. *212*, Article 104599.

Ishizaki, H. (1983). Wind profiles, turbulence intensities and gust factors for design in typhoon-prone regions. *13*(1–3), 55–66.

Jung, S., Ghaboussi, J., & Kwon, S. D. (2004). Estimation of aeroelastic parameters of bridge decks using neural networks. *Journal of Engineering Mechanics, 130*(11), 1356–1364.

Kareem, A. (1982). Acrosswind response of buildings. *Journal of the Structural Division, 108*(4), 869–887.

Kim, J., & Lee, C. (2020). Deep unsupervised learning of turbulence for inflow generation at various Reynolds numbers. *Journal of Computational Physics, 406*, Article 109216.

Ko, J. M., & Ni, Y. Q. (2005). Technology developments in structural health monitoring of large-scale bridges. *Engineering Structures, 27*(12), 1715–1725.

Kwok, K. C. S., & Melbourne, W. H. (1980). Cross-wind response of structures due to displacement dependent lock-in excitation. In *Wind Engineering*, pp. 699–708.

Laima, S., Feng, H., Li, H., Jin, Y., Han, F., & Xu, W. (2023). A Buffeting-Net for buffeting response prediction of full-scale bridges. *Engineering Structures, 275*, Article 115289.

Larsen, A. (1999, June). A 1999 view on aeroelastic stability-a driving force in bridge aerodynamics. In *Wind engineering into the 21st century: Proceedings of the 10th International Conference on Wind Engineering.*

Li, Q., & Y. X. (2000). Neural network method for aerodynamic characteristic analysis of bridge girder. *Central South Highway Engineering, 25*, 56–59

Li, L., Kareem, A., Xiao, Y., Song, L., & Zhou, C. (2015a). A comparative study of field measurements of the turbulence characteristics of typhoon and hurricane winds. *140*, 49–66.

Li, K., Ge, Y. J., Guo, Z. W., & Zhao, L. (2015b). Theoretical framework of feedback aerodynamic control of flutter oscillation for long-span suspension bridges by the twin-winglet system. *145*, 166–177.

Li, S., Laima, S., & Li, H. (2017). Cluster analysis of winds and wind-induced vibrations on a long-span bridge based on long-term field monitoring data. *Engineering Structures, 138*, 245–259.

Li, S., Li, S., Laima, S., & Li, H. (2021a). Data-driven modeling of bridge buffeting in the time domain using long short-term memory network based on structural health monitoring. *Structural Control & Health Monitoring, 28*(8), Article e2772.

Li, S. W., Laima, S. J., & Li, H. (2021b). Physics-guided deep learning framework for predictive modeling of bridge vortex-induced vibrations from field monitoring. *Physics of Fluids, 33*(3), 37113.

Liang, S. G., Wu, H. Y., & Chen, Z. Q. (2011). Investigation on model and responses of vortex-induced resonances of rectangular super high-rise buildings. *Journal of Vibration Engineering, 24*, 240–245.

Liao, H., Mei, H., Hu, G., Wu, B., & Wang, Q. (2021). Machine learning strategy for predicting flutter performance of streamlined box girders. *209*, Article 104493.

Lin, P. F., Ding F., Hu, G., Li, C., Xiao, Y.Q., Tse, K. T., Kwok, K. C. S., & Kareem, A. (2022). Machine learning-enabled estimation of crosswind load effect on tall buildings. *Journal of Wind Engineering and Industrial Aerodynamics, 220*, 104860. https://doi.org/10.1016/j.jweia.2021.104860

Lin, P., Hu, G., Li, C., Li, L., Xiao, Y., Tse, K. T., & Kwok, K. C. (2021). Machine learning-based prediction of crosswind vibrations of rectangular cylinders. *211*, Article 104549.

Liu, P., Zhao, L., Fang, G.S., & Ge Y.J. (2021) Explicit polynomial regression models of wind characteristics and structural effects on a long-span bridge utilizing onsite monitoring data. *Structural Control and Health Monitoring, 28*(5).

Liu, P., Chu, X., Cui, W., Zhao, L., & Ge, Y. (2022). Bayesian inference based parametric identification of vortex-excited force using on-site measured vibration data on a long-span bridge. *Engineering Structures, 266*, Article 114597.

Liu, X., Yu, J., Gong, L., Liu, M., & Xiang, X. (2024). A GCN-based adaptive generative adversarial network model for short-term wind speed scenario prediction. *Energy, 294*, Article 130931.

Liu, P. (2025). *Observation, statistics and AI-enhanced prediction of strong wind and typhoon field characteristics and structural effects on long-span bridges*. Shanghai, China.

Luo, X., & Kareem, A. (2020). Bayesian deep learning with hierarchical prior: Predictions from limited and noisy data. *Structural Safety, 84*, Article 101918.

Ma, T., Cui, W., Zhao, L., Ding, Y., Fang, G., & Ge, Y. (2022). Extreme wind speed prediction in mountainous area with mixed wind climates. *Stochastic Environmental Research and Risk Assessment, 37*(3), 1163–1181.

Ma, T., Chen, B., Cui, W., Chen, H., Song, S., & Zhao, L. (2025). Erection procedure optimization of a long-span suspension bridge considering flutter risk in mixed wind climates. *Journal of Tsinghua University, 65*(7), 1284–1290.

Ma, T., Cui W., Zhao, L., Ding, Y. J. Fang, G. S., Ge Y. J. Extreme wind speed prediction in mountainous area with mixed wind climates.

Marino, E., Lugni, C., Stabile, G., Borri, C., & Manuel, L. (2014). Coupled dynamic simulations of offshore wind turbines: Influence of wave modeling on the fatigue load assessment. In *Proceedings of the XIII Conference of the Italian Association for Wind Engineering.*

Marino, E., Giusti, A., & Manuel, L. (2017). Offshore wind turbine fatigue loads: The influence of alternative wave modeling for different turbulent and mean winds. *Renewable Engineering, 102*, 157–169.

Marino, E., Gkantou, M., Malekjafarian, A., Bali, S., et al. (2024). Offshore renewable energies: A review towards floating modular energy islands-monitoring, loads, modelling and control. *Ocean Engineering, 313*, 119251.

Matsumoto, M., Nakajima, N., Taniwaki, Y., & Shijo, R. (2001). Grating effect on flutter instability. *89*(14–15), 1487–1497.

Ohkuma, T. (1978). On the correlation between the shape of rectangular cylinders and characteristics of fluctuating lifts on them. In *Proceedings of the 5th Symposium on Wind Effects on Structures, Tokyo, Japan.*

Rocard, Y. (1965). Instabilité des ponts suspendus dans le vent—Experience sur modèle reduit. Pages 433–459 of: Wind effects on buildings and structures: Proceedings of the conference held at the National Physical Laboratory, Teddington, UK, Jun. 26–28, 1963. HMSO, London.

Scanlan, R. H. (1971). An examination of aerodynamic response theories and model testing relative to suspension bridges. In *Proceedings of the Third International Conference on Wind Effects on Buildings and Structures, Tokyo*, pp. 941–951.

Scruton, C. (1952). An experimental investigation of the aerodynamic stability of suspension bridges with special reference to the proposed Severn Bridge. *Proceedings of the Institution of Civil Engineers, 1*(2), 189–222.

Scruton, C. (1963). *Introductory review of wind effects on buildings and structures.* National Physical Laboratory.

Scruton, C. (1965). Introductory review of wind effects on buildings and structures. Pages 10–23 of: Wind effects on buildings and structures: Proceedings of the conference held at the National Physical Laboratory, Teddington, UK, Jun. 26–28, 1963. HMSO, London.

Solari, G. (1985). Mathematical model to predict 3-D wind loading on buildings. *Journal of Engineering Mechanics, 111*(2), 254–276.

Tamura, Y., Shimada, K., & Hibi, K. (1993). Wind response of a tower (Typhoon observation at the Nagasaki Huis Ten Bosch Domtoren). *50*, 309–318.

Tao, T., Xu, Y. L., Huang, Z., Zhan, S., & Wang, H. (2020). Buffeting analysis of long-span bridges under typhoon winds with time-varying spectra and coherences. *Journal of the Structural Engineering. American Society of Civil Engineers, 146*(12), 04020255.

Venanzi, I., & Materazzi, A. L. (2012). Acrosswind aeroelastic response of square tall buildings: A semi-analytical approach based of wind tunnel tests on rigid models. *Wind and Structures, 15*(6), 495–508.

Wang, H., Li, A., Guo, T., & Xie, J. (2009). Field measurement on wind characteristic and buffeting response of the Runyang Suspension Bridge during typhoon Matsa. *Science China E, 52*(5), 1354–1362.

Wang, H., Zhang, Y. M., Mao, J. X., & Wan, H. P. (2020). A probabilistic approach for short-term prediction of wind gust speed using ensemble learning. *202*, Article 104198.

Wang, Z., Zhao, L., Chen, H., Fang, G., Li, K., & Ge, Y. (2023). Flutter control of active aerodynamic flaps mounted on streamlined bridge deck fairing edges: An experimental study. *Structural Control & Health Monitoring, 2023*(1), 9970603.

Wang, Z., Zhao, L., Fu, Y., Fang, G., Cui, W., Li, K., & Ge, Y. (2024). Flutter control optimization for a 5000 m suspension bridge with active aerodynamic flaps: A CFD-enabled strategy. *Engineering Structures, 303*, Article 117457.

Wang, Z., Li, K., Wei, X., & Zhao, L. (2025). Optimization and mechanism of active aerodynamic control of combined dek flap system-PK box girder flutter control. *China Journal of Highway and Transport, 38*(8), 274–283.

Wilms, H., Cupelli, M., Monti, A., & Gross, T. (2019, March). Exploiting spatio-temporal dependencies for RNN-based wind power forecasts. In *2019 IEEE PES GTD grand international conference and Exposition Asia (GTD Asia)*, pp. 921–926.

Wold, J. W., Stadtmann, F., Rasheed, A., Tabib, M., San, O., & Horn, J. T. (2024). Enhancing wind field resolution in complex terrain through a knowledge-driven machine learning approach. *Engineering Applications of Artificial Intelligence, 137*, Article 109167.

Wu, B., Wang, Q., Liao, H., & Mei, H. (2020). Hysteresis response of nonlinear flutter of a truss girder: Experimental investigations and theoretical predictions. *Computers & Structures, 238*, Article 106267.

Wu, F., Wang, Z., Zhao, L., Pan, T., Xiao, H., & Ge, Y. (2023). Aerodynamic force distribution characteristics around a double slotted box girder of a long-span bridge during vortex-induced vibration. *Journal of Bridge Engineering, 28*(1), 04022132.

Wu, T., & Kareem, A. (2011). Modeling hysteretic nonlinear behavior of bridge aerodynamics via cellular automata nested neural network. *99*(4), 378–388.

Xie, Y., Li, C., Tang, G., & Liu, F. (2021). A novel deep interval prediction model with adaptive interval construction strategy and automatic hyperparameter tuning for wind speed forecasting. *Energy, 216*, Article 119179.

Yi, G., Pan, J., Zhao, L., Song, L., Fang, G., Cui, W., & Ge, Y. (2022). Profiles of mean wind and turbulence intensity during strong typhoon landfall. *228*, Article 105106.

Zhang, Y. M., Wang, H., Mao, J. X., Xu, Z. D., & Zhang, Y. F. (2021). Probabilistic framework with bayesian optimization for predicting typhoon-induced dynamic responses of a long-span bridge. *Journal of the Structural Engineering. American Society of Civil Engineers, 147*(1), 04020297.

Zhao, L. (2018). *The Wind Runner: Zhui feng che—ba shiyan shi kaizhe zou* [The Wind Runner: Wind chasing car-Driving the laboratory on the road]. WeChat. https://mp.weixin.qq.com/s/Pfn5ICkH2bJJWwcEH7LkWw

Zhao, L., Cui, W., & Ge, Y. (2019). Measurement, modeling and simulation of wind turbulence in typhoon outer region. *., 195*, Article 104021.

Zhao, L., Xie, X., Zhan, Y. Y., Cui, W., Ge, Y. J., Xia, Z. C., Xu, S. Q., & Zeng, M. (2020). A novel forced motion apparatus with potential applications in structural engineering. *Journal of Zhejiang University Science A, 21*(7), 593–608

Zhao, L., Cui, W., Shen, X., Xu, S., Ding, Y., & Ge, Y. (2022). A fast on-site measure-analyze-suppress response to control vortex-induced-vibration of a long-span bridge. *Structures, 35*, 192–201.

Zhao, L., Wang, Z., Chen, W., & Cui, W. (2024). Buffeting performance of long-span bridges with different span affected by parametric typhoon wind. *254*, Article 105903.

Zhao, L., Ding, Y., Cui, W., Fang, G., & Huang, M. (2025a). Nonstationary characteristics of short-rise-time gusts in a high-altitude deep-cut canyon. *Journal of the Structural Engineering. American Society of Civil Engineers, 151*(7), 04025091.

Zhao, L., Mei, Z., Cui, W., Hu, C., & Li, Y. (2025b). Nonlinear aerodynamic effects of a 5: 1 rectangular section in bending–torsional coupled flutter. *Journal of Fluids and Structures, 138*, Article 104387.

Zhou, Y., Kijewski, T., & Kareem, A. (2003). Aerodynamic loads on tall buildings: Interactive database. *Journal of the Structural Engineering. American Society of Civil Engineers, 129*(3), 394–404.

Chapter 2
Integrating Machine Learning with Computational Fluid Dynamics for Wind Energy Applications: From Turbulence Fundamentals to Intelligent Simulation

Hassan Hemida

Abstract This Chapter provides an overview of the integration between Machine Learning and Computational Fluid Dynamics in the analysis and optimization of wind energy systems, putting together fundamental fluid dynamics and emerging data-driven intelligence. It begins by introducing the physics and statistical characterisation of turbulence, the inherently chaotic, three-dimensional flow phenomenon that governs most wind and energy-related aerodynamics. By using such statistical parameters as mean velocity, variance, and turbulence kinetic energy, this Section outlines how turbulent motion and energy cascades, as described by Kolmogorov's theory, form the foundation for realistic CFD simulations. The Chapter then presents the governing equations of fluid dynamics and their turbulence modelling approaches-Direct Numerical Simulation, Reynolds-Averaged Navier–Stokes, and Large-Eddy Simulation-emphasising their application within offshore and wind turbine aerodynamics. It discusses how these techniques, along with hybrid models such as the Detached-Eddy Simulation, manage to capture the turbulence multi-scale structure, striking a balance between solution accuracy and computational cost. In the next Sections, attention is given to modeling the Atmospheric Boundary Layer inflow and the impact of inlet boundary conditions on the CFD predictions. Finally, how Artificial Intelligence, especially ML, Deep Learning, and Reinforcement Learning help CFD accelerate simulations, improve fidelity in predictions, and perform adaptive control is discussed. This Chapter draws on physics-based modeling combined with data-driven intelligence to create a framework for the next generation of AI-enhanced CFD tools that will revolutionise wind energy design, optimization, and real-time operational efficiency.

H. Hemida (✉)
School of Engineering, University of Birmingham, Birmingham, UK
e-mail: h.hemida@bham.ac.uk

C. Baniotopoulos and E. Marino (eds.), *Advances in Wind Energy in the Era of Artificial Intelligence*, CISM International Centre for Mechanical Sciences 625,
https://doi.org/10.1007/978-3-032-28815-8_2

2.1 Introduction

Wind energy has become a central pillar of global decarbonisation strategies, with modern wind turbines growing rapidly in rotor diameter, hub height, and rated power to improve energy yield and reduce levelised cost of electricity. Yet, the aerodynamic environment in which these machines operate remains inherently complex. A wind turbine rotor extracts momentum from a turbulent atmospheric boundary layer (ABL), while simultaneously generating strong shear layers, tip and root vortices, unsteady separation, and a highly three-dimensional wake that interacts with the nacelle, tower, terrain, and neighbouring turbines in wind farms. These physical processes occur over a wide range of spatial and temporal scales—from near-wall viscous layers on blade surfaces to wake meandering on the scale of multiple rotor diameters—making wind turbine aerodynamics an archetypal multiscale problem. Capturing the governing physics with sufficient fidelity is essential not only for accurate power prediction but also for assessing thrust, torque, fatigue loading, aeroelastic response, noise generation, and wake impacts that govern wind farm performance. Computational Fluid Dynamics (CFD) has therefore become an indispensable tool in wind energy research and design. Unlike purely empirical or low-order aerodynamic methods, CFD is built on the fundamental conservation laws of mass and momentum expressed through the Navier–Stokes equations. In principle, these governing equations contain the complete physics of incompressible and compressible flows, and thus provide a unified framework to model blade boundary layers, stall, wake development, and turbine-turbine interaction. In practice, however, the numerical solution of the Navier–Stokes equations at wind turbine Reynolds numbers is extremely demanding, primarily because of turbulence. Turbulent flows feature apparently random, three-dimensional vorticity and strongly unsteady motion; their instantaneous structure is too intricate to interpret without statistical tools. For this reason, turbulence is commonly described through averages and fluctuations, where the mean flow represents the organised component and the fluctuations represent the turbulent motions that drive mixing, transport, and unsteady loading. This chapter begins by establishing these foundations. It introduces the statistical decomposition of turbulent flow variables into mean and fluctuating components, defines key measures such as variance, standard deviation (r.m.s.), and turbulence kinetic energy, and links these quantities to turbulence intensity and the interpretation of fluctuating momentum transport. These definitions are not merely mathematical; they are directly connected to wind turbine aerodynamics, where atmospheric turbulence and shear determine inflow variability, and where Reynolds stresses influence wake recovery and mixing. The chapter also introduces the Reynolds number and the notion of turbulence scales, highlighting the cascade from large energy-containing eddies to small dissipative scales governed by viscosity and dissipation rate. This hierarchy explains why resolving all scales directly is rarely feasible for wind turbine flows, and it motivates the modelling strategies that underpin most practical CFD simulations. In wind energy applications, three main CFD approaches are widely discussed: Direct Numerical Simulation (DNS), Reynolds-Averaged Navier–Stokes

(RANS), and Large-Eddy Simulation (LES), along with hybrid approaches such as Detached-Eddy Simulation (DES). DNS resolves all turbulent scales down to Kolmogorov length and time scales; while it provides the most complete representation of turbulence, its computational cost scales prohibitively with Reynolds number, restricting DNS largely to canonical or fundamental studies. RANS, by contrast, time-averages the governing equations and models the effect of turbulence entirely through closure assumptions such as eddy viscosity; RANS is computationally efficient and widely used in industrial contexts, but it can struggle in flows dominated by separation, strong curvature, rotation, unsteadiness, and complex wake dynamics—features frequently encountered in wind turbine aerodynamics. LES offers a compromise by resolving the large, energy-containing eddies while modelling only the smaller subgrid-scale motions; it can capture unsteady separation and wake structures more faithfully than RANS but remains expensive, particularly if near-wall regions are treated with high resolution. DES and related hybrid methods bridge these regimes by applying RANS in near-wall regions while switching to LES-like behaviour in separated regions and wakes, offering improved fidelity over RANS at a lower cost than fully wall-resolved LES. Although CFD has enabled substantial advances in understanding and predicting wind turbine aerodynamics, its use at the scales required by modern wind farms—multiple turbines, complex terrain, transient atmospheric conditions, and repeated evaluations for design optimisation or control—still faces major barriers. High-fidelity simulations demand significant computational resources and specialist expertise in mesh generation, numerical stability, turbulence modelling, and validation. Moreover, uncertainties in inflow conditions, surface roughness, operational control settings, and atmospheric stability can substantially affect simulation outcomes. These challenges become even more pronounced when CFD is used not only for single-turbine aerodynamics but also for wind farm layout design, wake steering, load mitigation, and real-time or near-real-time operational decision-making. It is within this context that Artificial Intelligence (AI) and Machine Learning (ML) are emerging as transformative complementary technologies. The rapid growth of data availability—from high-fidelity CFD databases, wind tunnel experiments, field measurements, SCADA systems, LiDAR, and remote sensing—has created an opportunity to develop data-driven models that learn complex flow-load relationships directly from examples. At the same time, advances in learning algorithms, computing hardware, and scientific machine learning have enabled the development of surrogate models that can emulate expensive CFD simulations orders of magnitude faster. In wind energy, such models are attractive for tasks that require repeated evaluations: design-space exploration of blade geometries, rapid assessment of aerodynamic performance across operating conditions, wake prediction and farm optimisation, and control-oriented modelling for yaw and pitch strategies. This chapter is written with a practical goal: to explain, in a coherent flow, how fundamental fluid dynamics and turbulence modelling concepts connect to modern AI/ML opportunities in wind turbine CFD—without losing sight of physical interpretation, modelling assumptions, and validation requirements. A key theme is that ML should not be viewed as a replacement for physics-based modelling, but as a tool that can (i) accelerate computation via surrogates, (ii) improve model accuracy via data-informed corrections, (iii) automate or assist labour-intensive steps such as mesh generation or parameter selection, and (iv) enable decision-making frameworks for control and optimisation. These opportunities, however, come with risks: data bias, lack of generalisation beyond the training domain, limited interpretability of "black-box" models, and the possibility of violating physical constraints such as conservation laws or correct scaling behaviour. For wind energy, where design and operation are load- and safety-critical, these issues are not academic—they determine whether AI-enhanced CFD tools can be trusted by researchers, industry, and certification pathways. Accordingly, the chapter builds from first principles towards applications. It first establishes the governing equations and the statistical language of turbulence, then outlines the principal CFD strategies (DNS, RANS, LES, DES) and the closure problem that motivates turbulence models. It then introduces the aerodynamic forces and moments acting on wind turbine blades and rotors—lift, drag, thrust, torque, and unsteady load components—linking these directly to flow physics such as dynamic stall, rotational effects (centrifugal and Coriolis influences), yaw misalignment, tip vortices, and wake development. With that aerodynamic foundation in place, the chapter transitions to AI in wind energy systems, summarising key ML paradigms (supervised, unsupervised, reinforcement learning, and deep learning) and their relevance to CFD workflows. Particular attention is given to surrogate modelling as a "digital twin" concept: generating training data from a limited set of high-fidelity simulations, fitting learning-based models to capture input–output relationships, and deploying these surrogates for rapid predictions. Finally, the chapter highlights current research gaps and practical challenges that must be addressed for robust adoption: data quality and labelling, bridging scale gaps between mesoscale atmospheric models and turbine-resolved CFD, managing the cost of generating training datasets, ensuring interpretability and trustworthy validation, modelling turbulence and wake interactions under realistic atmospheric conditions, integrating AI tools into real-time control and industrial workflows, and propagating uncertainty through hybrid physics-data pipelines. By framing these challenges alongside the opportunities, the chapter aims to equip readers—particularly PhD students and advanced researchers—with both the conceptual tools and the critical perspective needed to develop and evaluate AI-enabled CFD methodologies for wind turbine aerodynamics. In summary, the chapter provides a structured pathway from the governing physics of turbulent flow to modern AI/ML techniques that can enhance CFD for wind energy applications. The emphasis throughout is on connecting models to physical meaning, understanding the limitations of each modelling choice, and identifying where data-driven methods can add genuine value—whether by improving fidelity, reducing computational cost, or enabling new capabilities in design, optimisation, and control.

2.2 Introduction to Turbulent Flow and Its Statistics

Turbulence is a state of fluid motion characterised by irregular, apparently random, and strongly three-dimensional fluctuations in velocity and vorticity. Unlike laminar flow, where fluid particles follow smooth and orderly trajectories, turbulent flow exhibits rapid variations in both space and time, accompanied by intense mixing and momentum transfer. Despite this apparent randomness, turbulence is not governed by stochastic equations; rather, it arises as a complex solution of the deterministic Navier–Stokes equations at sufficiently high Reynolds numbers. The challenge in turbulence lies not in the governing equations themselves, but in the extreme sensitivity of their solutions to initial and boundary conditions. This sensitivity leads to chaotic behaviour, making the instantaneous flow field exceedingly complex and practically unpredictable in detail. As a result, direct interpretation of instantaneous turbulent motion is rarely useful for engineering analysis. For this reason, the study of turbulence relies heavily on statistical and stochastic descriptions. By decomposing flow variables into mean and fluctuating components and analysing their statistical properties, it becomes possible to extract physically meaningful and reproducible information from an otherwise disordered flow. Quantities such as mean velocity, variance, turbulence intensity, Reynolds stresses, and energy spectra provide insight into the structure, transport mechanisms, and energetic content of turbulent motion. This statistical framework forms the foundation for turbulence modelling approaches used in CFD, including Reynolds-averaged, large-eddy, and hybrid simulation methods.

2.2.1 *The Mean (or Ensemble) Average*

Because turbulent flows exhibit strong, irregular fluctuations, meaningful analysis requires the use of averaging procedures that extract the organised, repeatable part of the motion. One of the most fundamental concepts in turbulence theory is the mean, also referred to as the ensemble average or true mean. This average represents the expected value of a flow quantity when all possible realisations of the turbulent motion are considered. The simplest form of averaging is the arithmetic mean. Given a finite set of N measurements x_n, the arithmetic mean is defined as:

$$X_N = \frac{1}{N}\sum_{n=1}^{N} x_n. \tag{2.1}$$

This definition is commonly used in experimental data analysis or numerical post-processing when only a limited number of samples is available. In turbulence theory, however, the concept of the mean is defined more rigorously through the ensemble average. An ensemble consists of a large number of statistically independent realisations of the same flow, each obtained under identical macroscopic conditions (such as geometry, boundary conditions, and inflow properties). The ensemble average of a quantity x is defined as

$$\bar{x} = \lim_{N\to\infty} \frac{1}{N} \sum_{n=1}^{N} x^{(n)}. \tag{2.2}$$

where x_n denotes the value of x in the n-th realisation of the ensemble. In the limit of an infinite number of realisations, this average converges to the true mean of the turbulent process. Applying this concept to a velocity field, the mean velocity at a spatial location x and time t is defined as:

$$\overline{u_i(\mathbf{x}, t)} = \lim_{N\to\infty} \frac{1}{N} \sum_{n=1}^{N} u_i^{(n)}(\mathbf{x}, t). \tag{2.3}$$

where $u_i^((n))(x, t)$ is the i-th component of the instantaneous velocity in the n-th realisation. The mean velocity represents the deterministic component of the flow, while deviations from this mean correspond to turbulent fluctuations. In practice, obtaining a true ensemble average is rarely possible, as it would require an infinite number of independent experiments or simulations. Under the assumption of statistical stationarity and ergodicity, the ensemble average can be replaced by a time average, which is more accessible experimentally and numerically. This equivalence between ensemble and time averaging underpins most practical turbulence analyses and forms the basis for Reynolds averaging and subsequent turbulence modelling approaches.

2.2.2 Fluctuations About the Mean

A central concept in turbulence analysis is the decomposition of any instantaneous flow quantity into a mean component and a fluctuating component. For the velocity field, this decomposition is written as

$$u(\mathbf{x}, t) = \bar{u}(\mathbf{x}) + u'(\mathbf{x}, t), \tag{2.4}$$

or equivalently,

$$u' = u - \bar{u} \tag{2.5}$$

Here, $u(x, t)$ is the instantaneous velocity, which varies in both space and time; $\bar{u}$ denotes the mean (or time-averaged) velocity; and u' represents the turbulent fluctuation about the mean. By definition, the mean velocity is obtained by averaging the instantaneous velocity over a sufficiently long time interval Δt,

$$\bar{u} = \frac{1}{\Delta t} \int_{0}^{\Delta t} u(\mathbf{x}, t)\, dt. \tag{2.6}$$

provided that the flow is statistically stationary. The fluctuating component u' captures the irregular, rapidly varying part of the motion associated with turbulence. A key property of this decomposition is that the average of the fluctuating component is zero,

$$\overline{u'} = 0. \tag{2.7}$$

This follows directly from the definition of the mean and ensures that all persistent or organised motion is contained in $\bar{u}$, while u' represents purely fluctuating behaviour. In the ideal case of an infinite number of statistically independent realisations, the ensemble average and the time average are equivalent. In practice, this equivalence is assumed when the turbulent flow is ergodic, allowing time averaging to be used in place of ensemble averaging in both experimental measurements and numerical simulations. This decomposition into mean and fluctuations forms the basis of Reynolds averaging and underpins the development of turbulence models in CFD.

2.2.3 *Variance and Standard Deviation*

While the mean of the velocity fluctuations is zero, the mean square of the fluctuations is not. This quantity is of fundamental importance in turbulence and is known as the *variance*. For a velocity component u, the variance is denoted by $\mathrm{var}[u]$ or $\overline{(u')^2}$ and is defined as

$$\mathrm{var}[u] = \overline{(u')^2} = \lim_{N \to \infty} \frac{1}{N} \sum_{n=1}^{N} (u - \bar{u})^2 . \tag{2.8}$$

In practice, assuming statistical stationarity, the variance is commonly evaluated using time averaging:

$$\mathrm{var}[u] = \overline{(u')^2} = \frac{1}{\Delta t} \int_{0}^{\Delta t} (u')^2\, \mathrm{d}t. \tag{2.9}$$

It should be noted that, like the ensemble average itself, the exact variance cannot be measured directly, as this would require an infinite number of realisations. Nevertheless, sufficiently long sampling times yield accurate statistical estimates in most practical applications.

The variance can also be written in an equivalent and often useful form:

$$\mathrm{var}[u] = \overline{u^2} - \bar{u}^2. \tag{2.10}$$

This expression shows that the variance is the second moment of the variable minus the square of the first moment (mean). For this reason, the variance is also referred to as the *second central moment* of the random variable.

Closely related to the variance is the standard deviation, or root-mean-square (r.m.s.) value, denoted by σ_u. It is defined as

$$\sigma_u = \sqrt{\mathrm{var}[u]} = \sqrt{\overline{(u')^2}} = \left(\frac{1}{\Delta t} \int_0^{\Delta t} (u')^2 \, \mathrm{d}t \right)^{1/2} . \tag{2.11}$$

The r.m.s. values of the velocity components are of particular importance in turbulent flows, as they provide a direct measure of the average magnitude of velocity fluctuations and are often the most accessible quantities in both experiments and numerical simulations.

2.2.4 *Turbulence Kinetic Energy and Turbulence Intensity*

In a three-dimensional turbulent flow, the instantaneous velocity field can be decomposed into three fluctuating components, u_1', u_2', and u_3', each associated with a corresponding variance $\overline{(u_1')^2}$, $\overline{(u_2')^2}$, and $\overline{(u_3')^2}$. One-half of the sum of these variances has a clear physical interpretation as the mean kinetic energy per unit mass contained in the turbulent fluctuations. This quantity is known as the *turbulence kinetic energy*, k, and is defined as

$$k = \frac{1}{2} \left(\overline{(u_1')^2} + \overline{(u_2')^2} + \overline{(u_3')^2} \right) . \tag{2.12}$$

Turbulence kinetic energy is a central variable in many turbulence models and provides a scalar measure of the intensity of turbulent motion at a given location. It represents the amount of energy available in the fluctuating velocity field to drive mixing, momentum transport, and unsteady aerodynamic loading.

A commonly used non-dimensional measure of turbulence is the *turbulence intensity*, T_i, defined as the ratio of a characteristic root-mean-square (r.m.s.) velocity fluctuation to a reference mean velocity U_{ref}. Assuming isotropic turbulence, the r.m.s. fluctuation can be related to the turbulence kinetic energy, leading to

$$T_i = \frac{\sqrt{2k/3}}{U_{\mathrm{ref}}} . \tag{2.13}$$

Turbulence intensity is widely used to characterise inflow conditions in wind engineering and wind turbine simulations, as it strongly influences aerodynamic loading, wake development, and fatigue behaviour.

2.2.5 Moments of Different Fluctuating Variables

The variance represents the second moment of a single fluctuating quantity. However, important information about turbulence structure and momentum transport is contained in second moments formed from pairs of different fluctuating variables. Consider two generic flow quantities,

$$\phi = \Phi + \phi', \qquad \psi = \Psi + \psi', \tag{2.14}$$

with

$$\overline{\phi'} = \overline{\psi'} = 0. \tag{2.15}$$

Their second moment, or covariance, is defined as

$$\overline{\phi'\psi'} = \frac{1}{\Delta t} \int_0^{\Delta t} \phi'(t)\, \psi'(t)\, \mathrm{d}t. \tag{2.16}$$

A particularly important example in turbulence is the covariance between velocity fluctuations in different directions, such as

$$\overline{u_1'u_2'} = \frac{1}{\Delta t} \int_0^{\Delta t} u_1'(t)\, u_2'(t)\, \mathrm{d}t. \tag{2.17}$$

These quantities represent the *Reynolds stresses*, which play a fundamental role in turbulent momentum transport and appear explicitly in the Reynolds-averaged Navier–Stokes equations.

If the velocity fluctuations in different directions were statistically independent, the cross moments $\overline{u_1'u_2'}$, $\overline{u_1'u_3'}$, and $\overline{u_2'u_3'}$ would be equal to zero. In real turbulent flows, however, these terms are generally non-zero, reflecting the anisotropic and correlated nature of turbulent motion.

2.2.6 Reynolds Number and the Nature of Turbulent Flow

In laminar flow, fluid particles follow smooth and orderly trajectories arranged in parallel layers, with no significant mixing between them. In turbulent flow, by contrast, velocity and pressure exhibit strong fluctuations in both space and time, leading to enhanced mixing and momentum transport.

The transition from laminar to turbulent flow depends on several factors, including the characteristic flow velocity U, a representative length scale l associated with the geometry or flow feature (such as boundary-layer thickness or wake width), and the kinematic viscosity of the fluid ν. These effects are combined into a single dimensionless parameter, the Reynolds number, defined as

$$\mathrm{Re} = \frac{Ul}{\nu}. \tag{2.18}$$

Although the precise Reynolds number at which transition occurs depends on the flow configuration and external disturbances, turbulent flow is generally associated with high Reynolds numbers. Turbulent motion is characterised by the following key features:

- Irregular, chaotic, and seemingly random behaviour,
- Strong three-dimensionality,
- A wide range of interacting length and time scales,
- Enhanced diffusion and mixing,
- Dissipation of kinetic energy through viscous effects.

It is important to emphasise that turbulence is a property of the flow, not of the fluid itself, and that it is a continuum phenomenon. Both laminar and turbulent flows are governed by the same conservation laws, but turbulence exhibits far richer dynamics due to the nonlinear interactions inherent in the Navier–Stokes equations.

2.2.7 *Turbulence Scales*

Turbulent flows are characterised by a wide range of interacting spatial and temporal scales, extending from the largest energy-containing motions down to the smallest dissipative structures. The largest scales are typically of the order of the characteristic size of the flow geometry, such as the diameter of a bluff body or the chord and span of a wind turbine blade. These large eddies extract their energy directly from the mean flow through shear and instability mechanisms and are strongly influenced by boundary conditions, geometry, and inflow characteristics.

As turbulence evolves, the large eddies become unstable and break down into progressively smaller eddies, transferring kinetic energy through a hierarchy of scales. This process, known as the *energy cascade*, continues until the eddies are sufficiently small that viscous effects dominate. At these smallest scales, turbulent kinetic energy is no longer transferred to smaller motions but is instead converted into internal energy (heat) through viscous dissipation. The rate at which this conversion occurs is quantified by the dissipation rate, denoted by ε. Figure 2.1 provides a schematic representation of the different turbulence scales in the wake of a wind turbine.

At sufficiently high Reynolds numbers, Kolmogorov's hypotheses state that the statistical properties of the smallest turbulent scales are universal and independent of

Fig. 2.1 Turbulence length scales around wind turbines

the large-scale flow geometry. These small-scale motions are governed solely by the kinematic viscosity of the fluid, ν, and the dissipation rate, ε. Dimensional analysis then yields the characteristic Kolmogorov scales, which define the smallest length, velocity, and time scales present in the turbulent flow:

$$\eta = \left(\frac{\nu^3}{\varepsilon}\right)^{1/4}, \qquad u_\eta = (\nu\varepsilon)^{1/4}, \qquad \tau_\eta = \left(\frac{\nu}{\varepsilon}\right)^{1/2}. \tag{2.19}$$

Here, η is the Kolmogorov length scale, u_η is the Kolmogorov velocity scale, and τ_η is the Kolmogorov time scale.

The relationship between the largest and smallest turbulent scales can be expressed in terms of the integral length scale, l, and the Reynolds number based on that scale, Re_l. The integral length scale represents the size of the largest energy-containing eddies and can be related to the mean flow velocity U and the dissipation rate as

$$l \sim \frac{U^3}{\varepsilon}. \tag{2.20}$$

The corresponding Reynolds number is defined as

$$\mathrm{Re}_l = \frac{Ul}{\nu}. \tag{2.21}$$

Under these conditions, the Kolmogorov length scale may be written as

$$\eta = l\,\mathrm{Re}_l^{-3/4}, \tag{2.22}$$

illustrating the rapid growth in scale separation as the Reynolds number increases.

These turbulence scales have direct implications for Computational Fluid Dynamics modelling. Resolving all scales down to η requires extremely fine spatial and temporal resolution, which explains the prohibitive computational cost of Direct Numerical Simulation at high Reynolds numbers. Practical turbulence models such as Reynolds-Averaged Navier–Stokes (RANS), Large Eddy Simulation (LES), and hybrid approaches exploit this scale separation by modelling part of the turbulent spectrum while resolving the most dynamically important motions.

2.2.8 *Turbulence Energy Spectrum*

The distribution of turbulent kinetic energy across different length scales is conveniently described in spectral space using the wavenumber κ, which is inversely proportional to the characteristic eddy size l, i.e.

$$\kappa \sim \frac{1}{l}. \tag{2.23}$$

Small wavenumbers therefore correspond to large-scale motions, while large wavenumbers represent small-scale eddies.

The turbulent energy spectrum, $E(\kappa)$, can be divided conceptually into three distinct regions. Figure 2.2 illustrates these regions in the energy spectrum.

Region I: Energy-Containing Range Region I is associated with the largest eddies in the flow and corresponds to small wavenumbers. These motions extract kinetic energy directly from the mean flow through shear and instability mechanisms. Their structure and magnitude depend strongly on the geometry, boundary conditions, and inflow characteristics of the flow.

Region II: Inertial Subrange Energy is transferred from the large eddies to progressively smaller ones through nonlinear interactions, a process known as the *energy cascade*. This transfer occurs in Region II, the inertial subrange, where the dynamics are assumed to be independent of viscosity and large-scale flow details. In this range, the energy transfer rate is constant and equal to the dissipation rate ε. Kolmogorov's theory predicts that the energy spectrum in the inertial subrange follows a universal power law:

$$E(\kappa) = C\,\varepsilon^{2/3}\kappa^{-5/3}, \tag{2.24}$$

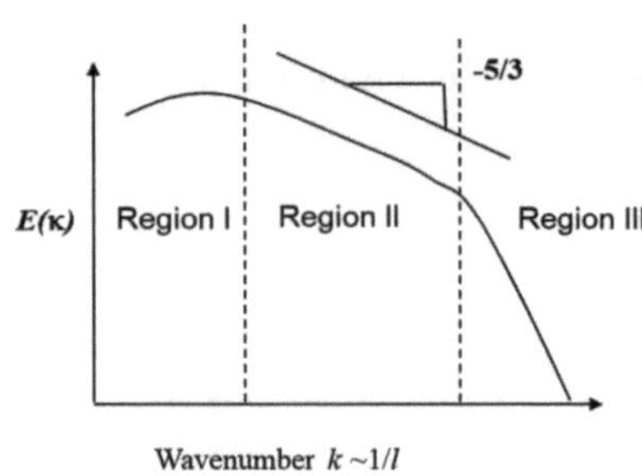

Fig. 2.2 The different regions in the turbulent energy spectrum

where C is a universal constant of order unity.

Region III: Dissipation Range At sufficiently large wavenumbers, corresponding to Region III, viscous effects dominate. In this range, turbulent kinetic energy is removed from the cascade and converted into internal energy (heat) at the smallest scales characterised by the Kolmogorov length, velocity, and time scales $(\eta, u_\eta, \tau_\eta)$. Beyond this region, the energy spectrum decays rapidly as viscosity suppresses further turbulent motion.

This spectral description provides a clear physical interpretation of turbulence as a multiscale process and underpins many modelling approaches in Computational Fluid Dynamics. In particular, it explains why Large Eddy Simulation (LES) resolves the energy-containing eddies while modelling the inertial and dissipation ranges, and why Direct Numerical Simulation (DNS) must resolve the full spectrum down to the highest wavenumbers associated with dissipation.

2.3 Governing Equations of Fluid Flow

The aerodynamic flows of interest in wind energy—ranging from the atmospheric boundary layer approaching the rotor to the strongly three-dimensional, unsteady wake downstream—are governed by the fundamental conservation laws of mass and momentum. When these laws are expressed mathematically, they naturally take the form of vector and tensor equations, because velocity, stress, and deformation are not single numbers but quantities with direction and, in many cases, multi-directional coupling. For this reason, it is convenient to introduce tensor notation before presenting the governing equations of fluid motion.

2.3.1 Tensor Definition

Tensor (or index) notation provides a compact and systematic way to write multi-dimensional equations without repeatedly expanding them into separate x-, y-, and z-components. In CFD and turbulence modelling, this compactness becomes especially valuable: the same governing equations are applied in two and three dimensions, in Cartesian or curvilinear coordinate systems, and in both instantaneous and averaged (RANS/LES-filtered) forms.

In fluid mechanics, many physical quantities—such as velocity, acceleration, and force—have both magnitude and direction and therefore require a vector or tensor description. In a Cartesian coordinate system (x, y, z), the velocity field is conventionally written in vector form as

$$\mathbf{u} = (u, v, w). \tag{2.25}$$

Using tensor (index) notation, the same velocity field can be written compactly as

$$\mathbf{u} \rightarrow u_i, \tag{2.26}$$

where the index $i = 1, 2, 3$ denotes the three Cartesian directions corresponding to (x, y, z). Thus, $u_1 = u$, $u_2 = v$, and $u_3 = w$. This notation allows vector equations to be written in a concise and general form that is independent of the coordinate direction.

As another example, consider Newton's second law. In vector form, it is written as

$$m\frac{\partial^2 \mathbf{x}}{\partial t^2} = \mathbf{F}, \tag{2.27}$$

which in component form becomes

$$m\frac{\partial^2 x}{\partial t^2} = F_1, \quad m\frac{\partial^2 y}{\partial t^2} = F_2, \quad m\frac{\partial^2 z}{\partial t^2} = F_3. \tag{2.28}$$

Using tensor notation, all three component equations can be written compactly as

$$m\frac{\partial^2 x_i}{\partial t^2} = F_i, \tag{2.29}$$

without explicitly listing the individual coordinate directions. This compactness is one of the key advantages of tensor notation in fluid dynamics and CFD.

2.3.2 Tensor Index Rules (Einstein Summation Convention)

Tensor notation follows the Einstein summation convention, which is based on the following rules:

- An index that appears once in a term is called a free index and indicates that the equation applies separately to each coordinate direction.
- An index that appears twice in a single term is called a dummy index, and summation over that index from 1 to 3 is implied.

For example, the tensor expression

$$\frac{\partial u_i}{\partial x_i} \tag{2.30}$$

represents a single scalar quantity, even though u_i has three components. Applying the summation convention, this term expands to

$$\frac{\partial u}{\partial x} + \frac{\partial v}{\partial y} + \frac{\partial w}{\partial z}, \tag{2.31}$$

which is the familiar divergence of the velocity field. It is important to note that $\partial u_i/\partial x_i$ is one term, whereas u_i represents three components.

More complex tensor expressions can be interpreted in a similar way. For instance, the nonlinear convective term in the Navier–Stokes equations is written in tensor form as

$$u_j \frac{\partial u_i}{\partial x_j}. \tag{2.32}$$

Expanding this expression yields three component equations:

$$\begin{aligned} i = 1 &: u\frac{\partial u}{\partial x} + v\frac{\partial u}{\partial y} + w\frac{\partial u}{\partial z}, \\ i = 2 &: u\frac{\partial v}{\partial x} + v\frac{\partial v}{\partial y} + w\frac{\partial v}{\partial z}, \\ i = 3 &: u\frac{\partial w}{\partial x} + v\frac{\partial w}{\partial y} + w\frac{\partial w}{\partial z}. \end{aligned} \tag{2.33}$$

Similarly, the viscous diffusion term in the momentum equation can be written in tensor form as

$$\frac{\partial}{\partial x_j}\left(\mu \frac{\partial u_i}{\partial x_j}\right). \tag{2.34}$$

When expanded in Cartesian coordinates, this produces:

$$\begin{aligned} i = 1 &: \mu\left(\frac{\partial^2 u}{\partial x^2} + \frac{\partial^2 u}{\partial y^2} + \frac{\partial^2 u}{\partial z^2}\right), \\ i = 2 &: \mu\left(\frac{\partial^2 v}{\partial x^2} + \frac{\partial^2 v}{\partial y^2} + \frac{\partial^2 v}{\partial z^2}\right), \\ i = 3 &: \mu\left(\frac{\partial^2 w}{\partial x^2} + \frac{\partial^2 w}{\partial y^2} + \frac{\partial^2 w}{\partial z^2}\right). \end{aligned} \tag{2.35}$$

This illustrates how tensor notation provides a clear, compact, and systematic way of writing and interpreting the governing equations of fluid flow. It is particularly powerful in turbulence modelling and CFD, where multi-dimensional, multi-component equations must be manipulated, averaged, and filtered repeatedly.

2.3.3 *Incompressible Flow*

A flow is termed incompressible if the density ρ of individual fluid particles remains constant as they move with the flow. This condition does not require the density to be uniform everywhere in space or time; rather, it requires that the material (or substantial) derivative of density vanishes:

$$\frac{D\rho}{Dt} = \frac{\partial \rho}{\partial t} + u_j \frac{\partial \rho}{\partial x_j} = 0, \tag{2.36}$$

where t denotes time, u_j the velocity components, and x_j the spatial coordinates. This expression states that each fluid particle retains its density as it is convected through the flow field.

In reality, no fluid is perfectly incompressible. However, many flows can be treated as effectively incompressible when density variations are sufficiently small that their influence on the momentum balance is negligible. In practical terms, this means that variations in density are much smaller than variations in velocity or pressure gradients within the flow,

$$\frac{\partial \rho}{\partial x} \ll \frac{\partial u}{\partial x}. \tag{2.37}$$

Under these conditions, the incompressible form of the governing equations provides an accurate and greatly simplified description of the flow.

2.3.4 Mach Number and the Validity of the Incompressible Assumption

The appropriateness of the incompressible-flow assumption is commonly assessed using the Mach number, M, a dimensionless parameter defined as the ratio of the characteristic flow velocity u to the local speed of sound a:

$$M = \frac{u}{a}. \tag{2.38}$$

The Mach number quantifies the importance of compressibility effects associated with pressure—density coupling and acoustic wave propagation.

For flows with $M < 0.3$, compressibility effects are typically small, and the flow can be considered both subsonic and incompressible for most engineering purposes. Consequently, the flow of liquids and many low-speed gas flows—such as ventilation flows and airflow around ground vehicles—can be accurately modelled using incompressible flow equations. Exceptions may arise at very high vehicle speeds, for example in the case of ultra-high-speed trains approaching or exceeding 400 km/h.

In wind turbine applications, the characteristic flow velocities are generally well below this threshold. As a result, the airflow around wind turbine blades, rotors, and wakes is normally treated as incompressible, allowing the use of incompressible Navier–Stokes equations in both analytical and numerical modelling without significant loss of accuracy.

2.3.5 Governing Equations in Fluid Dynamics

The motion of a fluid is governed by fundamental conservation laws, which express the principles of mass and momentum conservation. When applied to a continuum fluid and expressed in Cartesian tensor notation, these laws yield the governing equations of fluid dynamics that form the basis of analytical modelling and Computational Fluid Dynamics (CFD).

2.3.6 Conservation of Mass (Continuity Equation)

The conservation of mass requires that mass is neither created nor destroyed within a fluid element. In its most general form, the continuity equation is written as

$$\frac{\partial \rho}{\partial t} + \frac{\partial (\rho u_i)}{\partial x_i} = 0, \tag{2.39}$$

where ρ is the fluid density, u_i are the velocity components, and x_i denote the spatial coordinates. This equation is valid for both compressible and incompressible flows.

For an incompressible flow, where the density of each fluid particle remains constant, the continuity equation simplifies significantly. Since ρ is constant, it can be taken outside the derivatives, leading to the divergence-free condition

$$\frac{\partial u_i}{\partial x_i} = 0, \tag{2.40}$$

which states that the velocity field has zero divergence.

2.3.7 Conservation of Momentum (Navier–Stokes Equations)

The principle of momentum conservation, equivalent to Newton's second law applied to a fluid element, leads to the Navier–Stokes equations. For an incompressible, Newtonian fluid with constant viscosity, these equations can be written in tensor form as

$$\frac{\partial u_i}{\partial t} + u_j \frac{\partial u_i}{\partial x_j} = -\frac{1}{\rho}\frac{\partial p}{\partial x_i} + \nu \frac{\partial^2 u_i}{\partial x_j \partial x_j}, \tag{2.41}$$

where p is the instantaneous pressure field and ν is the kinematic viscosity of the fluid.

The left-hand side of the equation represents the local and convective acceleration of the fluid, while the right-hand side accounts for the forces acting on the fluid element: the pressure-gradient force and the viscous diffusion of momentum. The nonlinear convective term $u_j \partial u_i / \partial x_j$ is responsible for much of the complexity of fluid motion and plays a central role in the development of turbulence.

The kinematic viscosity ν is related to the dynamic viscosity μ and the fluid density ρ through the relation

$$\nu = \frac{\mu}{\rho}. \tag{2.42}$$

Together, the continuity and Navier–Stokes equations provide a complete mathematical description of incompressible Newtonian fluid flow. All turbulence modelling approaches—RANS, LES, and hybrid methods—are derived from these equations through averaging or filtering operations, making them the cornerstone of CFD-based analysis in wind energy and related applications.

2.4 Computational Fluid Dynamics (CFD)

Analytical solutions of the governing equations of fluid motion are limited to highly idealised cases, typically involving simple geometries and laminar flow conditions. For most practical engineering flows—particularly turbulent flows—exact analytical solutions are not available. Over the past five decades, Computational Fluid Dynamics (CFD) has therefore emerged as a powerful third pillar of fluid mechanics, complementing the classical theoretical and experimental approaches. By numerically solving the governing equations, CFD has enabled major advances across a wide range of disciplines, including aerodynamics, hydrodynamics, environmental flows, and energy engineering.

Today, CFD plays a central role in the design and analysis of systems involving fluid flow and heat transfer. In the context of wind energy, CFD provides detailed insight into blade aerodynamics, wake development, turbulencestructure interaction, and wind farm flow behaviour. However, the application of CFD to wind power problems has historically been constrained by the large computational resources required, particularly for high-Reynolds-number, three-dimensional, and unsteady flows. The rapid development of high-performance computing (HPC) and parallel computing architectures has significantly expanded the feasibility of CFD in wind energy research and industrial practice.

Figure 2.3 illustrates the standard workflow of a Computational Fluid Dynamics (CFD) simulation, highlighting the sequential stages from problem definition to solution validation. The process begins with the pre-processing stage, where the physical problem is translated into a numerical form. This involves creating the computational geometry and generating an appropriate mesh to discretise the flow domain.

Once the geometry and mesh are prepared, the simulation proceeds to the physics and model setup stage. At this stage, the relevant physical models are defined, including the selection of the solver, identification of the flow regime (e.g. laminar or

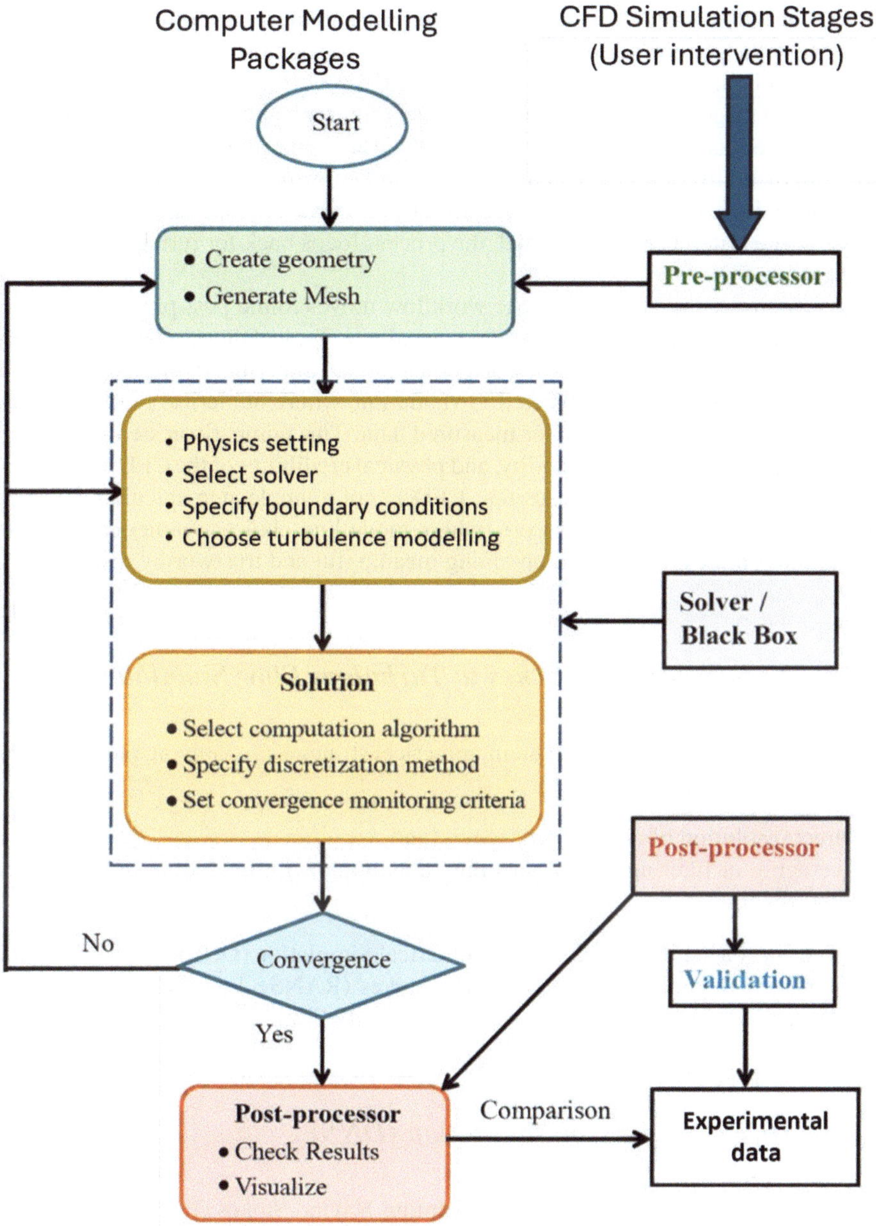

Fig. 2.3 Stages of a Computational Fluid Dynamics (CFD) simulation workflow

viscous/turbulent), and specification of boundary and initial conditions. These steps define how the governing equations are to be solved and how the flow interacts with the domain boundaries.

The next stage is the solution phase, which represents the core of the CFD solver (often treated as a "black box" by the user). Here, the numerical algorithms are selected, discretisation methods are applied, and convergence criteria are specified. The solver iteratively advances the solution until convergence is achieved. If the convergence criteria are not satisfied, the process loops back for further iterations or adjustments.

Once convergence is reached, the workflow moves to the post-processing stage, where results are checked, analysed, and visualised. Typical outputs include velocity fields, pressure distributions, forces, and other derived quantities of engineering interest. These results are then subjected to validation, where numerical predictions are compared against experimental or measured data. This comparison step is essential for assessing the accuracy, reliability, and physical credibility of the CFD simulation.

Overall, the figure emphasises that CFD is not a single-step calculation but an iterative and interconnected process, where modelling choices, numerical settings, and validation are all critical to obtaining meaningful and trustworthy results.

2.4.1 Numerical Approaches to Turbulent Flow Simulation

Broadly speaking, there are two main numerical strategies for solving the governing equations of turbulent flow:

- Direct resolution of all turbulent scales, and
- Averaging or filtering of the governing equations, with modelling of unresolved turbulence.

The first approach is known as Direct Numerical Simulation (DNS), while the second encompasses Reynolds-Averaged Navier–Stokes (RANS), Large-Eddy Simulation (LES), and hybrid methods such as Detached-Eddy Simulation (DES).

2.4.2 Direct Numerical Simulation (DNS)

In Direct Numerical Simulation, the governing Navier–Stokes equations are solved without any turbulence modelling. All relevant turbulent length and time scales—from the largest energy-containing eddies down to the smallest dissipative (Kolmogorov) scales—are fully resolved.

To achieve this, the computational grid spacing must be of the order of the Kolmogorov length scale, and the time step Δt must be sufficiently small to resolve the fastest turbulent motions. Two main constraints govern the choice of the time step. First, it must be smaller than the Kolmogorov time scale τ. Second, numerical stability and accuracy require that fluid particles do not travel more than one grid spacing in a single time step. In one dimension, this condition can be expressed as

$$\frac{U\Delta t}{\Delta x} \le 1, \tag{2.43}$$

where U is a characteristic velocity and Δx is the grid spacing.

Although DNS provides the most accurate and complete description of turbulent flow, its computational cost increases rapidly with Reynolds number. As a result, DNS is currently restricted to low-Reynolds-number flows and fundamental turbulence studies, and it is not feasible for realistic wind turbine or wind farm simulations.

2.4.3 Reynolds-Averaged Navier–Stokes (RANS)

In practical engineering applications, it is often more efficient to analyse turbulent flow by decomposing the instantaneous variables into mean and fluctuating components. For velocity and pressure, this decomposition is written as

$$u_i = \overline{u_i} + u_i', \qquad p = \bar{p} + p'. \tag{2.44}$$

Substituting this decomposition into the continuity and momentum equations and applying a time-averaging operation yields the Reynolds-Averaged Navier–Stokes (RANS) equations. For incompressible flow, the time-averaged momentum equation can be written as

$$\frac{\partial \overline{u_i}}{\partial t} + \overline{u}_j \frac{\partial \overline{u_i}}{\partial x_j} = -\frac{1}{\rho}\frac{\partial \bar{p}}{\partial x_i} + \frac{\partial}{\partial x_j}\left(\nu \frac{\partial \overline{u_i}}{\partial x_j} - \overline{u_i' u_j'}\right). \tag{2.45}$$

The additional term—$\rho \overline{u_i' u_j'}$ represents the Reynolds stress tensor, which accounts for the momentum transport due to turbulent fluctuations. These stresses introduce six new unknowns into the system.

2.4.4 The Turbulence Closure Problem

The appearance of the Reynolds stresses leads to a fundamental difficulty known as the turbulence closure problem. In three-dimensional flow, there are six independent Reynolds stress components:

$$\overline{u_1'^2},\ \overline{u_2'^2},\ \overline{u_3'^2},\ \overline{u_1' u_2'},\ \overline{u_1' u_3'},\ \overline{u_2' u_3'}. \tag{2.46}$$

These quantities must be expressed in terms of the mean flow variables in order to close the system of equations. Without additional modelling assumptions, the number of unknowns exceeds the number of equations, making the problem unsolvable.

2.4.5 *Eddy Viscosity and the Boussinesq Hypothesis*

A widely used approach to closing the RANS equations is based on the eddy viscosity concept, introduced through the Boussinesq eddy-viscosity assumption. This assumption postulates that the Reynolds stresses can be related to the mean rate of strain in a manner analogous to the viscous stresses in laminar flow:

$$-\rho\overline{u_i'u_j'} = \mu_t\overline{S_{ij}} - \frac{2}{3}\rho k\delta_{ij}, \tag{2.47}$$

where μ_t is the turbulent (eddy) viscosity, k is the turbulence kinetic energy, δ_{ij} is the Kronecker delta, and the mean strain-rate tensor $\overline{S_{ij}}$ is defined as

$$\overline{S_{ij}} = \frac{1}{2}\left(\frac{\partial\overline{u_i}}{\partial x_j} + \frac{\partial\overline{u_j}}{\partial x_i}\right). \tag{2.48}$$

Unlike molecular viscosity, the turbulent viscosity μ_t is not a property of the fluid; rather, it is a property of the flow and must be modelled. This requirement has led to the development of a wide range of turbulence models, including mixing-length models, one-equation models, and two-equation models such as the k–ε.

2.5 Turbulence Models

Because turbulent flows involve a wide range of interacting scales, direct resolution of all turbulent motions is often impractical for engineering applications. Turbulence models are therefore introduced to represent the effects of unresolved turbulent fluctuations on the mean or resolved flow field. Depending on how turbulence is treated, these models range from fully averaged approaches to scale-resolving methods.

2.5.1 *Mixing-Length Model*

One of the earliest and simplest turbulence models is the mixing-length model, which is based on the eddy-viscosity concept. In this approach, the turbulent (eddy) viscosity μ_t is assumed to be proportional to the square of a characteristic length scale—known as the mixing length, l_{mix}—and the magnitude of the local mean velocity gradient. It is expressed as

$$\mu_t = \rho \nu_t = \rho\, l_{\mathrm{mix}}^2 \, |\nabla \bar{u}|. \tag{2.49}$$

The key modelling assumption lies in prescribing the mixing length as a function of position within the flow. In unbounded shear flows, l_{mix} is typically taken to be of the order of 10% of the characteristic shear-layer thickness. Near solid walls, experimental observations suggest that the mixing length scales approximately with the distance from the wall,

$$l_{\mathrm{mix}} \approx 0.4\, y, \tag{2.50}$$

where y is the wall-normal distance. However, in the immediate vicinity of the wall, viscous effects dominate and the mixing length decreases more rapidly, eventually vanishing at the wall. While simple and computationally efficient, the mixing-length model is limited in its applicability and is generally restricted to simple shear flows.

2.5.2 The k–ε Model

A more widely used and robust turbulence closure is the two-equation k–ε model, which introduces transport equations for the turbulence kinetic energy k and its dissipation rate ε. In this model, the turbulent viscosity is defined as

$$\mu_t = \rho \nu_t = \rho C_\mu \frac{k^2}{\varepsilon}, \tag{2.51}$$

where C_μ is an empirical model constant.

The transport equation for the turbulence kinetic energy is

$$\frac{\partial(\rho k)}{\partial t} + \frac{\partial(\rho k u_i)}{\partial x_i} = \frac{\partial}{\partial x_j}\left[\left(\mu + \frac{\mu_t}{\sigma_k}\right)\frac{\partial k}{\partial x_j}\right] + 2\mu_t E_{ij} E_{ij} - \rho\varepsilon, \tag{2.52}$$

and the transport equation for the dissipation rate is

$$\frac{\partial(\rho \varepsilon)}{\partial t} + \frac{\partial(\rho \varepsilon u_i)}{\partial x_i} = \frac{\partial}{\partial x_j}\left[\left(\mu + \frac{\mu_t}{\sigma_\varepsilon}\right)\frac{\partial \varepsilon}{\partial x_j}\right] + C_{1\varepsilon}\, \frac{\varepsilon}{k}\, 2\mu_t E_{ij} E_{ij} - C_{2\varepsilon}\, \rho \frac{\varepsilon^2}{k}. \tag{2.53}$$

Here, u_i are the velocity components, and E_{ij} is the mean rate-of-deformation tensor. The model contains several empirical constants, typically taken as

$$\sigma_k = 1.0, \qquad \sigma_\varepsilon = 1.3, \qquad C_{1\varepsilon} = 1.44, \qquad C_{2\varepsilon} = 1.92. \tag{2.54}$$

The k–ε model offers a good compromise between accuracy and computational cost and has been widely applied in industrial CFD. However, it relies on isotropic turbulence assumptions and can struggle in flows with strong separation, curvature, rotation, or unsteadiness–features commonly encountered in wind turbine aerodynamics.

2.5.3 *Large-Eddy Simulation (LES)*

Large-Eddy Simulation (LES) represents a different modelling philosophy, in which the flow is decomposed into large, energy-containing eddies and small, subgrid-scale (SGS) motions. The large eddies, which are strongly dependent on geometry and boundary conditions, are resolved directly by the numerical simulation, while the influence of the smaller scales is modelled. LES therefore provides a compromise between the high accuracy of DNS and the low computational cost of RANS.

LES can be described conceptually through four main steps.

1. Filtering The first step in LES is the application of a spatial filter to the instantaneous velocity field $u_i(x,t)$, decomposing it into a resolved (filtered) component and a subgrid-scale component:

$$u_i(x,t) = \overline{U_i}(x,t) + U_i''(x,t). \tag{2.55}$$

The filtered velocity field is obtained through a convolution operation,

$$\overline{U_i}(x,t) = \int_{\Omega} G(x,r)\, u_i(x-r,t)\, \mathrm{d}r, \tag{2.56}$$

where Ω is the flow domain and $G(x,r)$ is the filter function. In practical CFD, the filter width is usually related to the local grid spacing.

2. Filtered Governing Equations Applying the filtering operation to the incompressible Navier–Stokes equations yields the LES governing equations:

$$\frac{\partial \overline{U_i}}{\partial x_i} = 0, \tag{2.57}$$

$$\frac{\partial \overline{U_i}}{\partial t} + \overline{U_j}\frac{\partial \overline{U_i}}{\partial x_j} = -\frac{1}{\rho}\frac{\partial \bar{p}}{\partial x_i} + \frac{\partial}{\partial x_j}\left(\nu\frac{\partial \overline{U_i}}{\partial x_j} - Q_{ij}\right). \tag{2.58}$$

The term Q_{ij} represents the subgrid-scale (SGS) stress tensor, defined as

$$Q_{ij} = \overline{u_i u_j} - \overline{U_i}\,\overline{U_j}, \tag{2.59}$$

which arises from the nonlinear interaction between resolved and unresolved scales.

3. Subgrid-Scale Modelling Because the SGS stresses cannot be computed directly, they must be modelled. The most widely used SGS model is the Smagorinsky model, which adopts an eddy-viscosity approach analogous to RANS:

$$Q_{ij} = 2\nu_{\mathrm{SGS}}\,\overline{S_{ij}}, \tag{2.60}$$

where the resolved strain-rate tensor is

$$\overline{S_{ij}} = \frac{1}{2}\left(\frac{\partial \overline{U_i}}{\partial x_j} + \frac{\partial \overline{U_j}}{\partial x_i}\right). \tag{2.61}$$

The SGS viscosity is defined as

$$\nu_{\mathrm{SGS}} = (C_S\Delta)^2\sqrt{2\,\overline{S_{ij}}\,\overline{S_{ij}}}, \tag{2.62}$$

where C_S is the Smagorinsky constant (typically in the range 0.06–0.2) and Δ is the filter width, commonly taken as the cubic root of the local cell volume.

4. Numerical Solution of the Resolved Flow In the final step, the filtered governing equations are solved numerically to obtain the resolved large-eddy velocity field. Because only the largest turbulent motions are resolved, the computational cost of LES is significantly lower than that of DNS. However, the smallest resolved eddies are of the order of the grid spacing, meaning that LES still requires fine spatial and temporal resolution, particularly in near-wall regions. Consequently, LES offers substantially improved physical fidelity compared to RANS—especially for unsteady separation and wake dynamics—while remaining computationally feasible for many wind energy applications, particularly when combined with wall-modelling or hybrid approaches.

2.5.4 Detached-Eddy Simulation (DES)

DES is a hybrid turbulence modelling approach that combines the strengths of Reynolds-Averaged Navier–Stokes (RANS) and Large Eddy Simulation (LES) to achieve an effective balance between accuracy and computational cost. In DES, the near-wall regions, where turbulent length scales are small and computationally expensive to resolve, are modelled using RANS formulations, while the separated flow regions and wakes, characterised by large, energy-containing eddies, are resolved using an LES-like approach. By avoiding full near-wall resolution, DES requires significantly fewer computational cells than wall-resolved LES, particularly for high-Reynolds-number flows typical of wind turbine blades and wakes. As a result, DES can capture unsteady flow separation, vortex shedding, and wake dynamics with substantially higher fidelity than conventional RANS models, while remaining computationally tractable for large-scale and industrial applications. Nevertheless, DES performance depends strongly on the underlying RANS turbulence model and grid design, as the transition between RANS and LES regions is implicitly controlled by the local grid spacing. Inappropriate mesh resolution may lead to modelled-stress depletion or delayed separation, necessitating careful turbulence-model selection and mesh-generation strategies to ensure reliable predictions.

2.6 Numerical Generation of Instantaneous Inlet Velocity

Accurate representation of the inflow conditions is a fundamental requirement for numerical simulations of wind energy systems, as the aerodynamic performance of wind turbines and the development of their wakes are strongly influenced by the characteristics of the incoming atmospheric flow. A widely adopted and practical framework for describing the atmospheric boundary layer is provided by Eurocode I (2005), which proposes a systematic and simplified representation suitable for engineering design. Within this framework, the vertical variation of the mean wind speed is described using a logarithmic (log-law) profile, while the turbulence intensity is represented through an inverse logarithmic relationship. The mean wind speed profile is given by

$$U(z) = V_b\, k_r\, \ln\left(\frac{z}{z_0}\right), \tag{2.63}$$

where $U(z)$ is the 10-min averaged wind speed at height z above the ground, z_0 is the ground roughness length, V_b is the regional base wind velocity, and k_r is a terrain-dependent parameter that depends only on the roughness length z_0. This formulation captures the essential effect of surface roughness on the vertical shear of the wind profile, with rougher terrain leading to lower wind speeds near the ground and stronger velocity gradients.

In addition to the mean wind profile, Eurocode I specifies the vertical variation of turbulence intensity, which quantifies the magnitude of velocity fluctuations relative to the mean flow. The mean turbulence intensity is defined as

$$I(z) = \frac{\sigma_u}{U(z)} = \frac{k_i}{\ln\left(\frac{z}{z_0}\right)}, \tag{2.64}$$

where σ_u is the standard deviation of the longitudinal velocity fluctuations evaluated over a 10-minute averaging window, and k_i is a parameter that, like k_r, depends solely on the ground roughness length. This inverse logarithmic relationship reflects the physical observation that turbulence intensity is generally higher near the ground, where shear and surface-induced turbulence are strongest, and decreases with height as the influence of the surface diminishes.

For CFD simulations of wind turbines and other wind-exposed structures, these profiles provide a consistent and standardised means of prescribing inlet boundary conditions that reflect realistic atmospheric conditions. By combining a logarithmic mean velocity profile with a height-dependent turbulence intensity, the Eurocode formulation enables the generation of inflow conditions that are compatible with both engineering practice and numerical modelling requirements. While this approach represents a simplification of the full complexity of the atmospheric boundary layer–neglecting, for example, thermal stratification and non-neutral stability effects–it offers a robust and widely accepted baseline for wind energy simulations, particularly in the context of design load assessment and compliance with engineering standards.

However, the structure of the atmospheric boundary layer is complex in which the wind field is inherently unsteady, turbulent, and spatially inhomogeneous. Temporal fluctuations in velocity, combined with vertical shear, turbulence intensity, length scales, and coherence, govern not only power production but also unsteady aerodynamic loading, fatigue, noise generation, and wake dynamics. Consequently, the numerical generation of realistic instantaneous inlet velocity fields is a critical component of high-fidelity CFD for wind energy applications.

In numerical modelling, prescribing a steady or uniform inlet velocity is often insufficient to capture the complex physics associated with turbulent inflow. Instead, time-resolved inlet conditions must be generated that reproduce key statistical properties of the atmospheric boundary layer, such as mean velocity profiles, turbulence intensities, integral length scales, spectral characteristics, and spatial and temporal correlations. Various numerical techniques have therefore been developed to synthesise turbulent inflow conditions, ranging from synthetic turbulence methods and spectral approaches to precursor simulations and data-driven techniques. These methods aim to generate instantaneous velocity fluctuations that are dynamically consistent with the governing equations and compatible with the turbulence model employed (RANS, LES, or hybrid approaches).

For wind energy simulations, the challenge lies in balancing physical realism with computational efficiency. High-fidelity approaches, such as precursor large-eddy simulations, can provide highly realistic inflow turbulence but at significant computational cost. Conversely, synthetic turbulence generators offer a more economical alternative, although care must be taken to ensure that the generated velocity field preserves correct turbulence statistics and does not introduce spurious numerical artefacts. Recent advances in data-driven and machine-learning-based inflow generation further expand the available toolkit, offering new possibilities for generating realistic, site-specific inflow conditions. This section introduces the key principles and numerical strategies used to generate instantaneous inlet velocity fields for wind energy applications, highlighting their relevance, capabilities, and limitations within modern CFD frameworks.

2.6.1 Generating Instantaneous Perturbations

Several numerical strategies have been developed to generate time-resolved turbulent inflow conditions in CFD, each with different levels of physical realism and computational cost. Below is a brief description of some of the existing methods, widely used in numerical simulations to create perturbations imposed on the mean velocity profile. More information about these methods can be found in the author's publications, such as Wang et al. (2022, 2025) and Vita et al. (2020a; 2020b).

Precursor-Based Methods In precursor approaches, a fully developed turbulent boundary layer is generated in a separate or auxiliary simulation, and the resulting velocity field is then used to prescribe the inlet condition for the main computational domain. In principle, this provides physically consistent turbulence that naturally satisfies the governing equations. However, a fully developed atmospheric boundary layer requires a long streamwise development distance. The thickness of the boundary-layer δ grows approximately as $O(\sqrt{x})$ and depends strongly on the thickness of the surface roughness z_0. As a result, generating a boundary layer of only 20 m height may require a precursor domain several hundred metres long, leading to a very high computational cost.

However, from a geometrical point of view, the flow domain is locally self-similar in the streamwise direction, with neighbouring streamwise segments differing mainly in their inlet conditions. This observation has motivated the development of periodic precursor methods in which specific periodic boundary conditions are applied in the streamwise direction to recycle turbulence. Variants of this approach include pre-library and simultaneous mapping techniques, where turbulent velocity fields are stored and reintroduced at the inlet. Although these methods can generate realistic turbulence, controlling the resulting flow statistics is often difficult, and significant tuning of the precursor domain is typically required to match target mean profiles, turbulence intensities, and length scales.

Random Perturbation Methods The most straightforward approach to generating unsteady inlet velocity is to superimpose random fluctuations onto a prescribed mean velocity profile. In this method, the instantaneous inlet velocity is updated at each time step according to

$$u_i^n = (1-\alpha)\, u_i^{(n-1)} + \alpha(U + sRU), \tag{2.65}$$

where u_i^n is the velocity at the current time step, $u_i^{(n-1)}$ is the velocity at the previous time step, U is the mean velocity, R is a random number, α is a weighting parameter controlling temporal correlation, and s is a fluctuation scale.

Although simple and computationally inexpensive, random perturbation methods generally fail to reproduce realistic turbulence spectra, spatial correlations, and Reynolds stress distributions. Consequently, they are mainly used for preliminary studies or low-fidelity simulations.

Recycling and Rescaling Method (Lund Method) A more advanced and widely used approach is the recycling and rescaling method, originally proposed by Lund et al. In this technique, the instantaneous velocity field is extracted from a downstream recycle plane within the computational domain, rescaled to match the desired inlet boundary-layer properties, and reintroduced at the inlet.

The rescaling is based on inner and outer layer similarity, using relationships between the inlet plane and the recycle plane. This method provides realistic turbulent inflow with good control of boundary-layer properties, but it still requires a sufficiently long domain to allow recycling and may be sensitive to grid resolution and numerical dissipation.

Synthetic Turbulence Generation Methods Synthetic turbulence generation methods are widely used in Computational Fluid Dynamics to prescribe realistic, time-dependent inlet velocity conditions without the need for expensive precursor simulations. The central idea of these methods is to construct an instantaneous turbulent velocity field by superimposing carefully designed fluctuations onto a prescribed mean flow profile, such that the resulting inflow reproduces key statistical properties of turbulence. These properties typically include the desired turbulence intensity, Reynolds stress components, integral length scales, and spatial coherence. Unlike simple random-noise approaches, synthetic methods are designed to introduce correlated fluctuations in space and time, allowing the generated turbulence to evolve naturally once it enters the computational domain.

In practice, synthetic turbulence is created by assembling a large number of artificial eddies or modes that are randomly distributed across the inlet plane and advected with the mean flow. The amplitude, orientation, and spatial extent of these structures are chosen to match target turbulence statistics, often derived from empirical models, atmospheric boundary-layer theory, or precursor simulations. This approach provides good control over inflow conditions while remaining computationally efficient, making it particularly attractive for Large-Eddy Simulation of wind energy flows, where realistic inflow turbulence is essential but computational resources are limited. Although synthetic turbulence does not initially satisfy the Navier–Stokes equations exactly, the generated fluctuations rapidly adjust and become dynamically consistent downstream of the inlet. As a result, synthetic turbulence generation methods offer a practical compromise between physical realism and computational efficiency for simulating atmospheric inflow conditions in wind energy applications.

Stored Turbulence Library/Mapping Method The stored turbulence library or mapping method is an approach for generating instantaneous inlet velocity fields that aims to combine the physical realism of precursor simulations with improved computational efficiency. In this method, a fully developed turbulent flow—typically representing an atmospheric boundary layer or a canonical turbulent boundary layer—is first generated in a separate high-fidelity simulation, often using Large-Eddy Simulation. The resulting time-resolved velocity fields are then stored in a database or "library." During the main CFD simulation, these precomputed turbulent fields are mapped or imposed at the inlet boundary, either directly or with appropriate spatial and temporal transformations to match the target inflow conditions.

The mapping process can involve scaling the velocity fluctuations to match the desired mean velocity profile, turbulence intensity, length scales, or Reynolds stress distributions at the inlet. Temporal shifting, interpolation, or cycling of the stored fields is often applied to avoid artificial periodicity and to maintain statistical stationarity over long simulation times. While the stored turbulence library method can produce highly realistic inflow conditions—closely resembling those obtained from live precursor simulations—it requires significant storage resources and careful preprocessing to ensure consistency between the library data and the main simulation grid and boundary conditions. As a result, this method is particularly attractive for repeated simulations under similar inflow conditions, such as parametric studies or wind farm layout optimisation, where the initial cost of generating the turbulence library can be amortised over many CFD runs.

Data-Driven/ML-Enhanced Methods Data-driven and machine-learning-enhanced methods represent an emerging class of techniques for generating instantaneous inlet velocity fields in CFD, offering a new balance between physical realism and computational efficiency. In these approaches, machine learning models are trained on data obtained from high-fidelity simulations, experiments, or field measurements—such as Large-Eddy Simulation of atmospheric boundary layers or full-scale wind measurements—to learn the statistical and structural characteristics of turbulent inflow. Once trained, the models can rapidly generate time-resolved, spatially correlated velocity fields at the inlet that reproduce key turbulence properties, including mean velocity profiles, turbulence intensities, spectra, coherence, and, in some cases, higher-order statistics.

A major advantage of data-driven inflow generation is that the computational cost is largely shifted to an offline training phase, while the online generation of instantaneous velocity fields during CFD simulations is extremely fast. This makes such methods particularly attractive for applications requiring repeated simulations, real-time or near-real-time predictions, or integration with control and optimisation frameworks in wind energy. Deep learning architectures, such as convolutional neural networks, recurrent neural networks, or generative models, can capture complex spatial and temporal correlations that are difficult to prescribe explicitly using traditional synthetic turbulence techniques. However, the realism and robustness of data-driven inflow methods depend strongly on the quality, representativeness, and diversity of the training data. Ensuring physical consistency, generalisation beyond the training conditions, and compatibility with different turbulence models remains an active area of research. Despite these challenges, data-driven methods show strong potential as a next-generation tool for realistic and efficient inflow generation in wind energy CFD simulations.

Each of the methods method prescribed above has advantage and disadvantage in terms of computational cost and physical realism. Table 2.1 shows the comparison between these methods for generating instantaneous inlet velocity fields.

Table 2.1 Comparison between different methods for generating instantaneous inlet velocity fields

Method	Computational cost	Physical realism	Remarks
Random noise/perturbation	Low	Low	Easiest to implement; adds uncorrelated fluctuations to a mean profile. Does not reproduce realistic turbulence spectra, spatial correlation, or Reynolds stresses
Synthetic turbulence generation (STG)	Low–Moderate	Moderate–High	Produces controlled turbulence statistics and spatial correlations without full precursor runs. Good compromise for LES inflow; requires parameter tuning
Recycling & rescaling (Lund)	Moderate	High	Uses internal domain flow to generate inlet turbulence with realistic boundary-layer structure. Cost arises from extended domain length and additional processing
Precursor simulation	High	Very high	Most physically accurate for turbulent boundary layers; generates self-consistent turbulence. Costly because a long upstream domain or separate run is required
Stored turbulence library/mapping	Moderate–high	High	Involves building a library of turbulent fields from precursor runs and mapping them to the inlet. Lower cost than live precursor but still needs extensive storage and setup
Data-driven/ ML-enhanced methods	Variable	Potentially high	Utilises learned models (GANs, DL) to generate inlet turbulence rapidly. Cost in training phase; inference is fast. Realism depends on training data quality

2.7 Wind Turbine Aerodynamics

Wind turbines operate by extracting kinetic energy from the atmospheric boundary layer through complex aerodynamic interactions between the incoming flow and the rotating blades. The aerodynamic forces generated on the rotor govern not only the power production but also the structural loading, fatigue life, control strategies, and wake behaviour of wind turbines. A clear understanding of these forces is therefore fundamental for both classical Computational Fluid Dynamics (CFD) approaches and emerging AI- and machine-learning-assisted modelling frameworks.

2.7.1 Lift and Drag Forces on Wind Turbine Blades

At the most fundamental level, thc acrodynamic forces acting on a wind turbine blade arise from lift and drag, generated by the relative motion between the blade sections and the surrounding airflow. Each blade can be discretised into a series of aerofoil sections along the span, with local flow conditions varying with radius due to rotation.

Lift is generated primarily due to the pressure difference between the suction and pressure sides of the aerofoil, resulting from flow acceleration and curvature around the blade profile. This force acts perpendicular to the local relative wind direction and is the dominant contributor to energy extraction in modern horizontal-axis wind turbines. Drag, acting parallel to the relative wind, arises from viscous shear stresses and pressure losses associated with flow separation. While drag is unavoidable, it represents an aerodynamic penalty and reduces the net torque produced by the rotor.

The magnitude of lift and drag depends on the local angle of attack, Reynolds number, surface roughness, and aerofoil geometry. Under normal operating conditions, blades are designed to operate in a high-lift, low-drag regime. However, during unsteady inflow, yaw misalignment, or high turbulence intensity, the local angle of attack can fluctuate significantly, leading to dynamic stall and rapid changes in aerodynamic loading. For instance the effect of the integral length scale of turbulence on performance of a wind turbine aerofoil can be found on Vita (2020c).

2.7.2 Normal and Tangential Force Components

For wind turbine analysis, lift and drag are often resolved into normal (thrust-related) and tangential (torque-producing) force components relative to the rotor plane. The tangential force component is directly responsible for generating aerodynamic torque and hence electrical power. The normal force component contributes primarily to axial thrust acting on the rotor, which is transmitted through the hub, nacelle, and tower.

Axial thrust is a critical design load, influencing tower bending moments, foundation design, and overall turbine stability. Excessive thrust loads can accelerate fatigue damage, particularly under turbulent inflow or extreme wind conditions. Accurate prediction of thrust is therefore a central objective in both low-fidelity models and high-fidelity CFD simulations.

2.7.3 Aerodynamic Torque and Power Extraction

The integration of tangential aerodynamic forces along the blade span produces the aerodynamic torque acting on the rotor. This torque is balanced by the generator load and drivetrain resistance, defining the operating rotational speed of the turbine. The relationship between torque, rotational speed, and power extraction is commonly represented through non-dimensional performance coefficients such as the power coefficient C_P and thrust coefficient C_T.

Aerodynamic torque is highly sensitive to inflow conditions, blade pitch angle, and tip-speed ratio. Even small errors in predicting local aerodynamic forces can lead to significant discrepancies in power estimation. This sensitivity motivates the use of advanced CFD and data-driven approaches to better capture nonlinear and unsteady aerodynamic effects.

2.7.4 Centrifugal and Coriolis Effects

Although not aerodynamic forces in the classical sense, centrifugal and Coriolis forces influence the effective flow experienced by rotating blades. Centrifugal effects can drive radial flow along the blade, modifying boundary-layer behaviour and delaying separation. Coriolis forces, arising due to rotation, alter the momentum balance within the boundary layer and can enhance lift compared to two-dimensional aerofoil predictions.

These rotational effects are particularly important near the blade root, where three-dimensional flow dominates and classical blade-element assumptions become less accurate. High-fidelity CFD is often required to capture these phenomena, while AI-based surrogate models increasingly aim to learn such effects directly from simulation or experimental data.

2.7.5 Unsteady Aerodynamic Forces and Dynamic Stall

Wind turbine blades operate in a highly unsteady environment characterised by atmospheric turbulence, wind shear, tower shadow, and wake interactions. These effects introduce time-varying aerodynamic forces that can significantly exceed steady-state loads. Dynamic stall, in particular, occurs when rapid changes in angle of attack lead to delayed flow separation, vortex formation, and sudden lift drop-off.

Unsteady aerodynamic forces are a major driver of fatigue loading and noise generation. They also pose significant challenges for CFD, as they require fine spatial and temporal resolution. Machine learning methods are increasingly explored to model unsteady force fluctuations, identify dominant flow features, and reduce the computational cost of resolving transient aerodynamics.

2.7.6 Yaw and Tilt-Induced Aerodynamic Loads

When a wind turbine operates under yaw misalignment, the inflow is no longer perpendicular to the rotor plane. This results in asymmetric loading across the rotor disk, generating cyclic blade loads and additional yaw moments. Similarly, rotor tilt introduces variations in inflow angle along the blade span.

These effects lead to periodic variations in aerodynamic forces at frequencies related to the rotational speed, contributing to fatigue damage. Accurate representation of yaw- and tilt-induced loads is essential for control system design and lifetime assessment.

2.7.7 Tip Losses and Three-Dimensional Effects

Near the blade tip, pressure equalisation between the suction and pressure sides gives rise to strong tip vortices, reducing the effective lift and increasing induced drag. These three-dimensional effects limit aerodynamic efficiency and strongly influence wake structure.

Tip-loss effects are traditionally accounted for through semi-empirical correction models, but high-fidelity CFD reveals a rich vortex dynamics that varies with operating condition. Data-driven models trained on CFD or experimental datasets offer new opportunities to represent tip effects more accurately within reduced-order frameworks.

2.7.8 Wake-Induced Aerodynamic Forces

The aerodynamic forces on a wind turbine are coupled to the development of its wake. Momentum extraction by the rotor generates a velocity deficit and enhanced turbulence downstream, which can feed back onto the rotor under certain conditions, especially in wind farms.

Wake-induced unsteadiness modifies the effective inflow seen by downstream turbines, altering lift, drag, thrust, and torque. Understanding this coupling between blade-scale forces and farm-scale aerodynamics is a key motivation for integrating CFD, reduced-order modelling, and machine learning techniques.

2.7.9 Implications for CFD and AI-Based Modelling

The aerodynamic forces acting on wind turbines are inherently nonlinear, multi-scale, and unsteady. While classical CFD provides detailed insight into these forces, it remains computationally expensive for design optimisation and real-time applications. AI and machine learning approaches increasingly aim to complement CFD by learning force-flow relationships, correcting model deficiencies, and enabling fast predictions of aerodynamic loads across a wide range of operating conditions.

A rigorous physical understanding of lift, drag, thrust, torque, and unsteady aerodynamic effects remains essential to ensure that data-driven models are physically consistent, interpretable, and robust.

2.8 AI in Wind Energy Systems

Computational Fluid Dynamics (CFD) has become an indispensable tool for the design, analysis, and optimisation of wind energy systems. It enables detailed modelling of airflow around wind turbines, including blade aerodynamics, wake formation and interaction, atmospheric boundary-layer effects, turbulence characteristics, and inflow variability. Despite its proven capability, CFD remains computationally expensive, particularly when applied to large-scale wind farms, high-fidelity turbulence simulations, or real-time control and optimisation tasks. These limitations motivate the integration of Artificial Intelligence (AI)—and in particular Machine Learning (ML)—as a complementary approach to enhance, accelerate, and extend CFD-based wind energy analysis.

AI techniques offer new pathways to overcome several long-standing challenges in wind energy modelling. These include the high computational cost of large-scale simulations, the geometric complexity of turbine blades and terrain, the inherent uncertainty and variability of atmospheric conditions, and the need for fast or near-real-time predictions in operational settings. When combined with CFD, AI can accelerate simulation workflows, automate labour-intensive processes such as mesh generation and parameter tuning, and improve predictive accuracy through data-driven representations of complex flow phenomena.

Figure 2.4 illustrates a typical data-driven workflow for wind power prediction using machine-learning techniques. The process begins with the acquisition of wind power-related data from a wind turbine or wind farm. These data include key operational and environmental variables such as wind speed, wind direction, active wind power, generator speed, and gearbox bearing temperature. Together, these variables represent the system state and external forcing acting on the turbine.

The collected wind power data are first subjected to a pre-processing stage, where tasks such as data cleaning, outlier removal, normalisation, and feature selection are performed to ensure data quality and consistency. The processed dataset is then divided into training data and testing data.

The training dataset is used to develop different machine-learning models (e.g. OGPR, OENS, OTREE, BOT), which learn the relationship between the input variables and the power output. Once trained, these models are applied to the testing dataset, where the input variables (x) are fed into the trained models to generate predicted power.

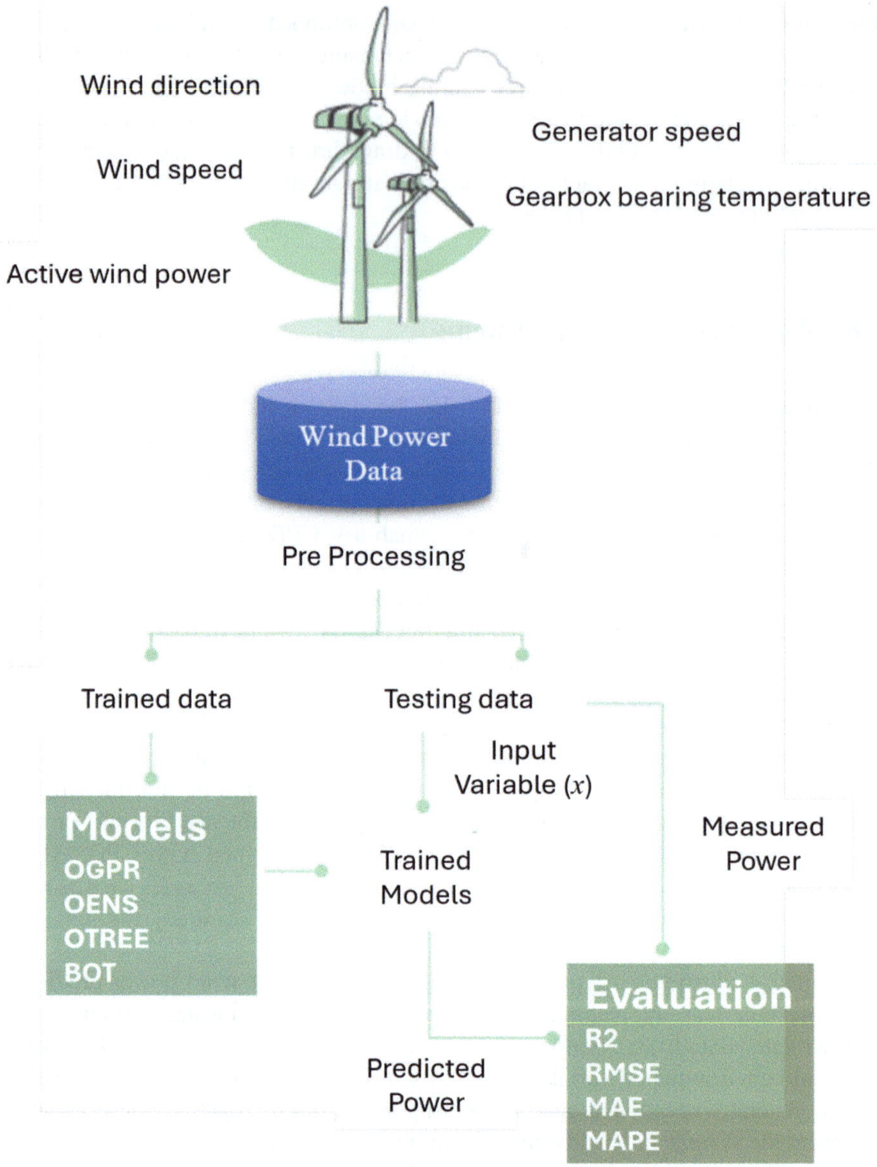

Fig. 2.4 Machine-learning workflow for wind power prediction

Finally, the predicted power is compared against the measured power to assess model performance. This evaluation step employs standard statistical metrics such as the coefficient of determination (R^2), root mean square error (RMSE), mean absolute error (MAE), and mean absolute percentage error (MAPE). The figure highlights the end-to-end pipeline from raw wind turbine data to validated machine-learning-based power prediction, demonstrating how data-driven models can support performance assessment and forecasting in wind energy systems.

2.8.1 Role of AI in CFD for Wind Energy

CFD simulations in wind energy are required to represent a broad range of strongly coupled physical processes, including wind resource assessment, rotor aerodynamics, wake evolution, turbulence transport, and atmospheric inflow modelling. Accurately resolving these phenomena often demands extremely high spatial and temporal resolution, leading to substantial computational cost, particularly for large wind farms and unsteady flow conditions. The problem is further complicated by the geometric complexity of wind turbine blades, support structures, and surrounding terrain, which poses challenges for mesh generation and numerical stability. In addition, atmospheric inflow conditions are inherently uncertain and highly variable in both space and time, introducing significant sensitivity in predicted aerodynamic loads, power production, and wake interactions. These difficulties are especially restrictive in operational contexts, where turbine control and optimisation tasks require fast or near-real-time predictions that are typically incompatible with high-fidelity CFD approaches.

Artificial Intelligence (AI) offers powerful tools to complement and extend CFD capabilities in wind energy applications by addressing these limitations. Data-driven and machine-learning-based methods can significantly accelerate CFD-based predictions through surrogate or reduced-order models, enabling rapid evaluation of complex flow scenarios. AI techniques can also automate labour-intensive pre-processing and post-processing tasks, such as mesh generation, parameter tuning, and feature extraction, thereby improving workflow efficiency and robustness. Moreover, by learning from high-dimensional simulation data or field measurements, AI models can uncover hidden patterns and relationships within turbulent flows that are difficult to capture using conventional modelling approaches alone. As a result, the integration of AI with CFD provides a promising pathway towards more efficient, scalable, and intelligent modelling frameworks for wind energy analysis, design, and operation.

2.8.2 AI Techniques Relevant to Wind Energy Applications

A broad range of artificial intelligence (AI) and data-driven techniques have been explored in wind energy research to support modelling, prediction, optimisation, and control tasks. These techniques include:

- Machine Learning (ML), encompassing supervised, unsupervised, and reinforcement learning methods,
- Computer Vision, for the analysis of flow-field images, wake visualisation, and condition monitoring,
- Natural Language Processing (NLP), for knowledge extraction, document analysis, and design automation,
- Evolutionary and metaheuristic optimisation methods, such as Genetic Algorithms (GA).

Genetic algorithms are not machine-learning models in the strict sense, as they do not learn explicit input–output mappings from data. Instead, they are population-based optimisation techniques inspired by natural evolution and are often used in combination with ML models for tasks such as feature selection, hyperparameter optimisation, model calibration, and aerodynamic or layout optimisation in wind energy applications.

2.8.3 Branches of Machine Learning in Wind Energy

1. **Supervised Learning** Supervised learning is one of the most widely used branches of machine learning in wind energy research and applications, owing to the availability of measured operational data and clearly defined input–output relationships. In supervised learning, models are trained using labelled datasets, in which each set of input variables is associated with a known output or target value. Through this process, the algorithm learns the underlying functional relationship between inputs and outputs, in a manner analogous to learning under guidance or supervision. Once trained, the model can be used to make predictions for new, unseen data, provided that the operating conditions remain within, or close to, the range represented in the training dataset. This paradigm is particularly well suited to wind energy systems, where large volumes of historical data are routinely collected from turbines and wind farms through supervisory control and data acquisition (SCADA) systems, meteorological masts, and remote sensing devices such as LiDAR. In wind energy applications, supervised learning has been extensively applied to power curve prediction, aerodynamic force and load estimation, and performance forecasting. Power curve prediction is a prominent example, where models are trained to map inflow conditions—such as wind speed, wind direction, turbulence intensity, and air density—to electrical power output. Accurate power curve models are essential for energy yield assessment, condition monitoring, and performance benchmarking. Supervised learning

methods often outperform traditional parametric power curve models by capturing nonlinear effects, wake-induced variability, and operational influences that are difficult to represent analytically. Similarly, supervised learning is increasingly used for aerodynamic force and load estimation, where the objective is to predict blade loads, thrust, or torque from measured or simulated flow variables. These predictions are crucial for fatigue analysis, control system design, and structural health monitoring. In performance forecasting, supervised models are employed to predict short-term or long-term power output based on weather forecasts, operational states, and historical trends, supporting grid integration, market participation, and maintenance planning. A range of supervised learning algorithms has been adopted in wind energy studies, each with distinct strengths and limitations. Support Vector Machines (SVMs), and their regression counterpart Support Vector Regression (SVR), are valued for their strong theoretical foundation and robustness in high-dimensional spaces. They are particularly effective for problems with limited training data and can provide good generalisation when appropriate kernel functions are selected. Random Forests, which are ensemble methods based on collections of decision trees, offer high predictive accuracy and resilience to noisy data. They are relatively easy to train and interpret, as they provide measures of feature importance that help identify the most influential input variables. Artificial Neural Networks (ANNs), including shallow and deep architectures, are amongst the most powerful supervised learning tools for wind energy applications. Their ability to approximate complex nonlinear relationships makes them well suited for modelling aerodynamic behaviour, power output, and load dynamics under varying operating conditions. However, ANNs typically require larger datasets and careful tuning to avoid overfitting and ensure stable performance. Overall, supervised learning provides a flexible and effective framework for modelling wind energy systems, particularly when combined with physical insight and high-quality data to ensure robustness and interpretability. For example, Pei and Li (2019) employed fuzzy c-means clustering to identify and remove outliers in wind data, followed by training multiple extreme learning machines. The resulting forecasts were combined using support vector regression to improve prediction accuracy, demonstrating the effectiveness of hybrid supervised-learning strategies.

2. **Unsupervised Learning** Unsupervised learning plays an increasingly important role in wind energy research, particularly in applications where large volumes of data are available but explicit labels or target outputs are not. Unlike supervised learning, unsupervised learning operates solely on unlabelled datasets and seeks to uncover hidden patterns, structures, or relationships within the data itself. This capability is especially valuable in Computational Fluid Dynamics (CFD) and wind energy applications, where simulations and measurements often generate high-dimensional datasets containing complex flow information that is difficult to interpret using conventional analysis techniques. By organising and reducing such data, unsupervised learning provides a powerful means of extracting physical insight and supporting both design and operational decision-making. In wind energy CFD, unsupervised methods are commonly applied to clustering flow regimes, identifying wake patterns, and reducing the dimensionality of CFD datasets using techniques such as

Principal Component Analysis (PCA). Clustering algorithms can be used to group similar flow states or operating conditions, for example separating attached-flow regimes from separated or stalled conditions on wind turbine blades, or distinguishing different wake interaction modes in wind farms. This enables researchers and engineers to classify complex flow behaviour into a small number of representative regimes, which can then be analysed in greater detail or used to guide reduced-order modelling. Wake pattern identification is another important application, where unsupervised learning helps reveal recurring wake structures, such as symmetric or asymmetric wakes, merged wakes in turbine arrays, or wake meandering under turbulent inflow. These insights are valuable for understanding wake dynamics, improving wind farm layout design, and developing control strategies for wake mitigation. Dimensionality reduction techniques, particularly PCA and related methods, are widely used to simplify high-dimensional CFD data by identifying a reduced set of dominant modes that capture most of the flow variability or energy content. By projecting complex velocity or pressure fields onto a low-dimensional basis, these methods make it possible to analyse, visualise, and model turbulent flows more efficiently. Such reduced representations can serve as the foundation for reduced-order models, surrogate modelling, or subsequent supervised learning tasks. Overall, unsupervised learning helps simplify and structure complex flow data, identify dominant aerodynamic behaviours, and expose underlying flow physics that may not be apparent through direct inspection. Because it does not rely on predefined output labels, unsupervised learning is particularly well suited for exploratory analysis, hypothesis generation, and the discovery of new flow phenomena in wind energy systems. As an example of the unsupervised learning, Ti and Yang (2020) proposed a novel CFD-machine-learning framework to improve wind turbine wake prediction accuracy and efficiency by coupling RANS simulations with an actuator disk model and an ANN trained to learn the spatial relationship between inflow conditions and three-dimensional wake characteristics. Validation against wind tunnel tests, LES, and field measurements demonstrates that the ANN-based wake model accurately predicts velocity deficits and turbulence levels for both single turbines and wind farm configurations while significantly reducing computational cost.

3. **Reinforcement Learning** Reinforcement Learning (RL) represents a distinct class of machine learning methods that is particularly well suited to decision-making and control problems, where actions taken by an agent influence the future state of a system. In RL, an agent interacts with an environment by observing its current state, selecting actions, and receiving feedback in the form of rewards or penalties. Through repeated interactions, the agent learns a control policy that maximises the cumulative reward over time. This trial-and-error learning paradigm makes RL especially attractive for complex, nonlinear systems where explicit control laws are difficult to derive, as is often the case in wind energy applications involving unsteady aerodynamics and coupled turbine-wake interactions. In wind energy, reinforcement learning has shown considerable promise for active flow control, blade pitch and flap optimisation, and real-time yaw and wake steering control. For active flow control, RL agents can learn how to adjust control devices–such as trailing-edge flaps or synthetic jets–to suppress

flow separation, delay stall, or reduce unsteady loads under varying inflow conditions. In blade pitch and flap optimisation, RL can identify control strategies that balance power maximisation with load alleviation, adapting dynamically to changes in wind speed, turbulence intensity, and operating constraints. At the wind farm level, RL-based yaw and wake steering control aims to mitigate wake losses by coordinating the behaviour of multiple turbines, intentionally misaligning upstream turbines to redirect wakes and increase overall farm power output. Such coordinated control strategies are difficult to design using conventional rule-based approaches but can emerge naturally through RL training. Despite its potential, the practical application of reinforcement learning in wind energy faces significant challenges, particularly related to training cost and robustness. Effective RL training typically requires millions of interactions with the environment to explore the state-action space and converge to an optimal policy. When high-fidelity Computational Fluid Dynamics (CFD) models are used to represent the environment, this requirement becomes computationally prohibitive. As a result, many studies rely on simplified flow models, reduced-order representations, or surrogate environments during training, which may limit the fidelity of the learned control policies. Bridging the gap between training in simplified environments and deployment in realistic, turbulent flow conditions remains an active area of research. Nevertheless, advances in surrogate modelling, parallel computing, and hybrid CFD-AI frameworks are steadily improving the feasibility of RL-based control in wind energy systems, particularly for real-time and adaptive applications. Portal-Porras and Berrizbeitia (2023), for example, demonstrated the use of RL coupled with CFD to optimise aerofoil flow control, achieving rapid and robust aerodynamic performance across a range of operating conditions. Abkar and Iosifidis (2023) reviewed the use of reinforcement learning in wind farm flow control. This review summaries recent advances in reinforcement-learning-based wind-farm flow control and highlights the key challenges and open research directions that must be addressed to enable the next generation of control strategies.

4. **Deep Learning** Deep learning is a specialised subset of machine learning that employs multi-layer artificial neural networks to learn complex, highly nonlinear relationships from large and often high-dimensional datasets. By stacking multiple hidden layers, deep learning models are able to extract hierarchical features, moving from low-level representations to increasingly abstract and informative patterns. This capability makes deep learning particularly well suited to problems where the underlying physics is complex, multi-scale, and difficult to capture using hand-crafted features alone. In wind energy applications, deep learning has gained increasing attention as the volume of available data—from high-fidelity CFD simulations, wind tunnel experiments, field measurements, and supervisory control and data acquisition (SCADA) systems—continues to grow. A key strength of deep learning lies in its ability to process unstructured data, such as images, flow-field snapshots, and time-series signals. Convolutional Neural Networks (CNNs), for example, are especially effective for analysing spatially distributed data and have been widely used for wake visualisation and flow-field reconstruction. When trained on CFD-generated

velocity or vorticity fields, CNNs can learn to identify coherent wake structures, estimate velocity deficits, and reconstruct full flow fields from sparse measurements. This capability is particularly valuable for wind farm applications, where direct measurement of the complete wake field is often impractical. CNN-based models can also support reduced-order modelling by providing fast approximations of complex flow features that would otherwise require expensive simulations. Recurrent Neural Networks (RNNs), including advanced variants such as Long Short-Term Memory (LSTM) and Gated Recurrent Unit (GRU) networks, are well suited to time-dependent problems and sequential data. In wind energy, RNNs have been successfully applied to the prediction of wind speed, power output, and unsteady load fluctuations, capturing temporal correlations and long-range dependencies that are difficult to represent with traditional regression models. These capabilities are essential for forecasting, condition monitoring, and control, where accurate prediction of future system behaviour can improve reliability and performance. Despite their strengths, deep learning models typically require large, high-quality datasets for training and careful validation to ensure generalisation beyond the training conditions. As such, their integration with CFD and physics-based constraints remains an important research direction, aiming to combine the expressive power of deep learning with the robustness and interpretability of physical models in wind energy applications. Wang and Bao (2025) applied a Soft Actor-Critic deep reinforcement learning framework, coupled with a surrogate neural-network model, to efficiently optimise wind turbine airfoil aerodynamics by formulating the design task as a Markov Decision Process. The results demonstrated high prediction accuracy and substantial improvements in lift-to-drag ratio across different airfoil classes, confirming the method's physical consistency, fast convergence, and strong potential for practical aerodynamic design optimisation. More work used deep learning could be found in Calzolari and Liu (2021) and Tanakal and Sone (2019).

2.8.4 Surrogate Modelling and Digital Twins

Surrogate modelling and digital twins represent one of the most impactful and rapidly developing applications of artificial intelligence within Computational Fluid Dynamics (CFD), particularly in the context of wind energy systems. The central idea of surrogate modelling is to replace, or augment, a computationally expensive high-fidelity CFD simulation with a fast, data-driven approximation that reproduces the essential input–output behaviour of the original model. This approach enables rapid evaluation of aerodynamic performance and flow responses across a wide range of operating conditions, making it especially attractive for design optimisation, uncertainty analysis, and real-time decision support.

The typical surrogate-modelling workflow begins with data generation, in which a carefully selected set of high-fidelity CFD simulations is performed. These simulations are designed to span the relevant parameter space, such as wind speed, turbulence intensity, yaw misalignment, blade pitch, or turbine spacing. Although each simulation may be computationally expensive, only a limited number is required, provided that the sampling strategy is chosen effectively. In the second stage, model training, a neural network or other machine learning model—such as Gaussian process regression, support vector regression, or deep learning architectures—is trained using the CFD-generated input–output data. Through this training process, the surrogate model learns the underlying relationships between operating conditions and aerodynamic responses, such as power output, thrust, load distributions, or wake characteristics. Once trained, the surrogate is deployed to provide fast predictions at a fraction of the computational cost of the original CFD model. In this deployed state, the surrogate effectively acts as a digital twin of the CFD simulation, capable of delivering near-instantaneous responses suitable for optimisation loops, sensitivity studies, or control applications.

In wind energy, surrogate models and digital twins are increasingly used to predict turbine performance under varying wind conditions, analyse wake interactions within wind farms, and support the optimisation of turbine layouts and control strategies. At the turbine scale, digital twins can assist in performance monitoring and load prediction by combining real-time measurements with surrogate-based flow predictions. At the wind farm scale, surrogate models enable rapid exploration of layout configurations and wake-steering strategies that would be infeasible using high-fidelity CFD alone. While challenges remain—particularly in ensuring robustness, generalisation, and physical consistency—the combination of CFD and AI-driven surrogate modelling offers a powerful framework for bridging the gap between high-fidelity simulation and practical, real-time wind energy applications. Tanaka and Sone (2019) demonstrated that ML models trained on CFD-generated flow-field images could reduce prediction time from minutes to milliseconds while preserving high accuracy.

2.8.5 Challenges and Research Gaps

Despite significant progress in both Computational Fluid Dynamics (CFD) and Artificial Intelligence (AI), their effective integration for wind energy applications continues to face a number of fundamental and practical challenges that span data, physics, computation, and industrial deployment. One of the most persistent barriers is data quality and availability. High-quality datasets are essential for training, validating, and deploying AI models, yet field measurements from wind turbines and wind farms are often noisy, incomplete, or affected by sensor faults, environmental contamination, and maintenance downtime. Moreover, much of the most valuable operational data is proprietary, owned by turbine manufacturers or wind farm operators, which severely limits data sharing, benchmarking, and reproducibility across the research community. This scarcity of open, standardised datasets hampers the development of robust, generalisable AI models and makes it difficult to rigorously compare AI-enhanced CFD approaches. Closely related to this issue is the challenge of scale and resolution mismatch. Atmospheric and mesoscale weather models typically operate at spatial resolutions of kilometres and temporal scales of minutes to hours, whereas wind turbine aerodynamics, wake interactions, and blade-scale flow physics occur at metre or sub-metre scales and over much shorter time horizons. CFD can bridge this gap by resolving fine-scale flow features, but at a very high computational cost. AI models trained at one scale often struggle to remain accurate when applied across scales, and coupling AI, CFD, and atmospheric models in a physically consistent and computationally efficient manner remains an open research problem.

Computational cost remains another central limitation. High-fidelity CFD approaches such as Large-Eddy Simulation (LES) or hybrid RANS–LES methods are essential for accurately capturing turbulence, unsteady separation, and wake dynamics, but they require substantial computational resources, long wall-clock times, and large memory footprints—particularly for full-scale wind farms or long-duration simulations. While AI-based surrogate models promise dramatic speed-ups, the generation of training data itself often relies on large ensembles of expensive CFD simulations. This creates a trade-off between up-front computational investment and downstream efficiency gains. Furthermore, ensuring that AI surrogates remain accurate outside their training envelope is non-trivial, especially under rare or extreme conditions such as gusts, storms, or emergency shutdown scenarios. Model interpretability and trust present additional obstacles, particularly for safety-critical and regulated applications. Many advanced AI models, especially deep learning architectures, function as black boxes, making it difficult to understand how predictions are generated or to ensure consistency with known physical laws. This lack of transparency complicates validation, uncertainty quantification, and regulatory acceptance, especially when AI-driven predictions are used to inform design decisions, control strategies, or lifetime assessments. Embedding physical constraints, conservation laws, and uncertainty estimates into AI models—through physics-informed learning or hybrid modelling—remains an active area of research but is not yet mature for widespread industrial use.

Beyond data and computation, the inherent complexity of wind energy flow physics poses a further challenge. Turbulence, wake-wake interactions, terrain effects, atmospheric stability, and extreme inflow conditions all interact in highly nonlinear and multiscale ways that are difficult to capture even with state-of-the-art CFD. AI models may identify correlations in data, but without careful physical grounding they risk misrepresenting causal relationships or failing under conditions not seen during training. Real-time or near-real-time deployment adds another layer of difficulty, as AI-CFD frameworks must interface reliably with sensors, actuators, and control systems while remaining robust to data latency, missing measurements, and model drift over time. In addition, wind energy systems are inherently multidisciplinary: aerodynamic performance is tightly coupled to structural loads, fatigue, acoustics, control strategies, cost, maintenance, and environmental impact. Many current AI or CFD studies focus on isolated aspects of the problem, whereas practical deployment requires integrated optimisation across multiple objectives and constraints. Ensuring that AI-enhanced CFD tools generalise from idealised test cases to real-world wind farms—with complex terrain, offshore conditions, aging turbines, and heterogeneous layouts—remains a major hurdle. Finally, industrial adoption is constrained by issues of certification, standardisation, and economic justification. Industry stakeholders require clear evidence of reliability, robustness, and return on investment before adopting AI-driven CFD tools in design or operation. Regulatory frameworks for certifying AI-assisted methods are still evolving, and standards for validation and uncertainty management are not yet well established. Addressing these challenges will require continued advances in physics-informed AI, scalable CFD-AI coupling strategies, open data initiatives, and close collaboration between academia, industry, and regulators to ensure that AI-enhanced CFD can be deployed with confidence in wind energy systems.

2.8.6 Concluding Remarks

AI-enhanced CFD is increasingly being used to explore large design spaces, optimise blade geometries, improve aerodynamic efficiency, and enhance annual energy production. In some studies, AI-assisted design workflows have reported substantial performance gains under nominal operating conditions. Beyond design, AI-CFD integration is already delivering value in forecasting, predictive maintenance, and turbine control, contributing to reduced operational costs and improved reliability. Nevertheless, many reported successes are demonstrated under controlled or idealised conditions. Ensuring robustness across diverse terrains, atmospheric regimes, and turbine types remains a key challenge. Balancing speed and fidelity, avoiding model bias, and maintaining physical consistency are central concerns. Emerging AI models trained on CFD or analytical wake representations—such as graph-based and neural wake models—show strong potential for accelerating wind farm layout design and wake steering while retaining much of the predictive power of CFD. Continued advances in physics-informed learning, uncertainty quantification, and validation frameworks will be essential to promote wider adoption of AI-driven CFD in wind energy systems.

References

Abkar, M., Zehtabiyan-Rezaie, N., & Iosifidis, A. (2023). Reinforcement learning for wind-farm flow control: Current state and future actions. *Theoretical and Applied Mechanics Letters, 13*.

Calzolari, G., & Liu, W. (2021). Deep learning to replace, improve, or aid CFD analysis in built environment applications: A review. *Building and Environment, 206*.

Eurocode 1. (2005). Actions on Structures—Part 1–4: General Actions—Wind Actions.

Pei, S., & Li, Y. (2019). Wind turbine power curve modeling with a hybrid machine learning technique. *Applied Science, 9*.

Portal-Porras, K., Fernandez-Gamiz, U., Zulueta, E., Garcia-Fernandez, R., & Berrizbeitia, S. E. (2023). Active flow control on airfoils by reinforcement learning. *Ocean Engineering, 287*.

Tanaka, H., Kawakami, T., Azegami, Y., Yamamoto, M., Ohtake, K., Matsuoka, Y., & Sone, T. (2019). Optimization calculations and machine learning aimed at reduction of wind forces act-

ing on tall buildings and mitigation of wind environment. *International Journal of High-Rise Buildings,8*, 291–302.
Ti, Z., Deng, X. W., & Yang, H. (2020). Wake modeling of wind turbines using machine learning. *Applied Energy,* 267.
Vita, G., Hemida, H., Andrianne, T., & Baniotopoulos, C. (2020a). The effect of the integral length scale of turbulence on a wind turbine aerofoil. *Journal of Wind Engineering and Industrial Aerodynamics, 204.*
Vita, G., Hemida, H., Andrianne, T., & Baniotopoulos, C. (2020b). The effect of the integral length scale of turbulence on a wind turbine aerofoil. *Journal of Wind Engineering and Industrial Aerodynamics, 204.*
Vita, G., Hemida, H., Andrianne, T., & Baniotopoulos, C. (2020c). The effect of the integral length scale of turbulence on a wind turbine aerofoil. *Journal of Wind Engineering and Industrial Aerodynamics, 204.*
Wang, Y., Vita, G., Fraga, B., Lyu, C., Wang, J., & Hemida, H. (2022). Influence of turbulent inlet boundary condition on large eddy simulation over a flat plate boundary layer. *International Journal of Computational Fluid Dynamics,3*, 232–259.
Wang, Y., Vita, G., Fraga, B., Wang, J.,& Hemida, H. (2021). Effect of the inlet boundary conditions on the flow over complex terrain using large eddy simulation. *Designs,2*, 5–34.
Wang, H., Baoyin, H., Han, S., Li, H., & Bao, C. (2025). Aerodynamic optimization of wind turbine blades via surrogate-assisted deep reinforcement learning. *Physics of Fluids, 37.*

Chapter 3
Dynamics of Offshore Wind Energy Systems: Fundamentals and Data-Driven Methods

Giulio Ferri and Enzo Marino

Abstract Floating Offshore Wind Turbines (FOWTs) enable exploitation of deep-water wind resources beyond the reach of fixed foundations, but their design and operation involve complex coupled dynamics, including aerodynamic, hydrodynamic, structural, and mooring interactions. Accurate modeling is essential to predict extreme and fatigue loads, ensure reliability, and optimize performance. This chapter reviews the fundamental principles of wind generation, wave kinematics, and wave–structure interactions and presents modeling strategies for FOWTs, covering rigid- and flexible-body dynamics, potential-flow and Morison-based hydrodynamics, and mooring system representations. Frequency-domain approaches are highlighted as efficient tools for computing hydrodynamic coefficients and steady-state responses, supporting preliminary design, structural optimization, and reliability assessment. The chapter also explores a few data-driven and surrogate modeling techniques, including machine learning and Gaussian process regression, which reduce computational costs while preserving fidelity, enabling long-term fatigue assessment and control-oriented applications. Overall, the chapter provides an integrated framework for designing robust, cost-effective, and reliable FOWTs.

3.1 Introduction

Offshore wind energy is a rapidly growing source of renewable power, driven by the need to decarbonize energy systems and to exploit wind resources available at sea. In particular, Floating Offshore Wind Turbines (FOWTs) enable access to deep-water sites beyond the limits of fixed-bottom foundations, significantly expanding the geographical and technical potential for offshore wind deployment. However, the transition from bottom-fixed to floating support structures introduces substantial challenges related to system dynamics, as FOWTs are exposed to the combined wind, wave, and current loading.

G. Ferri · E. Marino (✉)
Department of Civil and Environmental Engineering, University of Florence, Firenze, Italy
e-mail: enzo.marino@unifi.it

C. Baniotopoulos and E. Marino (eds.), *Advances in Wind Energy in the Era of Artificial Intelligence*, CISM International Centre for Mechanical Sciences 625,
https://doi.org/10.1007/978-3-032-28815-8_3

The dynamic behavior of offshore wind energy systems is governed by complex coupled interactions among aerodynamic loads acting on the rotor and tower, hydrodynamic forces induced by waves and currents, structural flexibility of the support structure, and nonlinear responses of the mooring system (Ferri et al., 2022; Otter et al., 2022; Ferri and Marino, 2023). Accurate representation of these interactions is essential because they directly affect fatigue life, extreme load responses, power production, and overall system reliability. Previous studies have highlighted the sensitivity of global dynamics to platform geometry, mass distribution, and mooring configuration, particularly for large-capacity turbines (Ferri et al., 2020; Bruschi et al., 2020).

Hydrodynamic modeling remains a critical source of uncertainty in offshore wind dynamics. The choice between linear and nonlinear wave-loading formulations, as well as the treatment of viscous effects, can strongly influence predicted responses, especially for floating platforms and slender structural members (Mockute et al., 2019; Marino et al., 2018). These challenges are further amplified by the increasing scale of modern wind turbines, whose tall towers and long rotor blades operate in highly nonlinear regimes, necessitating careful consideration of coupled hydro-aeroelastic effects.

Aero-hydro-servo-elastic simulation tools have become the standard for capturing the complex coupled behavior of offshore wind energy systems. However, many of these solvers rely on linear or weakly nonlinear formulations, which may limit their accuracy under strongly nonlinear operating conditions. High-fidelity models, on the other hand, can provide detailed insight into nonlinear system responses, but their high computational cost restricts their applicability in large-scale design optimization, uncertainty quantification, and real-time monitoring or control. To address these limitations, data-driven methods are increasingly integrated with physics-based formulations. These approaches enable efficient exploration of the design space, accelerate simulation workflows, and support digital twin for control-oriented applications, while retaining consistency with the underlying physical principles.

This chapter reviews the fundamental dynamic principles governing offshore wind energy systems, with a particular focus on floating platforms. It presents core modeling strategies for rigid-body and flexible dynamics, hydrodynamic interactions, and mooring system representations, before addressing multidisciplinary design and optimization challenges. Recent advances in data-driven techniques are then discussed as complementary tools that enhance accuracy and computational efficiency. By combining fundamental physics with modern data-driven approaches, this chapter provides a coherent framework for advancing the analysis, design, and reliability assessment of next-generation offshore wind energy systems.

3.2 Wind Energy

This section introduces the main sources of loads acting on Offshore Wind Turbines (OWTs). The fundamental physics underlying wind generation is first outlined, together with an overview of the offshore wind resource and the European and Italian wind potential. Subsequently, the basic concepts of wind energy converters are presented, with emphasis on the different operating principles of vertical- and horizontal-axis Wind Turbines (WTs).

Wind originates from the motion of air masses relative to the Earth's surface. Different wind scales can be identified, ranging from local breezes to global circulation patterns. Local winds typically persist for several hours and are primarily generated by pressure gradients arising from spatial variations in air temperature.

Local winds develop within the Atmospheric Boundary Layer (ABL), defined as the lowest portion of the atmosphere directly influenced by interaction with the planetary surface. The thickness of the ABL depends on surface roughness and typically ranges from about 200 m over open-sea conditions to approximately 1 km in densely built urban areas. Within the ABL, the mean wind speed increases with altitude, from values close to zero at the surface to a maximum that approaches the geostrophic velocity near the upper boundary of the layer.

Global (geostrophic) winds, by contrast, are generated by large-scale temperature differences between the equator and the poles, as well as by the rotation of the Earth through the Coriolis effect. These wind patterns are largely unaffected by local terrain or surface roughness.

In general, WTs operate within the local wind regime. However, new-generation Floating Offshore Wind Turbines (FOWTs), with rated power capacities reaching up to 15 MW, are characterized by extremely tall towers and rotor blades with very large spans. As a result, portions of the rotor may operate partially above the ABL.

The fundamental working principle of modern WTs, namely, the conversion of wind kinetic energy into mechanical work, has been adopted since ancient times. Early applications include windwheels and windmills, which can be traced back to approximately the ninth century (Hau, 2006).

Over the last decade, wind energy has emerged as one of the most significant resources in the renewable energy sector. Recent statistics on worldwide installed capacity indicate that offshore wind is a rapidly expanding segment, reaching a total installed capacity of 75.2 GW in 2024 (Fig. 3.1). Nearly half of global offshore installations are located in Europe, which—together with China—remains a leading actor in the sector. As of 2024, Europe's cumulative offshore wind capacity exceeds 35 GW, with annual installations fluctuating between 2.5 and 3.5 GW since 2017 (Fig. 3.2).

Within the offshore wind sector, FOWT installations currently constitute a relatively small portion of total operational capacity, with approximately 277 MW of floating wind deployed globally as of late 2025. However, the global floating wind fleet is expanding rapidly, with projections suggesting it could approach 2.5 GW

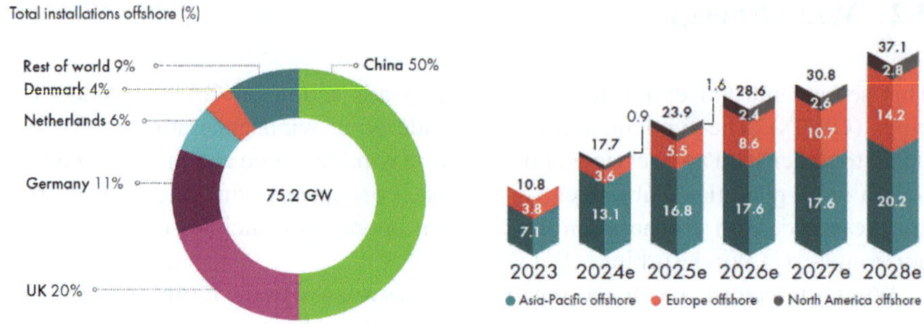

Fig. 3.1 Offshore installed capacity worldwide, courtesy of Williams et al. (2024)

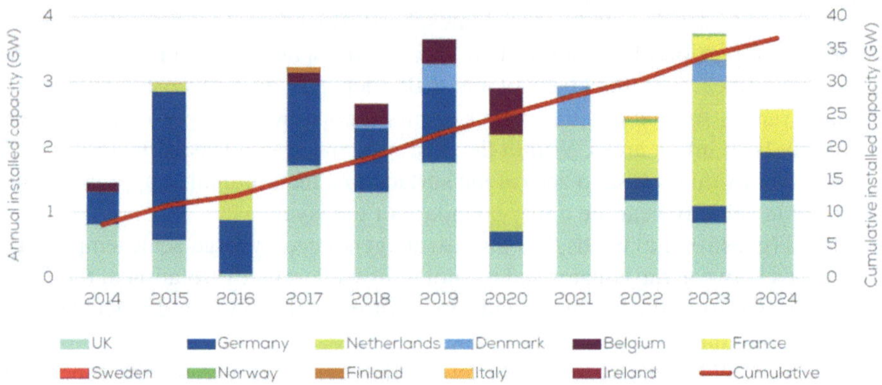

Fig. 3.2 Europe annual installed capacity, courtesy of Frémaux (2025)

of installed capacity by 2030, driven by numerous awarded and planned projects worldwide (Costanzo et al. 2025).

Higher average wind speeds, reduced turbulence intensity, the availability of vast deep-water sites, and a near absence of visual impact are among the main factors that make FOWTs competitive compared to fixed-bottom offshore and onshore wind technologies. Conversely, construction, installation, and maintenance costs increase substantially in open-sea environments. In addition, multiple fluid–structure interaction problems arise, involving very slender and highly flexible components such as rotor blades and mooring lines, as well as bluff bodies such as the floating platform. In the following, offshore wind turbines are described by introducing their main features, with particular attention devoted to support structures.

3.3 Floating Offshore Wind Turbines

A horizontal-axis Floating Offshore Wind Turbine (FOWT) consists of four interacting structural systems: rotor and nacelle, tower, floating platform, and anchoring system. The rotor and the nacelle are the components in which the kinetic energy of the rotating blades is converted into electrical energy. They are both located at the top of the turbine tower. All electronic devices and the control system are housed inside the nacelle. The control system consists of a set of devices and algorithms that adjust blade pitch, rotor speed, and turbine orientation to ensure correct energy production at each wind speed, from cut-in to cut-out. When the wind speed exceeds the cut-out value, the control system reduces the loads acting on the tower and helps prevent structural failures.

The tower is usually a steel beam with a tapered circular cross section. Its function is to place the rotor several meters above the Sea Water Level (SWL), where wind speed is higher and the effects of sea-surface roughness are reduced. The tower carries the aerodynamic and servo-dynamic loads transmitted by the rotor and nacelle.

The floating support system transfers loads from the tower base to the seabed through two subsystems: the platform and the mooring lines. The mooring system consists of a group of cables connected to the platform, providing the horizontal stiffness required to limit displacements within admissible bounds. Mooring lines are usually made of steel chains or steel wires, although polyester ropes have recently been adopted.

Slack catenary mooring systems are the most commonly used due to their simplicity in design and installation. They are very long (often exceeding four times the water depth) and, thanks to their dead weight, can compensate vertical loads. The cable length can be designed to ensure that a portion of the line always rests on the seabed. In this configuration, anchors are required to withstand only horizontal forces, allowing the use of simpler and more economical systems such as drag-embedded or gravity anchors. Slack catenary systems provide only a marginal contribution to the rotational stability of the structure.

Taut-line moorings consist of tensioned, inclined cables that contribute to both horizontal and rotational stability, enabling lighter platforms and stiffer systems. These moorings are significantly shorter than catenary ones and typically require a larger cross section. The design of the anchoring system is crucial, as the anchors are subjected to vertical loads. For this reason, suction piles or driven piles are generally preferred over gravity anchors.

Tension Leg (TL) moorings represent a particular case of taut-line systems characterized by vertical cables. A hybrid mooring system has recently been proposed for a 10 MW wind turbine (Bredmose et al., 2017). This configuration consists of two cable segments: the first, connected to the anchor, is a catenary chain that reduces anchor loads through its dead weight; the second is an inclined, tensioned taut-line wire. This solution combines the advantages of both technologies.

Based on the mechanism by which buoyancy stability is achieved, different platform technologies can be identified (Fig. 3.3).

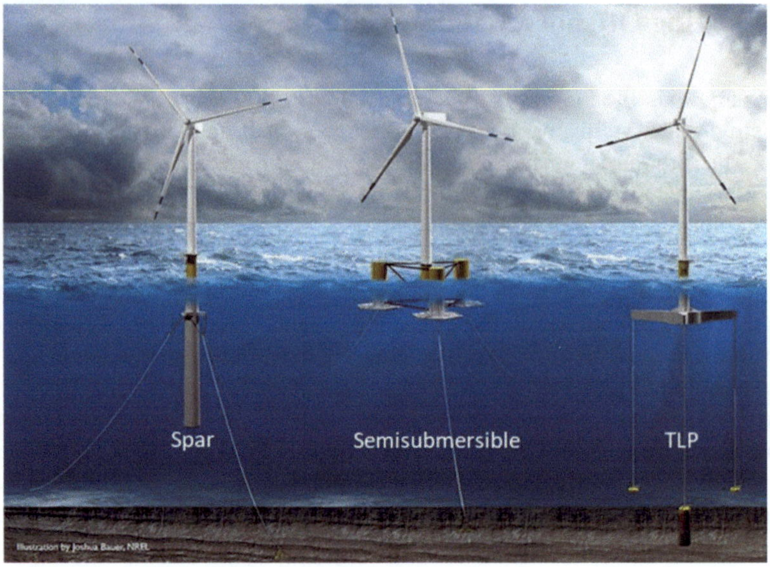

Fig. 3.3 FOWT principal concepts, courtesy of Barter et al. (2020)

- **Ballast-stabilized platforms**: Roll and pitch stabilities are achieved through ballast weights that lower the overall center of mass of the FOWT, resulting in larger restoring moments. Surge, sway, and yaw motions must be constrained by the mooring system. The most common concept of this type is the spar platform (Jonkman, 2010). It consists of a single cylindrical structure with a very large draft. Owing to its simplicity, it was among the first concepts to be developed and remains one of the most widely installed FOWT platforms worldwide. However, its large dimensions—typically between one and one and a half times the turbine tower height—may make it unsuitable for 10 MW wind turbines and may restrict its applicability to specific water-depth ranges.
- **Buoyancy-stabilized platforms**: Roll and pitch stabilities are almost entirely provided by the platform itself through the righting moment associated with the inertia of the waterplane area. The mooring system is primarily responsible for constraining horizontal displacements and, to a lesser extent, yaw rotation. Compared to ballast-stabilized platforms, these systems are characterized by reduced draft and a large footprint, resulting in lighter but stiffer structures. Consequently, particular attention must be paid to the dynamic response to avoid excessive amplifications that could negatively affect turbine performance (Robertson and Jonkman, 2011).
- **Mooring-stabilized platforms**: Roll and pitch stabilities are achieved through a tensioned mooring system capable of providing restoring moments. This approach allows reduced platform size and weight, as well as shorter mooring lines. However, the cables are subjected to higher tensile stresses than in other concepts. Special care must be taken in the design of the anchoring system, which must

withstand vertical loads, as well as in the dynamic design. These concepts are particularly sensitive to roll and pitch motions, which may lead to loss of cable tension. For this reason, resonance under combined wind and wave conditions must be avoided. One of the most common examples is the Tension Leg Platform (TLP) (Matha, 2010), although hybrid concepts such as taut-line and semi-taut-line platforms have also been developed (Bredmose et al., 2017).

By combining the characteristics of these three main concepts, several hybrid solutions have been developed. The most common of these is the semi-submersible platform, which achieves stability through both ballast and buoyancy effects (Butterfield et al., 2007).

3.3.1 Floating Platforms Modeling

From a mechanical point of view, a FOWT is a complex multi-body system exposed to a multi-physics fluid–structure interaction that involves wind and water (waves and currents). To properly reproduce its dynamic response, nonlinear aero-hydro-servo-elastic time-domain simulations are usually employed. A common approach is to combine modal- and multi-body dynamics formulations (Jonkman and Buhl, 2007). As far as the WT is concerned, both blades and tower are modeled as flexible elements adopting a linear modal representation, considering three DoFs for the blades (two flapwise and one edgewise bending mode) and four DoFs for the tower (two fore-aft and two side-to-side bending modes). Nacelle and hub are modeled as rigid bodies, with appropriate lumped mass and inertia.

Since the floating platform is stiffer than blades and tower, hydro-elasticity is generally neglected, and its kinematics is represented by a six-DoF rigid-body model: three translations (surge, sway, heave) and three rotations (roll, pitch, yaw). A small-angle assumption is also commonly adopted.

Different representations can be used for the mooring lines. The simplest employs quasi-static models that solve the catenary system of equations at each time step to update, according to platform motions, the stiffness and restoring forces at the platform fairleads. More advanced computational models divide the cable into elements with lumped mass connected by springs and dampers or discretize it using finite elements with appropriate elastic or viscoelastic properties. These approaches allow loads to be distributed along the cable length and enable hydrodynamic loads due to waves and currents to be accounted for. Recent developments also allow the localized rotational stiffness arising from contact and friction between chain links to be considered.

3.4 Review of the Wave–Body Interaction Problem

In this section, we briefly review fundamental concepts of fluid mechanics that are essential for the hydrodynamic characterization of floating structures supporting FOWTs.

3.4.1 The Potential-Flow Model

For many offshore engineering applications, the assumptions of an incompressible, inviscid fluid and irrotational flow are acceptable. Under these assumptions, the gravity-wave kinematics can be fully described by the velocity potential $\Phi(t, q)$, i.e., an Eulerian scalar field such that

$$\boldsymbol{v}(t, q) = \nabla\Phi(t, q)\,, \tag{3.1}$$

where $\boldsymbol{v}(t, q)$ is the Eulerian velocity at time t and at spatial position q within the domain Ω. Together with the incompressibility condition, this leads to the Laplace equation

$$\nabla^2\Phi(t, q) = 0\,. \tag{3.2}$$

The translational momentum balance is expressed through the unsteady Bernoulli equation:

$$\frac{\partial\Phi(t, q)}{\partial t} + \frac{1}{2}\left(\nabla\Phi(t, q)\right)^2 + \frac{\mathrm{p}(t, q)}{\rho} + gz(t, q) = \text{constant in space}\,, \tag{3.3}$$

where g is the gravitational acceleration, and $z(t, q)$ and $\mathrm{p}(t, q)$ denote the vertical coordinate and the hydrostatic pressure at q, respectively.

3.4.2 Linear Waves Theory

We provide a brief review of the linear waves theory. For additional details, the reader is referred to classical textbooks, e.g., (Sarpkaya and Isaacson 1981; Chakrabarti 1987; Journèe and Massie 2012).

The Laplace equation is enforced throughout the fluid domain, while suitable boundary conditions (BCs) are imposed at the domain boundaries to reproduce the relevant physics.

Without loss of generality, we consider the two-dimensional fluid domain. The fluid domain $\Omega(t)$ is bounded by $\partial\Omega(t) \equiv \Gamma_f(t) \cup \Gamma_i(t) \cup \Gamma_b \cup \Gamma_o(t)$. The governing equations for this 2D problem are as follows:

- Laplace equation in $\Omega(t)$: $\nabla^2\Phi(t,q) = 0,\ t \in T,\ q \in \Omega(t)$.
- Impermeability at the seabed (Γ_b): no fluid crosses the bottom boundary: $\nabla\Phi(t,q) \cdot \mathbf{n}(t,q) = 0,\ t \in T,\ q \in \Gamma_b$, where $\mathbf{n}(t,q)$ is the outward normal unit vector at the boundary.
- Far-field conditions on $\Gamma_i(t)$ and $\Gamma_o(t)$: appropriate inlet and outlet velocities or fluxes.
- Free-surface dynamic BC: the pressure at the free surface equals atmospheric pressure. From the Bernoulli equation:

$$\frac{\partial\Phi(t,q)}{\partial t} = -\frac{\mathrm{p}_{atm}(t,q)}{\rho} - g\zeta(t,q) - \frac{1}{2}\left(\nabla\Phi(t,q)\right)^2, t \in T,\ q \in \Gamma_f(t),$$

  where ρ is the fluid density, g is the gravitational acceleration, $\mathrm{p}_{\mathrm{atm}}$ is the atmospheric pressure, and ζ denotes the free-surface elevation. Note that in the following, the position q is defined by the Cartesian coordinates (x, z).
- Free-surface kinematic BC: fluid particles at the free surface remain on the surface during motion.

This formulation represents a fully nonlinear boundary-value problem (BVP), which generally requires numerical methods. However, in many practical applications, linearization allows for analytical solutions of the velocity potential, as in the case of linear waves.

Linear Regular Waves To review the linear theory of regular waves, we first introduce some preliminary notions:

- Still Water Level (SWL): average water level in the absence of waves (see Fig. 3.4).
- Water depth (h): vertical distance from the seabed to the SWL.
- Crest: highest point of the wave surface.
- Trough: lowest point of the wave surface.
- Wave amplitude (ζ_a): vertical distance from SWL to crest/trough.
- Wave height (H): crest-to-trough vertical distance; for linear waves, $H = 2\zeta_a$.
- Wavelength (λ): horizontal distance between two successive crests (or troughs).
- Wave period (T): time interval between two successive crests (or troughs) at a fixed location.
- Wave number $k = 2\pi/\lambda$.
- Circular wave frequency $\omega = 2\pi/\mathrm{T}$.

The wave steepness is defined as H/λ. For small steepness, higher-order nonlinear terms in the governing equations can be neglected, and displacements, velocities, accelerations, and pressures can be considered harmonic functions linearly related to the wave elevation. Moreover, particle motion depends primarily on the depth below the SWL. This leads to the linear expression for the free-surface elevation:

$$\zeta(t,x) = \zeta_a \cos\left(kx - \omega t\right),$$

where x is the horizontal coordinate along the still water level.

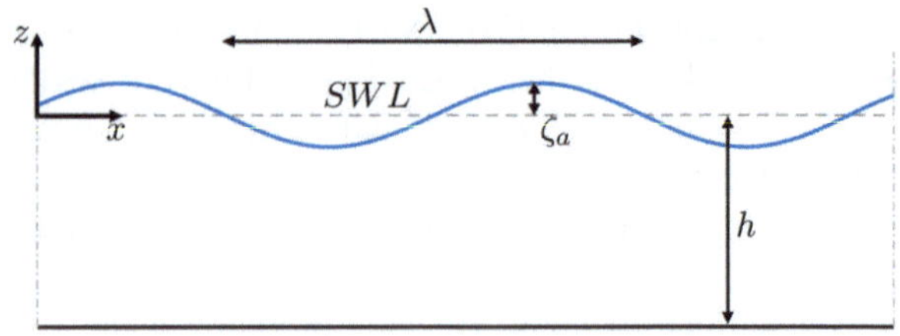

Fig. 3.4 Linear regular wave

The linearized boundary conditions are as follows:

- Seabed impermeability: $\frac{\partial \Phi(t,q)}{\partial z} = 0, \; t \in T, \; z = -h.$
- Free-surface dynamic BC: $\frac{\partial \Phi(t,q)}{\partial t} + g\zeta(t, q) = 0, \; t \in T, \; z = 0.$
- Free-surface kinematic BC: $\frac{\partial^2 \Phi(t,q)}{\partial t^2} + g\frac{\partial \Phi(t,q)}{\partial z} = 0, \; t \in T, \; z = 0.$

The kinematic BC leads to the well-known dispersion relation:

$$\omega^2 = kg \tanh(kh),$$

linking the wave number to the wave frequency.

Assuming a separable form for the velocity potential,

$$\Phi(t, q) = \mathrm{P}(z) \sin(kx - \omega t),$$

where $\mathrm{P}(z)$ is an unknown function, and applying the BCs, the velocity potential for a linear regular wave can be obtained as

$$\Phi(t, q) = \frac{\zeta_a g}{\omega} \frac{\cosh k(z + h)}{\cosh kh} \sin(kx - \omega t).$$

From this velocity potential, velocities, accelerations, and dynamic pressures can be computed at any point within the fluid domain.

Irregular Waves Ocean waves are not purely harmonic. The wave elevation is commonly modeled as a stationary, zero-mean Gaussian process, characterized by its variance,

$$\sigma_\eta^2 = \int_0^\infty S_{\eta\eta}(\omega)\, d\omega,$$

where $S_{\eta\eta}(\omega)$ is the one-sided wave spectrum.

Irregular waves are represented as a superposition of N regular-wave components, each with amplitude

$$\zeta_i = \sqrt{2 S_{\eta\eta}(\omega_i)\, \Delta\omega_i}.$$

The resulting wave elevation along the x-axis is

$$\zeta(t, x) = \sum_{i=1}^{N} \zeta_i \sin\left(k_i x - \omega_i t + \epsilon_i\right),$$

where k_i and ϵ_i denote the wave number and random phase shift of the ith component.

3.5 Dynamic Response

3.5.1 Hydrodynamic Modeling of Offshore Structures

In this section, we provide a concise introduction to the hydrodynamics of offshore structures, with particular emphasis on floating systems. For a more exhaustive treatment, the reader is referred to Faltinsen (1990); Journèe & Massie (2012).

First, the potential-flow problem for a freely floating structure is presented, highlighting the linearization and decoupling of the BVP. Then, the equations of motion of the floating system are introduced.

Floating platforms are relatively large structures exposed to wave excitation. To study the hydrodynamic problem, the system is often modeled as a rigid body with six DoFs, $\boldsymbol{q} = [q_1, q_2, q_3, q_4, q_5, q_6]^{\mathrm{T}}$ (see Fig. 3.5), where

- q_1: surge, displacement in the x-direction;
- q_2: sway, displacement in the y-direction;
- q_3: heave, displacement in the z-direction;
- q_4: roll, rotation about the x-axis (surge);
- q_5: pitch, rotation about the y-axis (sway);
- q_6: yaw, rotation about the z-axis (heave).

Potential-flow theory generally applies to elements of the floating body that are "bluff" with respect to the incident wave. We define $\alpha = \lambda/D$ as the element slenderness, where λ is the wavelength and D is a characteristic length (e.g., cylinder diameter). For $\alpha < 5$, the element is considered bluff and its hydrodynamics can be modeled using potential-flow theory, accounting for its influence on the surrounding flow. For $\alpha > 5$, the element is "slender," viscous effects may dominate, and the hydrodynamic behavior cannot be accurately represented by potential-flow theory.

In the following sections, these two approaches for hydrodynamic characterization are presented. Here, attention is restricted to first-order problems; for second- and higher-order effects, see Faltinsen (1990); Journèe & Massie (2012).

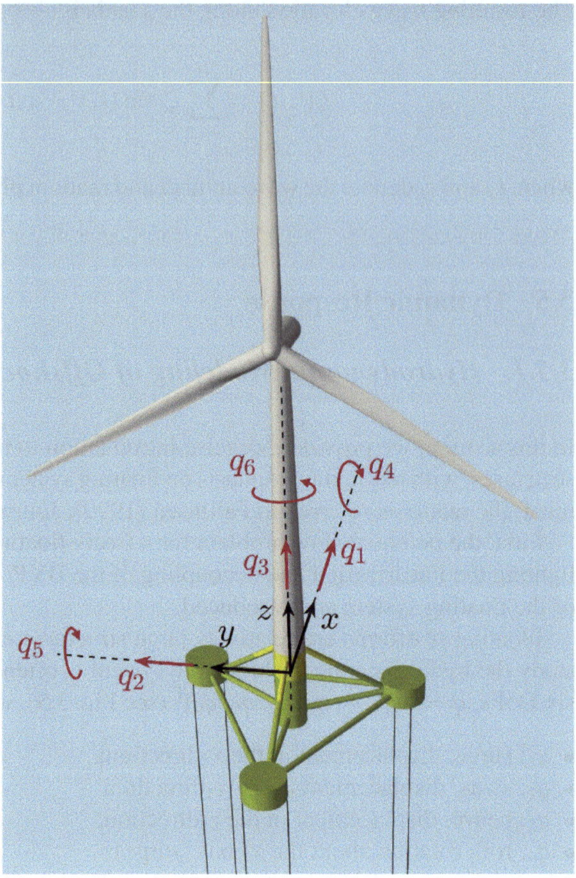

Fig. 3.5 Floating platform rigid-body model

3.5.1.1 The Potential-Flow BVP of a Floating Body

Considering a floating body exposed to waves (Fig. 3.6), we assume small displacements and rotations, allowing the linearized free-surface BCs from the regular-wave case to be adopted

$$\frac{\partial^2 \Phi(t,q)}{\partial t^2} + g\frac{\partial \Phi(t,q)}{\partial z} = 0,\ t \in T,\ z = 0\,.$$

Under stationary conditions, the spatial and temporal dependencies of the velocity potential can be separated

$$\Phi(t,q) = \phi(q)e^{-i\omega t}\,,$$

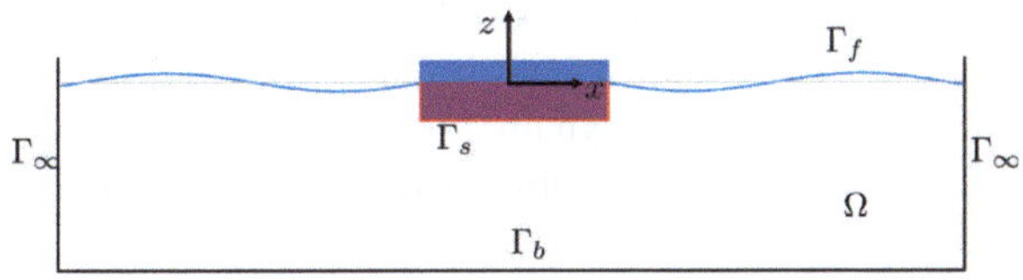

Fig. 3.6 Schematic of the BVP of a floating body

where $\phi(q)$ is the stationary potential, depending only on the geometry of the domain. Substituting into the linearized free-surface BC yields

$$\omega^2\phi(q) + g\frac{\partial\phi(q)}{\partial z} = 0, \quad z = 0\,.$$

The spatial potential can be split into diffraction and radiation contributions

$$\phi(q) = \phi_D(q) + \phi_R(q)\,,$$

leading to two separate BVPs. In the diffraction problem, the body is fixed and subjected to the incident wave. The diffraction potential $\phi_D(q)$ is the sum of the incident potential $\phi_I(q)$ and the scattering potential $\phi_S(q)$, which enforces the impermeability condition at the wetted surface Γ_s.

In the radiation problem, the body oscillates in one DoF at a time while the surrounding fluid is at rest. Since $\phi_I(q)$ is known analytically, only the scattering and radiation BVPs must be solved numerically.

Diffraction + Scattering BVP

- $\partial\phi_S(q)/\partial z = 0,\ q \in \Gamma_b$;
- $\omega^2\phi_S(q) + g\partial\phi_S(q)/\partial z = 0,\ q \in \Gamma_f$;
- Far-field condition on Γ_∞;
- $\partial\phi_S(q)/\partial\mathbf{n}(q) = -\partial\phi_I(q)/\partial\mathbf{n}(q),\ q \in \Gamma_s$.

Radiation BVP

- $\partial\phi_{Rj}(q)/\partial z = 0,\ q \in \Gamma_b \quad j = 1, \ldots, 6$;
- $\omega^2\phi_{Rj}(q) + g\partial\phi_{Rj}(q)/\partial z = 0,\ q \in \Gamma_f \quad j = 1, \ldots, 6$;
- Far-field condition on $\Gamma_\infty \quad j = 1, \ldots, 6$;
- $\partial\phi_{Rj}(q)/\partial\mathbf{n}(q) = \mathrm{n}_j(q),\ q \in \Gamma_s \quad j = 1, \ldots, 6$

with

$$\mathrm{n}_j(q) = \begin{cases} \mathrm{n}_j(q) & \text{if } j = 1, 2, 3 \\ (\boldsymbol{r}(q) \times \mathbf{n}(q))_j & \text{if } j = 4, 5, 6 \end{cases}$$

where $\boldsymbol{r}(q)$ is the position vector of q relative to the reference frame in Fig. 3.6.

These problems are generally solved numerically for different frequencies ω, yielding the velocity potential at the wetted surface for unit wave amplitude.

3.5.1.2 Hydrodynamic Coefficients

The first-order hydrodynamic loads are obtained by integrating the dynamic pressure over Γ_s. Using the linearized Bernoulli equation (see Eq.(3.3)) and removing hydrostatic pressure, we obtain

$$
\begin{aligned}
\mathrm{q}(t, q) &= -\rho \frac{\partial \Phi(t, q)}{\partial t}, \\
\mathrm{q}(t, q) &= -\rho \frac{\partial \phi(q) e^{-i\omega t}}{\partial t} = i\omega\rho\Big(\phi_D(q) + \sum_{j=1}^{6} \phi_{Rj}(q)\Big) e^{-i\omega t}.
\end{aligned}
$$

The kth component of the total hydrodynamic force is

$$
\begin{aligned}
F_k(\omega) &= -\int_{\Gamma_s} \mathrm{q}(\omega, q) \mathrm{n}_k \, d\Gamma \\
&= -i\omega\rho \int_{\Gamma_s} \Big(\phi_D(q) + \sum_{j=1}^{6} \phi_{Rj}(q)\Big) \mathrm{n}_k \, d\Gamma \\
&= -i\omega\rho\Big(X_k(\omega) + \sum_{j=1}^{6} X_{jk}^{R}(\omega)\Big),
\end{aligned}
$$

where $X_k(\omega)$ is the diffraction force per unit wave amplitude, often expressed in amplitude and phase. The radiation forces X_{jk}^R are split into added-mass and damping contributions:

$$
X_{jk}^{R}(\omega) = \omega^2 A_{jk}(\omega) + i\omega B_{jk}(\omega),
$$

where $\boldsymbol{A}(\omega)$ and $\boldsymbol{B}(\omega)$ are the added-mass and radiation-damping matrices, respectively, depending only on the geometry of the floating body.

3.5.1.3 Hydrodynamics of Slender Elements with the Morison Equation

For hydrodynamically slender elements ($\alpha > 5$), viscous effects may dominate. The Morison equation Morison et al. (1950) estimates the unit-length wave-induced forces as the sum of inertial and quadratic drag contributions

$$
d\mathbf{F}(t) = d\mathbf{F}_{in}(t) + d\mathbf{F}_{vis}(t),
$$

given by

$$
\begin{aligned}
d\mathbf{F}_{in}(t) &= \rho A_c \left[(1 + C_A)\dot{\boldsymbol{v}}_n(t) - C_A \ddot{\mathbf{x}}_n(t)\right], \\
d\mathbf{F}_{vis}(t) &= \frac{1}{2}\rho C_D D \left[\boldsymbol{v}_n(t) - \dot{\mathbf{x}}_n(t)\right] \|\boldsymbol{v}_n(t) - \dot{\mathbf{x}}_n(t)\| \ ,
\end{aligned}
$$

where A_c and D are the cylinder cross-sectional area and diameter, C_A and C_D are the added-mass and drag coefficients, $\mathbf{x}_n$ is the displacement of the structural member centerline normal to the member axis, $\boldsymbol{v}_n$ is the component of the undisturbed fluid particle velocity normal to the member axis, evaluated at the member location, and $(\dot{\circ})$ indicates standard time derivative.

The Keulegan–Carpenter number

$$
KC = \frac{v_A T}{D} \ ,
$$

where v_A is the amplitude of the oscillatory fluid particle velocity, T is the oscillation or wave period, and D, the characteristic diameter of the structural member, is a dimensionless parameter used to assess the relative importance of inertia and drag forces in oscillatory flow, with guidance on applicable regimes commonly reported in the literature, see, e.g., (Journèe and Massie 2012).

3.5.1.4 Floating Rigid-Body Equation of Motion

For a moored floating body under a 0°-heading regular wave $\zeta(t, q) = \zeta_a \cos(kx - \omega_k t)$, the linearized six-DoF equation of motion is

$$
\begin{aligned}
&[\boldsymbol{A}(\omega_k) + \boldsymbol{M}_{float}]\ddot{\boldsymbol{q}}(t) + \boldsymbol{B}(\omega_k)\dot{\boldsymbol{q}}(t) + [\boldsymbol{K}_{hydro} + \boldsymbol{K}_{mooring}]\boldsymbol{q}(t) \\
&= \boldsymbol{F}_{wave}(\omega_k, t) + \boldsymbol{F}_{MOR}(t) = ||\boldsymbol{X}(\omega_k)||\zeta_a \cos\big(-\omega_k t + \varphi(\omega_k)\big) + \boldsymbol{F}_{MOR}(t) \ ,
\end{aligned}
$$

where $\boldsymbol{M}_{float}$ is the mass matrix, $\boldsymbol{F}_{MOR}(t)$ the Morison forces, and $\boldsymbol{K}_{hydro}$, $\boldsymbol{K}_{mooring}$ the hydrostatic and mooring stiffness matrices.

3.5.1.5 FOWTs Equation of Motion in the Frequency Domain

The FOWT can be modeled as a seven-DoF system by adding a tower-top fore-aft flexible mode to the six rigid-body DoFs (Ferri and Marino 2023). Around the steady-state operating point, the frequency-domain equation reads as follows:

$$
\begin{aligned}
&\Big[-\omega^2\big(\boldsymbol{A}(\omega) + \boldsymbol{M}_{float} + \boldsymbol{M}_{turb}\big) + i\omega\big(\boldsymbol{B}(\omega) + \boldsymbol{B}_{turb}\big) \\
&\qquad + \boldsymbol{K}_{hydro} + \boldsymbol{K}_{mooring} + \boldsymbol{K}_{turb}\Big]\hat{\boldsymbol{q}}(\omega)e^{i\omega t} \\
&= \Big[\boldsymbol{X}(\omega)\sqrt{2S_{wave}(\omega)\Delta\omega} + \boldsymbol{F}_d\sqrt{S_{wind}(\omega)\Delta\omega}\Big]e^{i\omega t} \ ,
\end{aligned}
$$

where $\hat{\boldsymbol{q}}(\omega)$ collects the seven-DoF response amplitudes, and the matrices account for hydrodynamics, turbine mass, stiffness, damping, gyroscopic effects, and aerodynamic forces. The linearized Morison forces can be included as additional damping.

3.6 Optimization and Data-Driven Methods for FOWTs

This chapter presents applications of frequency-domain models for the optimization and reliability assessment of FOWTs. Furthermore, some data-driven methods are reviewed, as they are becoming increasingly attractive for predicting the response of FOWTs, particularly when informed by physics-based models like the one presented here. Note that this short review of data-driven approaches is limited to cases that make use of synthetic "data."

3.6.1 Optimization Processes for FOWTs

Optimization procedures are widely employed to reduce the Levelized Cost of Electricity (LCOE) and/or improve structural performance (Karimi et al. 2017; Ferri et al. 2022; Ferri and Marino 2023). Frequency-domain models, as discussed above, provide an appealing trade-off between accuracy and computational efficiency and can be used in preliminary design phases to identify optimal floating substructures (platform + moorings) for a given installation site (Ferri and Marino 2023).

3.6.1.1 Environmental Actions

This type of analysis requires defining site-specific environmental actions using long-term metocean records (IEC 2009), which are essential to statistically characterize wind and wave conditions. Typically, three variables are considered:

- 10-min average wind speed, U, measured at 10m above sea level.
- Significant wave height, H_S.
- Peak spectral period, T_p.

Once the turbulence model and wind profile are selected, U defines the turbulence spectrum, while H_S and T_p define the wave spectrum. Inverse First Order Reliability Methods (IFORM) (Li et al. 2013; Ferri and Marino 2023) can then be applied to generate T_R-year return-period Environmental Contour Surfaces (ECSs), i.e., ensembles of (U, H_S, T_p) events corresponding to a return period T_R (Fig. 3.7). To construct these ECSs, marginal and conditional distributions of the variables are fitted to historical wind and wave records.

With the environmental conditions defined, wind and wave loadings can be fully characterized.

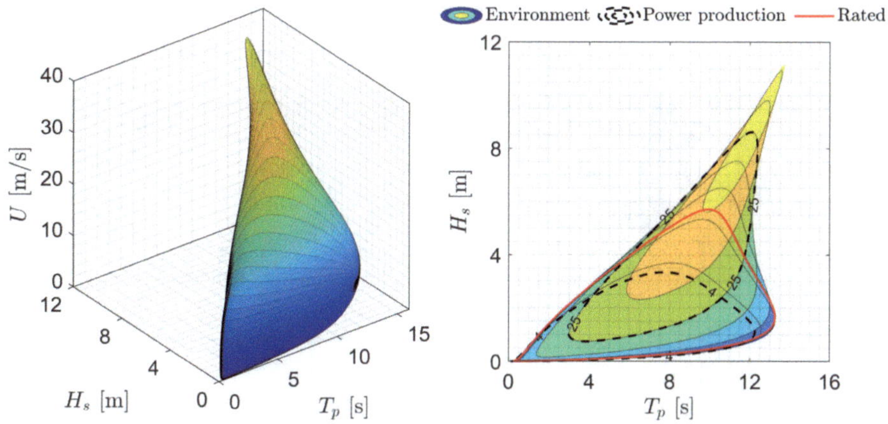

Fig. 3.7 Example of an ECS for a Mediterranean Sea installation site with a 50-year return period: 3D plot (left) and 2D contour plot (right)

3.6.1.2 The Optimization Problem

Optimization procedures have been developed for the design of FOWTs to reduce the cost of the floating substructure while maintaining adequate structural performance. Early-stage design strategies for platforms and mooring lines often relied on rational upscaling approaches, see Fig. 3.8, which do not guarantee globally optimal solutions, particularly in terms of cost efficiency.

To formulate the optimization problem, design variables, constraints, and objective functions must first be defined. Design variables are typically associated with the floater geometry, such as platform radius, draft, and column diameters. Constraints are selected to satisfy design standards, power-production requirements, cost targets, and other system-level criteria. Objective functions are then constructed to quantify optimization goals, such as minimizing stresses in critical structural components, manufacturing costs, or rotor and blade accelerations.

Considering multiple objective functions with potentially competing targets (Ferri and Marino 2023) enables the identification of trade-off solutions between structural performance and cost.

Results of a representative multi-objective optimization process are shown in Figs. 3.9–3.13. The design variables include parameters of both the platform and mooring lines, while constraints incorporate wind-turbine power performance (Ferri and Marino 2023). The objective functions considered are structural performance, $f_{obj,1}$, and manufacturing cost, $f_{obj,2}$. Both objectives are normalized relative to the results of a single-objective optimization that minimizes only the tower-base bending moment under an ultimate load state condition. Further methodological details can be found in (Ferri and Marino 2023).

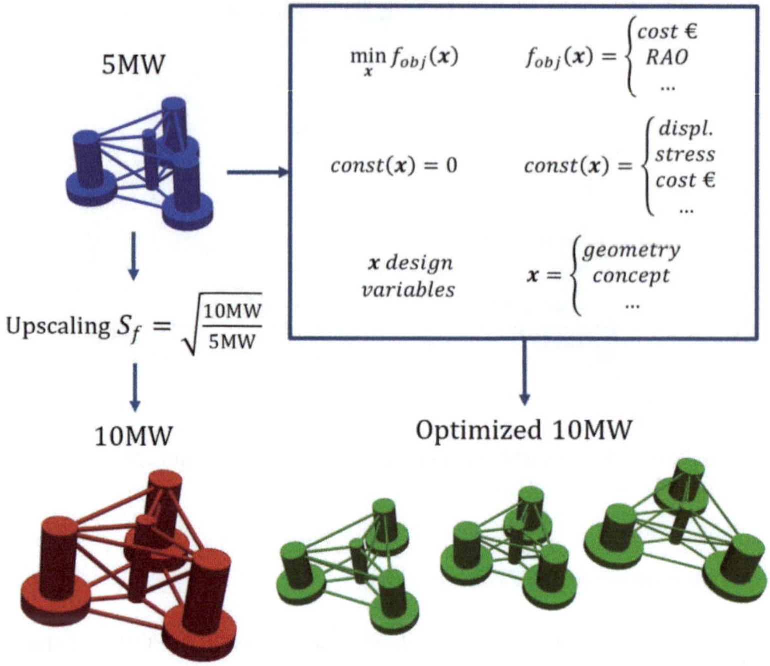

Fig. 3.8 Optimization process

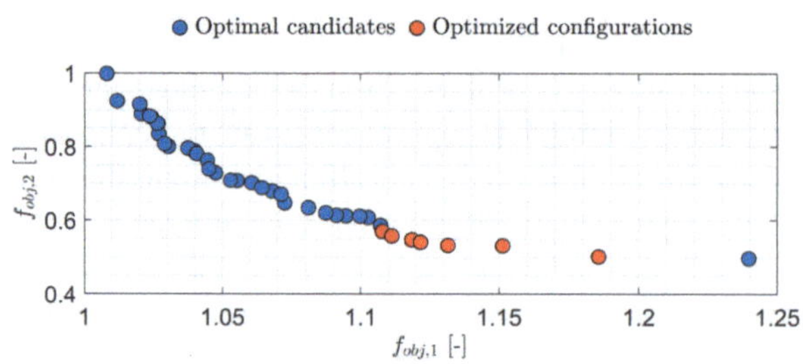

Fig. 3.9 Pareto front of a multi-objective optimization of a 10 MW semi-submersible FOWT. The x-axis shows the normalized bending moment at the tower base, while the y-axis shows the normalized manufacturing cost

The Pareto front in Fig. 3.9 demonstrates that optimized configurations (orange dots) achieve significant reductions in manufacturing cost while inducing only moderate increases in tower-base stresses.

The optimized floater concepts are characterized by relatively small column diameters (Fig. 3.10), large platform radii (Fig. 3.11), and moderate draft (Fig. 3.12). These configurations enhance the bending stiffness of the substructure while controlling costs through reduced column diameters and draft. Consequently, pitch rotations are

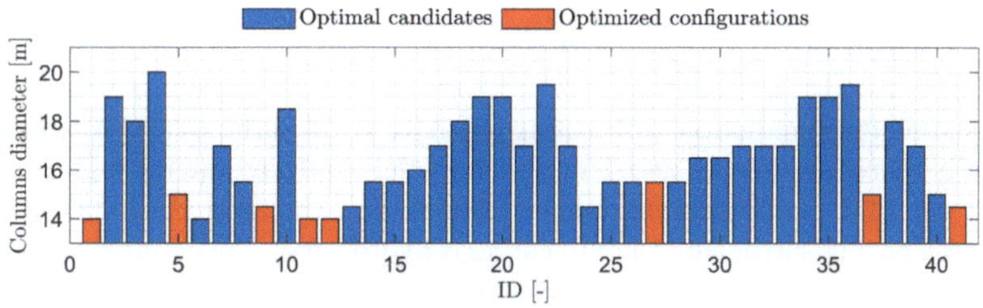

Fig. 3.10 Optimized side column diameters

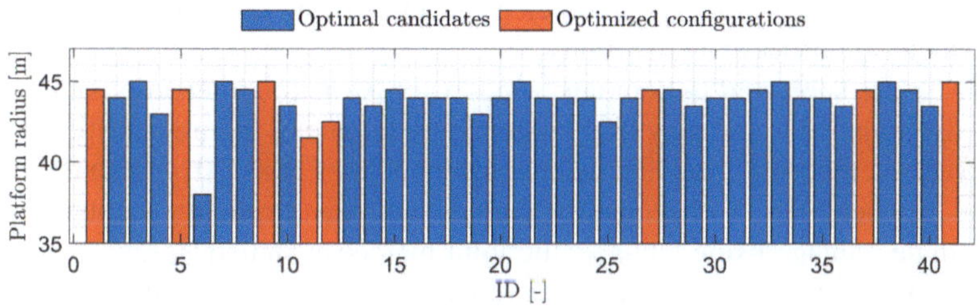

Fig. 3.11 Optimized platform radii

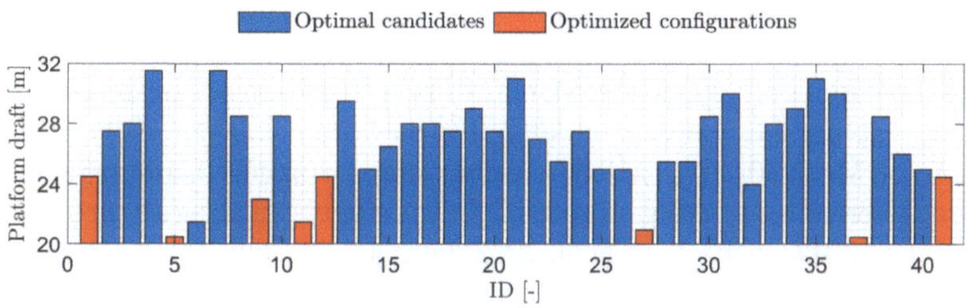

Fig. 3.12 Optimized platform drafts

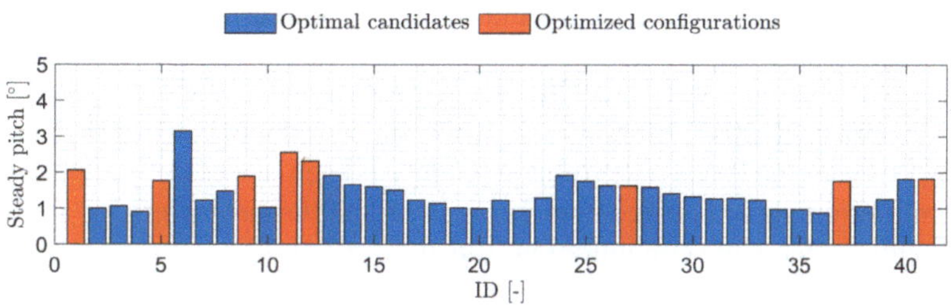

Fig. 3.13 Steady pitch at rated wind speed for the optimized platform concepts

minimized (Fig. 3.13), which leads to reduced tower-base bending moments (Ferri and Marino 2023).

3.6.2 Data-Driven Methods for FOWTs

Analysis and load prediction for Floating Offshore Wind Turbines (FOWTs) increasingly rely on reduced-order and data-driven models to overcome the high computational cost of fully coupled time-domain simulations. Ahmad et al. (2023) propose a machine-learning surrogate of nonlinear aero-hydro-servo-elastic models, enabling fast and accurate response predictions suitable for reliability analysis and control-oriented applications. Similarly, Medina-Manuel et al. (2024) investigate surrogate-based strategies for fatigue and reliability assessment, demonstrating that physics-informed data-driven models can efficiently capture long-term structural responses under stochastic metocean conditions.

FOWTs are subjected to complex wind–wave–structure interactions that induce highly dynamic loads and may lead to premature fatigue or failure of critical components, such as mooring lines and turbine towers (Thies et al. 2014; Hsu et al. 2017; Liu and Manuel 2018). Consequently, their design must explicitly account for long-term fatigue damage using site-specific joint metocean statistics, together with the selection of appropriate load cases for reliability assessment (Manuel et al. 2018; Marino et al. 2017; Ziegler et al. 2015).

For site-specific fatigue analysis, representative sea states are selected based on their probability of occurrence, severity, and interaction with the floating system, forming the basis for so-called short-term fatigue assessments. High-fidelity time-domain simulations (Jonkman and Matha 2011) or CFD-based approaches (Liu et al. 2017) can capture these effects in detail, but they are computationally demanding (Kvittem and Moan 2015). Coupled frequency-domain models therefore represent an appealing alternative, offering a favorable balance between accuracy and efficiency (Karimi et al. 2017, 2019; Ferri et al. 2022; Ferri and Marino 2023). Computational costs can be further reduced by introducing surrogate models.

Surrogate modeling techniques have been developed for computationally intensive components such as structural finite-element models (Zheng et al. 2023), CFD simulations (Lim and Manuel 2021), and condition monitoring applications (Zhang et al. 2022). Common approaches include machine-learning techniques (Müller and Cheng 2018; Richmond et al. 2020) and regression-based methods (Singh et al. 2022; Teixeira et al. 2017; Murcia et al. 2018; Shi and Manuel 2019). In the context of FOWTs, Gaussian Process Regression (GPR) has been successfully applied to estimate long-term fatigue damage in mooring lines and turbine towers for a reference 5 MW semi-submersible configuration (Liu et al. 2024).

In the proposed framework, frequency-domain simulations are used to generate the training data for the surrogate model. Critical sea states are selected based on the joint metocean probability distribution at the installation site and on the sensitivity of the structural response to wind and wave characteristics. The surrogate predictions are validated by computing the exact short-term 1-Hz damage-equivalent loads ($DEL_{1\text{-Hz}}$) for each sea state, as shown in Fig. 3.14.

The developed surrogate framework shows excellent computational efficiency when compared to Monte Carlo simulations (MCSs). While MCS randomly sam-

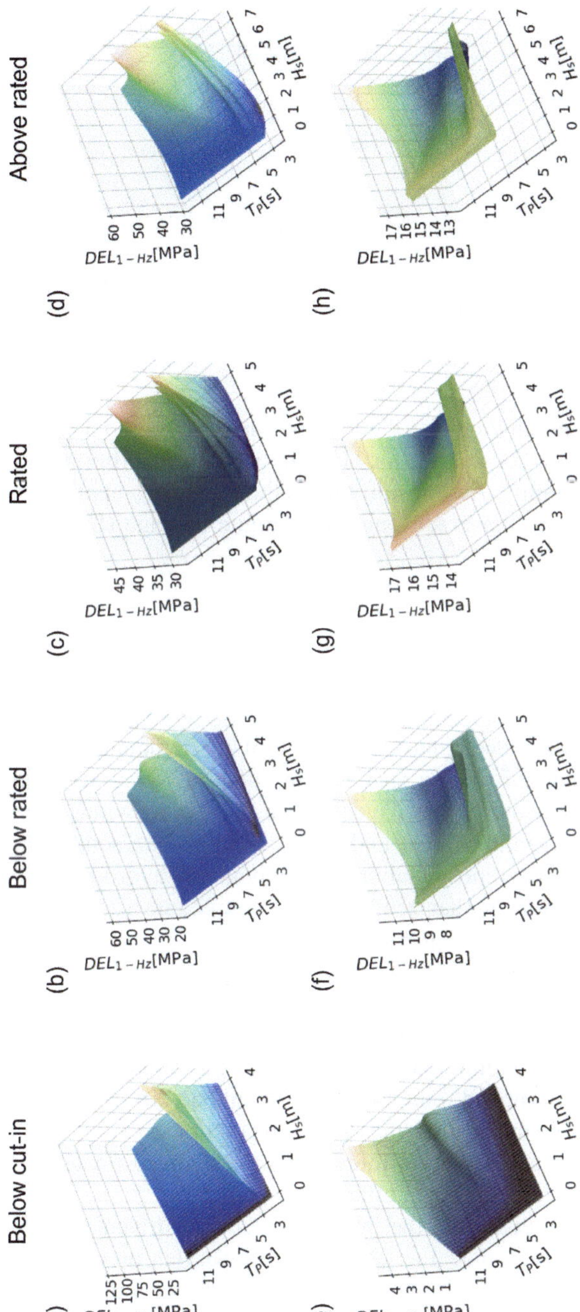

Fig. 3.14 Exact $DEL_{1\text{-Hz}}$ from full-grid simulation for (a–d) the tower base and (e–h) the mooring line. Courtesy of Liu et al. (2024)

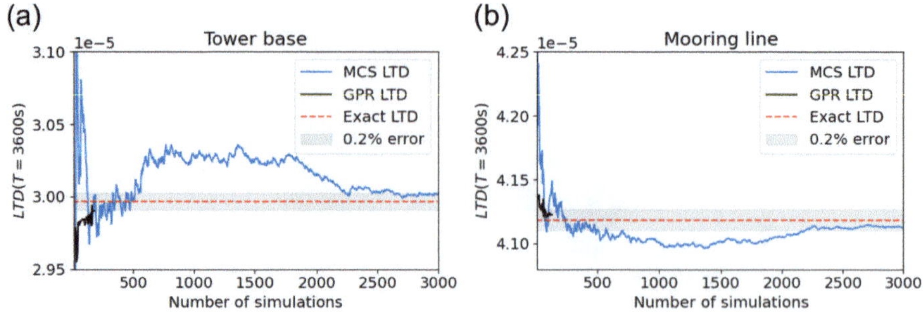

Fig. 3.15 Comparison of convergence rates for $LTD(T = 3600\text{s})$ estimation between MCS and GPR approaches for **a** the tower-base bending moment and **b** the mooring line fairlead tension. Courtesy of Liu et al. (2024)

ples sea states from the joint distribution, the surrogate-based approach sequentially selects sea states by accounting for both the joint probability distribution and the GPR model uncertainty. As a result, the method achieves approximately a tenfold faster convergence to the exact long-term fatigue damage (LTD), as illustrated in Fig. 3.15.

3.7 Conclusions

This chapter has provided a basic overview of the dynamic behavior of offshore wind energy systems, focusing on Floating Offshore Wind Turbines (FOWTs). It reviewed the fundamental physical mechanisms governing wind generation, offshore wind resources, wave kinematics, and wave–structure interactions, highlighting the importance of accurate representation of aerodynamic, hydrodynamic, and mooring effects. The discussion emphasized how these coupled interactions influence system response, fatigue, and overall reliability.

Modeling strategies for FOWTs were presented, including rigid- and flexible-body dynamics, potential-flow formulations, Morison-based viscous corrections, and mooring system representations. Frequency-domain approaches were highlighted as efficient tools for estimating steady-state responses and enabling preliminary design, structural optimization, and reliability assessments under stochastic wind and wave conditions. These methods allow the systematic computation of hydrodynamic coefficients, diffraction and radiation forces, and the coupled floater–turbine response while maintaining computational efficiency.

The chapter also discussed the integration of data-driven and surrogate modeling techniques, such as machine learning and Gaussian process regression, which enable rapid prediction of FOWT responses while preserving physical fidelity. Such approaches are particularly valuable for fatigue assessment, long-term reliability evaluation, and control-oriented applications, reducing the need for extensive time-domain simulations.

In summary, combining fundamental physics, frequency-domain modeling, and data-driven strategies provides a powerful framework for designing robust, cost-effective, and reliable FOWTs, supporting the sustainable expansion of offshore wind energy in deep-water environments.

References

IEC 61400-3. (2009). Wind Turbines-Part 3: Design requirements for offshore wind turbines. Technical report, International Electrotechnical Commission.

Ahmad, I., M'zoughi, F., Aboutalebi, P., Garrido, I., & Garrido, A. J. (2023). Machine-learning surrogate modeling for hybrid floating offshore wind turbines with integrated oscillating water columns. *Scientific Reports*, *13*(1), 1–19. https://doi.org/10.1038/s41598-023-28703-z

Barter, G. E., Robertson, A., & Musial, W. (2020). A systems engineering vision for floating offshore wind cost optimization. *Renewable Energy Focus*, *34*, 1–16. https://doi.org/10.1016/j.ref.2020.03.002

Bredmose, H., et al. (2017). The triple spar campaign: Model tests of a 10mw floating wind turbine with waves, wind and pitch control. *Energy Procedia*, *137*, 58–76. https://doi.org/10.1016/j.egypro.2017.10.334

Bruschi, N., Ferri, G., Marino, E., & Borri, C. (2020). Influence of clumps-weighted moorings on a spar buoy offshore wind turbine. *Energies*, *13*(23). https://doi.org/10.3390/en13236407. All Open Access; Gold Open Access; Green Accepted Open Access.

Butterfield, S., Musial, W., Jonkman, J., & Sclavounos, P. (2007). Engineering challenges for floating offshore wind turbines. Golden, CO, USA. NREL/CP-500-38776 Technical Report.

Chakrabarti, S. K. (1987). *Hydrodynamics of Offshore Structures*. WIT Press.

Costanzo, G., Brindley, G., & Tardieu, P. (2025). Wind energy in Europe: 2024 statistics and the outlook for 2025–2030. WindEurope: Technical report.

Faltinsen, O. M. (1990). *Sea Loads on Ships and Offshore Structures*.

Ferri, G., Marino, E., & Borri, C. (2020). Optimal dimensions of a semisubmersible floating platform for a 10 mw wind turbine. *Energies*.

Ferri, G., & Marino, E. (2023). Site-specific optimizations of a 10 MW floating offshore wind turbine for the Mediterranean Sea. *Renewable Energy*, *202*, 921–941. ISSN 09601481. https://doi.org/10.1016/j.renene.2022.11.116

Ferri, G., Marino, E., Bruschi, N., & Borri, C. Platform and mooring system optimization of a 10 MW semisubmersible offshore wind turbine. *Renewable Energy*, 182: 1152–1170, jan 2022. ISSN 0960-1481. https://doi.org/10.1016/J.RENENE.2021.10.060.

Frémaux, A. (2025). Offshore wind energy statistics 2024. Technical report, WindEurope. www.windeurope.org

Hau, E. (2006). *Wind Turbines: Fundamentals, Technologies, Application Economics* (2nd ed.). Springer.

Hsu, W. T., Thiagarajan, K. P., & Manuel, L. (2017). Extreme mooring tensions due to snap loads on a floating offshore wind turbine system. *Marine Structures*, *55*, 182–199. ISSN 0951-8339. https://doi.org/10.1016/j.marstruc.2017.05.005

Jonkman, J., & Buhl, M. L. (2007). *FAST user's guide*. Technical report. National Renewable Energy Laboratory.

Jonkman, J. (2010). *Definition of the floating system for phase IV of OC3*. National Renewable Energy Laboratory (NREL), Golden, CO. Technical report.

Jonkman, J. M., & Matha, D. (2011). Dynamics of offshore floating wind turbines-analysis of three concepts. *Wind Energy*, *14*(4), 557–569.

Journèe, J. M. J., & Massie, W. W. (2012). Offshore hydromechanics (January). https://doi.org/10.1109/ICSENS.2012.6411159

Karimi, M., Hall, M., Buckham, B., & Crawford, C. (2017). A multi-objective design optimization approach for floating offshore wind turbine support structures. *Journal of Ocean Engineering and Marine Energy, 3*, 69–87.

Karimi, M., Buckham, B., & Crawford, C. (2019). A fully coupled frequency domain model for floating offshore wind turbines. *Journal of Ocean Engineering and Marine Energy, 5*, 135–158.

Kvittem, M. I., & Moan, T. (2015). Frequency versus time domain fatigue analysis of a semisubmersible wind turbine tower. *Journal of Offshore Mechanics and Arctic Engineering,* (1), 137.

Li, L., Gao, Z., & Moan, T. (2013). Joint environmental data at five European offshore sites for design of combined wind and wave energy devices. In: *Proceedings of the international conference on offshore mechanics and arctic engineering*. https://doi.org/10.1115/OMAE2013-10156

Lim, H. U., & Manuel, L. (2021). Distribution-free polynomial chaos expansion surrogate models for efficient structural reliability analysis. *Reliability Engineering & System Safety*, *205*, Article 107256. https://doi.org/10.1016/j.ress.2020.107256

Liu, D. P., Ferri, G., Heo, T., Marino, E., & Manuel, L. (2024). On long-term fatigue damage estimation for a floating offshore wind turbine using a surrogate model. *Renewable Energy*, *225*, 120238. ISSN 18790682. https://doi.org/10.1016/j.renene.2024.120238

Liu, J., & Manuel, L. (2018). Alternative mooring systems for a very large offshore wind turbine supported by a semisubmersible floating platform. *Journal of Solar Energy Engineering*, *140*(5), 051003. ISSN 0199-6231. https://doi.org/10.1115/1.4039984

Liu, Y., Xiao, Q., Incecik, A., Peyrard, C., & Wan, D. (2017). Establishing a fully coupled CFD analysis tool for floating offshore wind turbines. *Renewable Energy, 112*, 280–301.

Manuel, L., Nguyen, P. T. T., Canning, J., Coe, R. G., Eckert-Gallup, A. C., & Martin, N. (2018). Alternative approaches to develop environmental contours from metocean data. *Journal of Ocean Engineering & Marine Energy*, *4*(4), 293–310.

Marino, E., Giusti, A., & Manuel, L. (2017). Offshore wind turbine fatigue loads: The influence of alternative wave modeling for different turbulent and mean winds. *Renewable Energy*, *102*, 157–169. ISSN 0960-1481. https://doi.org/10.1016/j.renene.2016.10.023

Marino, E., Mockute, A., Borri, C., & Lugni, C. (2018). Wave loads on monopile-supported offshore wind turbines: Current methods and future challenges. In: Structural control and fault detection of wind turbine systems (pp. 9–44).

Matha, D. (2010). *Model development and loads analysis of a wind turbine on a floating offshore tension leg platform* [PhD thesis]. Stuttgart, Germany.

Medina-Manuel, A., Sánchez, R. M., & Souto-Iglesias, A. (2024). Surrogate modeling framework integrating reduced-order hydrodynamics and artificial neural networks for floating platforms under stochastic wave loading. *Journal of Marine Science and Engineering*, *12*(9). https://doi.org/10.3390/jmse12091464

Mockute, A., Marino, E., Lugni, C., & Borri, C. (2019). Comparison of nonlinear wave-loading models on rigid cylinders in regular waves. *Energies, 12*(21), 1–22. https://doi.org/10.3390/en12214022

Morison, J. R., Johnson, J. W., & Schaaf, S. A. (1950). The force exerted by surface waves on piles. *Journal of Petroleum Technology*, *2*(5), 149–154. ISSN 0149-2136.

Müller, K., Cheng, P. W. (2018). Application of a Monte Carlo procedure for probabilistic fatigue design of floating offshore wind turbines. *Wind Energy Science,3*(1), 149–162.

Murcia, J. P., Réthoré, P. E., Dimitrov, N., Natarajan, A., Sørensen, J. D., Graf, P., & Kim, T. (2018). Uncertainty propagation through an aeroelastic wind turbine model using polynomial surrogates. *Renewable Energy*, *119*, 910–922.

Otter, A., Murphy, J., Pakrashi, V., Robertson, A., & Desmond, C. (2022). A review of modelling techniques for floating offshore wind turbines. *Wind Energy, 25*(5), 831–857. https://doi.org/10.1002/we.2701

Richmond, M., Sobey, A., Pandit, R., & Kolios, A. (2020). Stochastic assessment of aerodynamics within offshore wind farms based on machine-learning. *Renewable Energy*, *161*, 650–661.

Robertson, A. N., & Jonkman, J. M. (2011). Loads analysis of several offshore floating wind turbine concepts. In: *Proceedings of the International Offshore and Polar Engineering Conference*, pp. 443–450.

Sarpkaya, T., & Isaacson, M. (1981). *Mechanics of Wave Forces on Offshore Structures*. Van Nostrand Reinhold Company.

Shi, C., & Manuel, L. (2019). Non-parametric prediction of the long-term fatigue damage for an instrumented top-tensioned riser. *Applied Ocean Research, 82*, 245–258. https://doi.org/10.1016/j.apor.2018.11.001

Singh, D., Dwight, R. P., Laugesen, K., Beaudet, L., & Viré, A. (2022). Probabilistic surrogate modeling of offshore wind-turbine loads with chained Gaussian processes. *Journal of Physics: Conference Series*, *2265*, 032070. IOP Publishing.

Teixeira, R., O'Connor, A., Nogal, M., Krishnan, N., & Nichols, J. (2017). Analysis of the design of experiments of offshore wind turbine fatigue reliability design with kriging surfaces. *Procedia Structural Integrity, 5*, 951–958.

Thies, P. R., Johanning, L., Harnois, V., Smith, H. C. M., & Parish, D. N. (2014). Mooring line fatigue damage evaluation for floating marine energy converters: Field measurements and prediction. *Renewable Energy*, *63*, 133–144. ISSN 0960-1481. https://doi.org/10.1016/j.renene.2013.08.050

Williams, R., Zhao, F., Backwell, B., Lee, J., Patel, A., Hutchinson, M., Qiao, L., Nguyen, T., Lathigara, A., Liang, W., Fang, E., Cheong, J., Tan, H., Francisco, A. M., Bui, V., Nguyen, T., Fiestas, R., Cox, R., Ruas, M., Lima, J. P. D., Rabie, H., Gitobu, J., Ladwa, R., Madan, K., Shardul, M., Muchiri, W., Weekes, N., Ogilvie, C., Khinda, N., Yamamura, J., Mhi, T. S., & Vestas Offshore Wind. (2024). Global offshore wind report 2024. Technical report, Global Wind Energy Council. www.gwec.net

Zhang, S., Robinson, E., & Basu, M. (2022). Hybrid Gaussian process regression and fuzzy inference system based approach for condition monitoring at the rotor side of a doubly fed induction generator. *Renewable Energy, 198*, 936–946.

Zheng, S., Li, C., & Xiao, Y. (2023). Efficient optimization design method of jacket structures for offshore wind turbines. *Marine Structures,89*, Article 103372.

Ziegler, L., Voormeeren, S., Schafhirt, S., & Muskulus, M. (2015). Sensitivity of wave fatigue loads on offshore wind turbines under varying site conditions. *Energy Procedia*, *80*, 193–200. ISSN 1876-6102. https://doi.org/10.1016/j.egypro.2015.11.422

Chapter 4
Digital Twins and Data-Driven Methods for Intelligent Operation and Interceptive Maintenance of Wind Energy Structures

Ahmed Elgammal, Yasmin Ali, Junlin Heng, Kaoshan Dai, and Charalampos Baniotopoulos

Abstract Wind energy constitutes a critical component of sustainable transitions, yet turbine maintenance in harsh environments poses considerable challenges. This chapter examines how digital twins (DTs) and data-driven methodologies establish intelligent operation and maintenance (O&M) frameworks. The evolution of DT technology is detailed, encompassing Internet of Things (IoT) integration, simulation capabilities, artificial intelligence (AI) applications and the roles of model calibration and uncertainty quantification. Specific applications are subsequently addressed: predictive maintenance strategies, virtual sensing techniques, lifecycle management approaches and integrated aero-structural control systems. Key challenges including

A. Elgammal · Y. Ali · J. Heng (✉) · K. Dai
Department of Civil Engineering, Sichuan University, Chengdu, China
e-mail: junlin.heng@scu.edu.cn

A. Elgammal
e-mail: ahmed_elgammal@stu.scu.edu.cn

Y. Ali
e-mail: yasmin_ali@stu.scu.edu.cn

K. Dai
e-mail: kdai@scu.edu.cn

Y. Ali
Department of Civil Engineering, Delta University for Science and Technology, Gamasa, Egypt

K. Dai
State Key Laboratory of Intelligent Construction and Healthy Operation and Maintenance of Deep Underground Engineering, Sichuan University, Chengdu, China

C. Baniotopoulos
Department of Civil Engineering, University of Birmingham, Birmingham, UK

Institute for Steel Construction, Faculty of Civil Engineering and Geodetic Science, Leibniz University Hannover, Hannover, Germany

C. Baniotopoulos
e-mail: c.baniotopoulos@bham.ac.uk

C. Baniotopoulos and E. Marino (eds.), *Advances in Wind Energy in the Era of Artificial Intelligence*, CISM International Centre for Mechanical Sciences 625,
https://doi.org/10.1007/978-3-032-28815-8_4

environmental factors and cybersecurity vulnerabilities are discussed alongside emerging innovations such as generative AI. A detailed case study illustrates a DT framework for corrosion-fatigue monitoring in offshore towers. The proactive DT-informed maintenance demonstrates a ninefold return on investment and extends remaining useful life (RUL) by 2.75 years. This analysis establishes how DTs transform intelligent O&M practices in the wind energy sector.

4.1 Introduction

Wind energy stands as a cornerstone in the global pursuit of renewable energy targets and the fundamental shift towards sustainable power generation systems. Its rapid ascent has positioned it among the most competitively viable and swiftly expanding renewable technologies, holding substantial potential to contribute significantly to emissions reduction objectives outlined in the Paris Agreement (Summerfield-Ryan & Park, 2023). The trajectory of growth has been remarkable, with global installed wind capacity effectively doubling roughly every three years over recent decades (Singh et al., 2022). Indeed, projections indicate that renewable sources, prominently featuring wind, could potentially fulfil up to two-thirds of the global primary energy supply by the middle of this century (Hassan et al., 2024). This continued worldwide expansion is propelled by a confluence of factors, including persistent technological innovation, enhanced financial viability and supportive governmental policies and incentives (Haces-Fernandez et al., 2022).

Recent industry analyses, particularly insights from the Global Wind Energy Council's (GWEC) 2025 report (Global Wind Energy Council, 2025), underscore this momentum, portraying wind energy not merely as a niche alternative but as a mature technology delivering tangible value across multiple societal dimensions: governmental strategic goals, economic development and public welfare. It has firmly established itself as a key pillar supporting the broader energy transition, achieving significant cost-competitiveness relative to traditional energy sources globally. The scaling of wind power deployment has demonstrably contributed to lower consumer electricity prices, stimulated job creation, particularly in manufacturing and maintenance sectors, and aided in the regeneration of communities, often in rural or coastal areas. Furthermore, a robust domestic wind sector enhances national energy resilience by mitigating dependence on volatile international fossil fuel markets. According to the GWEC report (Global Wind Energy Council, 2025), wind energy contributed to 20% of the growth within the renewables sector in 2024, with wind and solar installations collectively dominating new power capacity additions, although solar deployment currently proceeds at a faster pace. Ensuring an accelerated return to growth for wind energy is considered crucial for maintaining a balanced and orderly progression towards decarbonization goals (Global Wind Energy Council, 2025). Economies leveraging their wind resources benefit from increased energy independence and supply security, while large-scale investments foster local industrial capabilities and provide affordable power, stimulating wider economic activity.

The sector's societal contribution is also significant, supporting an estimated 1.5 million direct jobs globally and yielding considerable environmental benefits through reduced carbon emissions and improved air quality. Despite facing macroeconomic headwinds, trade complexities and supply chain pressures, the wind industry demonstrates notable resilience, though further progress hinges on strengthened international cooperation, streamlined permitting and stable policy frameworks (Global Wind Energy Council, 2025).

Quantitative benchmarks reported for 2024 (Global Wind Energy Council, 2025) further illustrate the scale of recent achievements. A record 117 GW of new wind capacity was installed worldwide during that year, elevating the global cumulative installed capacity to 1136 GW, an 11% increase from the previous year (see Fig. 4.1 and Table 4.1). Onshore wind installations achieved a historic peak, reaching 109 GW and surpassing the 100 GW annual installation mark for the second consecutive year, with cumulative onshore capacity exceeding 1000 GW for the first time. Offshore wind contributed 8 GW of new capacity in 2024, although this represented a decrease compared to the prior year and marked the lowest annual addition since 2021, bringing total global offshore capacity to 83.2 GW. Regionally, the Asia Pacific led deployment, accounting for 75% of the global market share in new installations for 2024, followed by Europe (14%), North America (5%), Latin America (4%) and Africa and Middle East (2%). The leading five national markets for new installations in 2024 were China (constituting 68% of global additions), the USA (12%), Germany (3%), India (3%) and Brazil (3%), representing 81% of the global total together. These same five countries also led in terms of total cumulative installed capacity by the end of 2024, accounting for 73% of the world's installations (Global Wind Energy Council, 2025).

Looking ahead, market intelligence forecasts from the GWEC 2025 report (Global Wind Energy Council, 2025) suggest continued growth, with new installations projected to reach 138 GW in 2025. Under existing policy scenarios, an estimated 982 GW of new capacity is anticipated between 2025 and 2030, equating to roughly 164 GW of new installations annually and reflecting a projected compound annual growth rate of 8.8% for this period. Significant acceleration in deployment is expected particularly in emerging markets across Southeast Asia, Central Asia and the Middle East and North Africa (MENA) region, with forecasts indicating potentially record

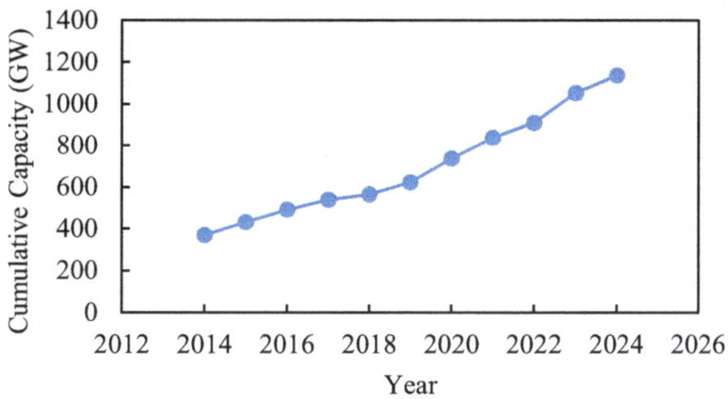

Fig. 4.1 Cumulative wind power capacity worldwide (Global Wind Energy Council, 2025)

Table 4.1 Key global wind energy statistics and insights (Global Wind Energy Council, 2025)

Metric	Value (GW)	Insight
New global installations (2024)	117	2024 saw a historical peak in new installations, making it the best year ever for the wind industry globally
New onshore installations (2024)	109	Onshore wind continued to dominate new capacity additions, surpassing the 100 GW milestone for the second consecutive year and reaching a historical peak
New offshore installations (2024)	8	While a smaller portion of new capacity compared to onshore, offshore wind continues to be a vital part of the energy transition, though installations saw a decrease from the previous year
Cumulative global capacity (end of 2024)	1136	Global cumulative capacity reached 1136 GW, representing an 11% increase compared to the end of 2023. This signifies significant growth beyond the 1 TW milestone achieved in 2023
Projected new installations (2025–2030 total)	982	It is forecasted that nearly 1 terawatt of new wind power capacity would be added over the next six years
Estimated annual installations (2025–2030 average)	164	This average annual projection is significantly higher than the 117 GW installed in the record year of 2024, indicating an anticipated acceleration in the rate of deployment

installations each year through 2030. Industry assessments within the report (Global Wind Energy Council, 2025) suggest a readiness to potentially triple annual growth to 320 GW, aligning with the COP28 goal of tripling renewable energy capacity by 2030, contingent upon the establishment of more favourable regulatory and investment conditions. This substantial and ongoing expansion underscores the critical importance of ensuring the long-term reliability and efficiency of wind energy assets.

Despite the notable advancements and expanding deployment of wind energy, ensuring the operational integrity and longevity of wind turbine structures presents substantial and multifaceted challenges. These challenges are particularly acute due to the diverse and often harsh operational environments encountered, spanning onshore, offshore and increasingly deep-water floating installations. Turbines are persistently subjected to dynamic and aggressive conditions, including strong winds, significant wave loading in marine environments and wide-ranging thermal cycles, all of which contribute to accelerate structural fatigue and amplify maintenance requirements. The structural integrity of these large-scale systems is a primary concern; for instance, turbines, particularly those sited offshore, endure extreme environmental loads that induce vibrations and impose heavy stresses, directly contributing to material fatigue and potential structural failure (Salic et al., 2019). Furthermore, although advanced fibre-reinforced composite materials enable lightweight turbine blade designs with favourable aerodynamic performance, their widespread use introduces challenges in maintaining adequate stiffness and preventing buckling under complex, variable loading scenarios (Hayman et al., 2008).

The difficulties associated with maintenance are significantly magnified in offshore settings. Accessing offshore wind turbines for inspection, repair or component replacement is inherently complex and costly, frequently necessitating specialized vessels and sophisticated logistical planning, thereby escalating operational expenditures (Santos et al., 2015). Moreover, the degradation processes affecting materials and components tend to occur more rapidly in corrosive marine environments compared to their onshore counterparts, demanding more frequent and intensive maintenance interventions to ensure continued operational safety and efficiency (Santos et al., 2015). Floating offshore wind turbines unlock access to deeper waters with stronger wind resources, but they also introduce additional complexity. These platforms face particularly severe environmental dynamics, where the interplay of strong winds and waves can lead to pronounced vibrations, intensified fatigue accumulation and substantial loads on the entire structure, including mooring systems, thereby increasing both maintenance needs and the risk of unexpected failures (Salic et al., 2019). The intricate installation procedures and inherently high capital investment associated with floating wind farms also present considerable hurdles to their widespread adoption (Hong et al., 2024), with operation and maintenance (O&M) in such deep-water locations posing distinct and demanding problems that call for innovative solutions (Ambarita et al., 2023).

Consequently, these operational and environmental challenges culminate in significant O&M costs, which typically constitute a substantial portion, estimated between 20 and 30%, of the levelized cost of electricity for wind power systems; these costs are generally higher for offshore installations due to the aforementioned logistical constraints (Costa et al., 2021; Fu et al., 2015). While research explores avenues to mitigate these expenses, such as comparative studies indicating potentially lower operational costs for direct-drive turbines versus medium-speed configurations (Donnelly et al., 2024), the overall financial burden of O&M remains a critical factor. The complex operational environments also render wind turbines susceptible to a variety of component failures, which can markedly affect their reliability and further inflate operational costs (Peng et al., 2023). Although effective condition monitoring systems are recognized as essential for the early detection of incipient faults and for optimizing maintenance schedules, existing technologies often encounter technical and commercial limitations, including challenges related to their sustained reliability and overall cost-effectiveness (Yang et al., 2014). The confluence of these pervasive structural, environmental, accessibility and economic challenges underscores a pressing need for more advanced, integrated and intelligent approaches to operation and maintenance. Such strategies are paramount not only for enhancing the reliability and efficiency of existing and future wind energy assets but also for ensuring their continued economic viability and contribution to sustainable energy goals.

Addressing these pervasive structural, environmental, accessibility and economic challenges effectively necessitates a paradigm shift towards more sophisticated O&M frameworks, and in this context, digital twin (DT) technology is emerging as a particularly potent enabler for advancing intelligent O&M strategies. DTs offer a transformative approach by creating comprehensive, dynamic digital representations of physical wind turbine assets. This virtual mirroring facilitates continuous, real-time monitoring of crucial components, such as blades and towers, by reflecting their current operational status and performance characteristics. Such capabilities allow for the accurate and timely detection of incipient anomalies or deviations from optimal performance, thereby supporting proactive predictive maintenance strategies designed to pre-empt failures and potentially extend the operational lifespan of critical systems (Jia & Jiang, 2023; Luo et al., 2023). By simulating the complex behaviour of turbines under a spectrum of operational and environmental conditions, DTs can also be employed to forecast potential failure modes and optimize maintenance schedules, which contributes to enhanced overall reliability and operational safety (Ibrahim et al., 2023).

Beyond predictive capabilities, DTs provide a platform for a suite of intelligent services, including advanced data analysis, sophisticated fault diagnosis and comprehensive performance evaluation. These services are instrumental in optimizing the operational efficiency of individual turbines as well as entire wind farms, leading to improved management practices, enhanced control levels and tangible reductions in O&M expenditures (Zhang et al., 2024a). The effective integration of DTs with high-fidelity physical modelling techniques and advanced big data analytics further allows for the optimization of resource utilization and power output, thereby contributing significantly to the overall systemic efficiency of wind energy generation (Siddiqui et al., 2023). Moreover, DTs can function as virtual sensors, augmenting or even supplanting physical sensor platforms to overcome inherent limitations such as data inaccuracies, sensor degradation or outright failures. This virtual sensing capability enhances the precision of critical parameter measurements, including wind speed and structural loads, which is vital for ensuring efficient and safe turbine operation (Ibrahim et al., 2023). The deployment of DTs within web-based platforms can also foster multi-party collaborative monitoring, further bolstering the prospects for long-term safe and reliable operation of wind turbine fleets (Jia & Jiang, 2023). Crucially, this digital paradigm allows for the sophisticated integration of aero- and structural control strategies directly within the O&M framework. By enabling detailed scenario modelling, real-time performance assessment and predictive insights into structural health, DTs empower operators to make more informed decisions regarding turbine management, actively optimizing performance while concurrently mitigating operational costs and structural risks (Moghadam et al., 2024; Zhao et al., 2022).

Building on this understanding of the pivotal role that DTs can play in transforming wind turbine O&M, this chapter aims to comprehensively explore the multifaceted applications of DT technology across the full spectrum of wind energy systems. The ensuing discussion will encompass onshore, fixed-bottom offshore and the increasingly prevalent floating offshore wind turbine typologies, recognizing the distinct operational challenges and opportunities inherent to each. A vital focus will be placed on demonstrating how DTs serve as foundational enablers for advanced predictive maintenance strategies, holistic lifecycle management from design through decommissioning and sophisticated intelligent decision-making processes. This will be achieved by examining the development and implementation of integrated operational and maintenance frameworks, wherein DTs facilitate a seamless synergy between data acquisition, advanced analytics, simulation and control, ultimately striving for enhanced reliability, optimized performance and improved economic viability of wind energy assets.

4.2 Evolution and Fundamentals of Digital Twins

Historical Development. The concept of the DT, denoting a dynamic virtual representation of a physical asset, system or process, has undergone a significant evolutionary trajectory over recent decades, maturing from a nascent idea into a transformative technology with broad industrial applicability. Understanding this historical development (illustrated in Fig. 4.2) provides essential context for its current and future role in complex engineering domains such as wind energy. Although the foundational principles can be traced to early applications of mirroring physical systems with computational models, notably in NASA's Apollo programme during the 1960s where initial concepts of creating digital counterparts for physical assets were explored in engineering (Rosen et al., 2015), the formal conceptualization of the DT is more recent. In 2003, Michael Grieves first proposed a model that is now widely regarded as the prototype of the DT concept, initially developed within the context of product lifecycle management to simulate product behaviour and optimize performance (Grieves, 2014; Abayadeera & Ganegoda, 2024). During this early period, from 2003 to 2010, the idea was articulated through terms such as the mirror space model or the information mirror model (Githens, 2007; Grieves, 2005) before the term DT was officially introduced in NASA's 2010 technology roadmap (Shafto et al., 2012). Grieves subsequently adopted and popularized the DT terminology in 2011, solidifying its place in academic and industry discourse (Grieves, 2011). At its core, this early vision already encompassed the creation of a virtual replica to enable comprehensive analysis, understanding and optimization of physical entities, facilitating bidirectional data exchange and interaction to establish a closed-loop management system.

The period following 2010 marked an acceleration in the development and recognition of DT technology. A pivotal moment occurred around 2012 when the United States Air Force Research Laboratory (AFRL) and NASA jointly proposed a DT model specifically for future aircraft (Glaessgen & Stargel, 2012; Tuegel et al., 2011), a development that garnered widespread attention and signalled the technology's expanding potential beyond manufacturing. This spurred significant interest across various industries, rapidly positioning DTs as a promising technological frontier (Rasheed et al., 2020; Itxaro Errandonea et al., 2020; Van Dinter et al., 2022). Consequently, major technology providers such as ANSYS, Siemens and Bentley began launching cloud platforms and commercial software solutions tailored for DT implementation, thereby catalysing its broader application and accessibility (Rasheed et al., 2020), a trend evident from around 2014 onwards (as depicted in Fig. 4.2). Despite this proliferation, the diverse application requirements across different sectors have

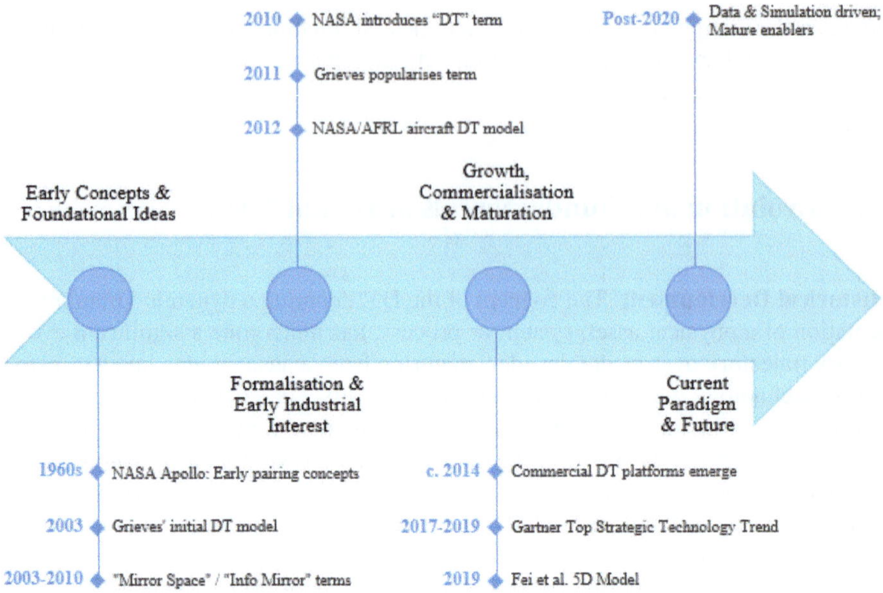

Fig. 4.2 Developmental phases and key events in the history of DTs

led to varied interpretations and implementations of the DT concept, and a universally standardized definition remains elusive (Errandonea et al., 2020). Nevertheless, most technical frameworks retain fidelity to Grieves's original three-component model: the physical entity in real space, its virtual counterpart in digital space and the data interface connecting them (Errandonea et al., 2020; Semeraro et al., 2021). Further conceptual refinement has also occurred, notably with (Fei et al., 2019) proposing an expanded five-dimensional DT model, offering a more granular descriptive framework with significant implications for future development.

Over nearly two decades of evolution, the foundational technologies underpinning DTs have matured considerably, creating a robust platform for rapid advancements. The developmental paradigm has notably shifted from an early emphasis on simulation-driven approaches (pre-2010) to a more integrated mode where data and simulation now jointly drive progress, particularly evident from 2020 onwards (Fig. 4.2). This maturation has been substantially supported by the concurrent advancement and convergence of enabling technologies. Innovations in the Internet of Things (IoT) for pervasive sensing and data acquisition, big data analytics for processing vast datasets, machine learning (ML) for pattern recognition and predictive modelling and cloud computing for scalable computational resources have been crucial. These technologies provide critical support for deeper technological integration and the deployment of DTs at scale, in addition to shaping the trajectory of future expansion. The strategic importance of DTs has been consistently recognized, for instance, by Gartner, which listed DTs among the top ten strategic technology trends for several consecutive years between 2017 and 2019 (Fig. 4.2).

The adoption of DT technology within wind energy systems represents a particularly significant milestone in this developmental narrative, driven by the pressing need to enhance the operational efficiency, reliability and cost-effectiveness of wind turbines, which are critical assets in the renewable energy landscape (Poddar and VP 2024; Kandemir et al., 2024). In this sector, DTs are increasingly utilized for sophisticated applications such as predictive maintenance, enabling the anticipation of component failures; for overall energy optimization, ensuring turbines operates at peak performance under varying environmental conditions; and for advanced fault diagnosis, allowing for quicker and more accurate identification of operational issues. This transition towards data-driven, intelligent O&M, facilitated by DTs, is pivotal for managing the complex lifecycle of wind energy structures, especially as they are deployed in increasingly challenging environments.

Core Components and Frameworks. Following the historical evolution of DT technology, a deeper understanding necessitates an examination of its fundamental architecture, specifically the core components and operational frameworks that enable its functionality. At its essence, a DT is more than a static model; it is a dynamic, interconnected system built upon several key pillars that facilitate its ability to accurately reflect and predict the behaviour of its physical counterpart. The efficacy of any DT depends on the seamless and continuous exchange of information between the physical asset and its digital representation, a bidirectional flow that forms the lifeblood of the system. This intricate interplay is further powered by the synergistic integration of diverse technologies, including the IoT for data acquisition, sophisticated simulation tools for emulating physical processes, artificial intelligence (AI) for advanced analytics and learning and control systems for implementing optimized operational strategies. The nature and fidelity of the models embedded within the DT are also critical. Rigorous calibration and validation processes are equally important, as they ensure trustworthy outputs and support robust real-time monitoring and operational decision-making.

This fundamental bidirectional data exchange is indeed the cornerstone of DT's operation, underpinning its capacity to create and sustain an accurate, dynamic virtual representation of a physical object or system. It is through this continuous, real-time transmission and integration of data that the digital model effectively mirrors its real-world counterpart, thereby enabling sophisticated functionalities such as comprehensive monitoring, detailed simulation, predictive analytics and ultimately the optimized control of physical entities within a virtual (Saracco, 2019; Segovia & Garcia-Alfaro, 2022). The successful actualization of this physical-to-digital linkage, where data from the physical asset informs the digital model, and the subsequent digital-to-physical connection, where insights or commands from the DT influence the physical asset, relies heavily on the principles of big data. The requisite volume, velocity and variety of information, often sourced from an array of sensors and IoT devices embedded within or monitoring the physical system, are crucial for maintaining the fidelity of the DT, ensuring its real-time responsiveness and facilitating robust analysis (Abdulhayan & Ali Abd, 2024).

The intricate integration of physical and digital domains is typically achieved through persistent, real-time data connections that allow for the fluid sharing of operational inputs, performance outputs and systemic processes. This high level of interconnectedness enhances user efficiency by providing actionable insights and also plays a significant role in mitigating operational risks through early detection of anomalies or deviations from desired performance (Mazumder et al., 2023). In this context, the DT often serves a dual conceptual purpose: it acts as a digital shadow, meticulously reflecting the current operational status and condition of its physical counterpart, while simultaneously functioning as a digital thread, diligently recording the asset's historical evolution and accumulating a comprehensive lifecycle record (Saracco, 2019). The fulfilment of these roles is critically dependent upon the availability and utilization of well-formed, high-quality data that accurately captures the salient features, characteristics and behaviours of the physical object. Such data integrity ensures that the DT can effectively monitor, simulate and interpret the physical system's status and predict its future behaviour with a high degree of confidence (Minerva et al., 2022b). Consequently, the robust synergy achieved through this continuous physical-to-digital and digital-to-physical data exchange is essential to unlocking the transformative potential and utility of DTs across a diverse array of applications and industries, including manufacturing, health care and smart cities (Minerva et al., 2022b).

The robust synergy achieved through this continuous physical-to-digital and digital-to-physical data exchange is further translated into actionable intelligence and optimised performance by the sophisticated integration of several core technologies within the DT framework. Principally, the IoT, advanced simulation tools, AI and embedded control systems work in concert to achieve comprehensive real-time monitoring and facilitate effective operational optimization across diverse and complex systems.

IoT devices, particularly sensors deployed throughout the physical asset or its operational environment, play a crucial role by providing a continuous stream of high-resolution data. This affords an unprecedented level of visibility into ongoing operations, enabling immediate and informed responses to disruptions and allowing for the dynamic optimisation of processes, a capability proving invaluable in sectors such as energy and construction (Agho et al., 2025; Pandit, 2025). The data acquired through IoT networks then fuels advanced analytical engines.

Specifically, the combination of AI algorithms with sophisticated simulation tools allows for in-depth, real-time data analysis and the exploration of various operational scenarios. AI contributes by identifying patterns, predicting future states and diagnosing anomalies, while simulation tools provide a virtual testbed for evaluating the potential impact of different strategies or conditions (Ferrigno & Barsola, 2023). This powerful pairing is fundamental to enabling predictive maintenance regimes, which can significantly minimize operational downtime and extend the effective lifespan of assets, thereby enhancing overall operational efficiency (Agho et al., 2025; Lachvajderová et al., 2024). Furthermore, extensive scenario testing in the digital environment enables proactive performance optimization, which is crucial for managing the inherent complexities of modern industrial environments.

Eventually, this convergence of technologies within the DT empowers enhanced decision-making processes and supports the implementation of refined control strategies. The insights generated can also be leveraged to advance sustainability objectives, for instance, by enabling the meticulous real-time monitoring of resource utilisation and environmental impacts (Agho et al., 2025). The synergistic operation of these components thus redefines operational capabilities, fostering greater efficiency and driving innovation.

The operational capabilities and analytical depth of DTs, stemming from the integration of prementioned technologies, are fundamentally reinforced by the diverse array of mathematical and computational models employed within their frameworks. These models are crucial for accurately replicating, analysing and predicting the behaviour of the physical systems they represent. Generally, the modelling approaches utilized can be categorized into three principal types: physics-based models, data-driven (or AI-based) models and hybrid models, each offering distinct advantages depending on the application and available information.

Physics-based models are constructed upon fundamental scientific principles, established physical laws and engineering equations that govern the behaviour of the system. These models aim to capture the intrinsic mechanisms and dynamics of the physical asset, as exemplified in applications such as the modelling of battery energy storage systems or electrolysers where equivalent circuit models are used to simulate complex electrochemical dynamics based on first principles (Chang et al., 2024). Such models provide a strong theoretical foundation and can offer insights even in scenarios with limited operational data.

In contrast, data-driven models primarily leverage ML algorithms and statistical techniques to learn patterns and relationships directly from historical and real-time operational data, often sourced from extensive sensor networks. These models excel in identifying complex, nonlinear correlations that might be difficult to formulate from first principles and are frequently employed in industrial settings for tasks such as intelligent process monitoring, anomaly detection and predictive control based on observed system behaviour (Raşit Cesur et al., 2024). Their predictive power is directly related to the quality and quantity of the data used for their training.

Recognizing the complementary strengths and limitations of the aforementioned approaches, hybrid models seek to combine the interpretability and generalizability of physics-based understanding with the adaptive learning capabilities of data-driven techniques. This integration, for instance by incorporating first-principles physics to inform or constrain ML algorithms, can lead to enhanced accuracy, improved robustness and better extrapolation capabilities, particularly in complex systems where purely physics-based or data-driven models might fall short, as demonstrated in areas like heat diffusion problems (Blakseth et al., 2022). This multifaceted modelling strategy, allowing for the selection or combination of these distinct approaches, enables the development of tailored and effective DT solutions capable of addressing the intricate complexities inherent in modern industrial systems (Munko et al., 2024).

Regardless of the specific modelling approach, the trustworthiness and practical utility of any DT are critically dependent upon rigorous model calibration and validation processes. These stages are indispensable for ensuring that the digital representation accurately and reliably mirrors the behaviour and characteristics of its real-world physical counterpart, thereby forming a dependable basis for analysis, prediction and decision-making.

Model calibration is the systematic process of refining and optimizing the parameters and structure of the chosen model (or models) within the DT. This involves systematically comparing model outputs with actual observational data collected from the physical asset and adjusting the model to minimize discrepancies. The objective is to align the digital representation as closely as possible with real-world performance, a step that is particularly essential in complex operational scenarios such as those encountered in advanced manufacturing paradigms like Industry 4.0 or in demanding applications where precision is paramount (Fakhraian et al., 2024; Sung & Tuo, 2024). Effective calibration ensures that the DT is not merely a theoretical construct but a faithful reflection of the physical system's current state and behaviour.

Following calibration, model validation serves to formally assess the extent to which the tuned DT accurately represents the physical system across a range of conditions, often using datasets not employed during the calibration phase. Validation addresses the fundamental question of the DT's fidelity and, consequently, its trustworthiness as a tool for informed decision-making (Hua et al., 2022; Vanderhorn & Valluri, 2024). This process quantifies the model's predictive accuracy and its ability to generalize to new, unseen operational states or environmental inputs. Advanced methodologies, such as the integration of model-based reinforcement learning, can further enhance calibration and validation efforts by explicitly accounting for inherent uncertainties within the system and improving the robustness of control policies derived from the DT, especially in highly dynamic environments.

Despite their critical importance, establishing universally robust and adaptable validation frameworks tailored to the unique, dynamic nature of DTs presents ongoing challenges. Addressing these often necessitates a multifaceted approach that judiciously combines domain-specific expert knowledge with the rich, real-time data streams provided by IoT devices and other sensing technologies (Hua et al., 2022). Ultimately, diligent and continuous calibration and validation are not merely procedural steps but are paramount for building confidence in the DT and fully leveraging its potential to deliver tangible benefits across diverse industrial applications.

Beyond rigorous calibration and validation, constructing an effective DT necessitates careful consideration of model fidelity, as choices among physics-based, data-driven and hybrid approaches involve distinct trade-offs concerning accuracy, computational load, interpretability and adaptability. Coupled with this, the explicit quantification of uncertainty in predictions is fundamental for reliable decision-making.

Physics-based models offer high interpretability and reliability, especially for well-understood systems (Kapteyn et al., 2022; Kochunas & Huan, 2021), but can be computationally intensive and less adaptive to novel data or unmodelled complexities (Kapteyn et al., 2022; Liu et al., 2024). Conversely, data-driven models excel at learning from operational data, enabling rapid adaptation and management of complex, nonlinear systems (Chen et al., 2022; Cohen & Huan, 2024; Kessels et al., 2025; Yu et al., 2021), though they may lack physical transparency, and their reliability can be compromised by data scarcity or when extrapolating beyond observed conditions (Kapteyn et al., 2022; Kochunas & Huan, 2021). Hybrid models strive to amalgamate these strengths, for instance, by using data to refine physics-based frameworks or by integrating multi-fidelity models to balance accuracy and cost (Chen et al., 2022; Kapteyn et al., 2022). While promising, these can introduce integration complexities (Chen et al., 2022; Kapteyn et al., 2022). The selection thus balances the strong interpretability of physics-based methods against the adaptive capabilities of data-driven approaches, with hybrids offering a synergistic, albeit potentially more intricate, path.

Irrespective of the chosen model, uncertainty quantification is critical for trustworthy DT applications (Cohen & Huan, 2024; De Angelis et al., 2023; Kapteyn et al., 2022; Kessels et al., 2025; Thelen et al., 2023). Uncertainty quantification methods, such as Bayesian inference or probabilistic neural networks (Cohen & Huan, 2024; De Angelis et al., 2023; Kapteyn et al., 2022; Kessels et al., 2025; Thelen et al., 2023; Yu et al., 2021), provide essential confidence measures for model outputs. This allows users to assess prediction reliability and make informed decisions, supporting robust online updating, real-time health monitoring and optimization under uncertainty; all vital for high-stakes applications where decision consequences are significant (Cohen & Huan, 2024; De Angelis et al., 2023; Kapteyn et al., 2022; Kessels et al., 2025; Thelen et al., 2023; Yu et al., 2021).

Applications Across Wind Turbine Types. The application and design of DTs for wind energy systems are not monolithic; rather, they must be carefully tailored to address the distinct operational environments and specific technical demands associated with different turbine typologies, notably onshore, fixed-bottom offshore and floating offshore installations. Each category presents unique requirements for data acquisition, modelling complexity and maintenance strategies, alongside characteristic challenges that DTs aim to mitigate.

For onshore wind turbines, effective DT implementation often necessitates the seamless integration of real-time supervisory control and data acquisition (SCADA) system data to continuously monitor turbine performance and prevailing environmental conditions (Stadtmann et al., 2023c). Accurate wind resource assessment and operational planning, particularly in topographically complex landscapes, can be significantly enhanced by incorporating high-resolution terrain modelling to visualize wind flow patterns (Stadtmann et al., 2023c). Furthermore, predictive analytics, leveraging detailed weather forecasts in conjunction with historical operational data, are crucial for optimizing power output and scheduling maintenance (Siddiqui et al., 2023; Stadtmann et al., 2023c). However, onshore applications face challenges such as inherent data variability induced by terrain-specific turbulence, which can complicate the development of robust predictive models (Stadtmann et al., 2023c).

Older turbine fleets may also suffer from limited sensor coverage, thereby hindering comprehensive real-time data acquisition for the DT (Sakr & Sadhu, 2024), and the retrofitting of existing infrastructure with the necessary IoT-enabled devices to support advanced DT functionalities can entail considerable investment (Siddiqui et al., 2023).

Transitioning to offshore wind turbines, typically fixed-bottom structures, introduces a more demanding set of requirements. Robust and reliable communication systems are paramount for transmitting vast quantities of data from remote offshore locations to onshore control and analysis centres (Chen et al., 2024; Kale & Gawali, 2023). The harsh marine environment dictates the use of corrosion-resistant sensors and components capable of withstanding prolonged exposure to saltwater and dynamic weather conditions (Chen et al., 2024; Sakr & Sadhu, 2024). Consequently, predictive maintenance frameworks within the DT become even more critical, specifically addressing accelerated wear and tear induced by saltwater corrosion and heightened mechanical stresses on the structure and machinery (Chen et al., 2024; Kale & Gawali, 2023). The challenges in this domain are significant, with severe environmental conditions such as storms and high salinity accelerating component degradation and increasing failure rates (Chen et al., 2024; Sakr & Sadhu, 2024). High latency in data transmission from distant offshore sites can also delay real-time operational adjustments or interventions (Ambarita et al., 2024), and the effective integration of heterogeneous data streams from diverse sources (including individual turbines, offshore substations and grid connection points) presents a considerable technical hurdle (Ambarita et al., 2024; Siddiqui et al., 2023).

Floating offshore wind turbines represent a further escalation in complexity, imposing even more specialized requirements on DT systems. Continuous, real-time monitoring of dynamic mooring lines, including parameters such as axial tension and fatigue accumulation, is essential for structural integrity and safety (Branlard et al., 2024; Daraki et al., 2024; Walker et al., 2022). The DTs for floating offshore wind turbines must incorporate advanced, high-fidelity physics-based models capable of accurately simulating the complex, coupled interactions between the floating platform, waves, wind and mooring system (Branlard et al., 2024; Siddiqui et al., 2023). Sophisticated motion compensation algorithms are also necessary to accurately interpret sensor data, accounting for platform tilting, heaving and potential drifts in GPS data (Branlard et al., 2024). The challenges are correspondingly amplified: the intricate structural dynamics arising from strong wind and wave coupling necessitate computationally intensive, high-fidelity simulations for accurate representation within the DT (Branlard et al., 2024; Daraki et al., 2024). Ensuring sensor reliability and accuracy in a constantly moving and harsh marine environment complicates precise load estimation and condition monitoring (Branlard et al., 2024; Walker et al., 2022). Furthermore, the relative novelty of many floating wind designs means there is often limited historical operational data available for thoroughly validating predictive models and training data-driven components of the DT (Chen et al., 2024; Siddiqui et al., 2023).

The practical implementation of DT concepts is increasingly evident across various wind turbine structural setups, with a growing body of research demonstrating tailored solutions that often place significant emphasis on capturing complex aero-structural interactions. These interactions, which govern how aerodynamic forces influence structural responses and vice-versa, are critical for understanding turbine performance, predicting fatigue and optimizing control strategies.

In the onshore wind turbine domain, for example, DT development has yielded specific applications focusing on critical components. One such implementation is a DT designed for the aerodynamic monitoring of rotor blades. This system utilizes data from micro-electro-mechanical systems sensors strategically placed on the blades, coupled with cloud-based analytics, to provide detailed insights into the aerodynamic performance and condition of these vital components (Marykovskiy et al., 2024). This allows for the detection of issues such as leading-edge erosion or changes in aerodynamic efficiency, which directly impact power production and can induce undesirable structural loads.

Transitioning to offshore wind turbines, a notable DT solution has been developed and validated using full-scale prototype data. This approach effectively integrates Kalman filters with physics-based models to achieve reliable fatigue load estimation, a crucial aspect for managing the long-term structural integrity of turbines operating in demanding marine environments (Branlard et al., 2024). Another pertinent example for offshore applications involves the creation of a computationally efficient DT for a turbine situated on a monopile foundation. This was achieved through the application of reduced-order modelling techniques, which successfully balanced high accuracy in representing the system's behaviour with the computational speed necessary for practical real-time applications (Zhao et al., 2023b). Such models are vital for understanding the dynamic interplay between wind loads, wave actions and the structural response of the support structure and tower.

4.3 Applications in Wind Turbine Structures

Predictive Maintenance. Wind turbine structures, particularly their primary components such as blades, towers and support structures, are subjected to demanding operational and environmental conditions that necessitate periodic and effective maintenance to ensure their longevity and optimal performance. Blades, for instance, are highly susceptible to issues like leading-edge erosion from particle impact and fatigue damage accumulated from fluctuating aerodynamic loads, both of which can significantly degrade turbine efficiency and shorten operational lifespan (Kaewbumrung et al., 2024). Similarly, wind turbine support structures, including towers and foundations, experience substantial fatigue under cyclic wind loads and, where applicable, wave loads, alongside the persistent risk of corrosion (Rangel-Ramírez & Sørensen, 2008). Given the exceptionally high number of load cycles a typical wind turbine endures throughout its design life, ensuring adequate fatigue resistance is paramount for achieving secure service duration and an economically viable design (Schaumann et al., 2011). Consequently, regular monitoring, often guided by methods such as computational fluid dynamics to investigate performance-affecting factors (Kaewbumrung et al., 2024), and preventive maintenance, potentially optimized through risk-based inspection planning, especially for less accessible offshore turbines (Rangel-Ramírez & Sørensen, 2008), are essential. While various damage detection methods, including the use of advanced sensor technologies like fibre optic and piezoelectric sensors, are continuously being developed to improve maintenance practices, particularly for blades (Li et al., 2015), the scale and complexity of modern wind energy systems call for more integrated and predictive approaches.

It is precisely in addressing these inherent maintenance challenges that the established capabilities of DTs (encompassing detailed simulation, continuous data integration and advanced modelling) offer significant advancements for predictive maintenance strategies. By enabling real-time condition monitoring and sophisticated predictive analytics for these critical components, DTs allow operators to move beyond reactive or merely scheduled maintenance towards a more proactive and optimized regime. This capability to anticipate potential failures and refine maintenance schedules, based on integrated sensor data and advanced modelling, is particularly valuable for high-stress components such as gearboxes and main bearings, as well as for the aforementioned blades, towers and support structures. The ultimate aim is to predict the remaining useful life (RUL) of these parts with greater accuracy, thereby substantially reducing unplanned downtime and associated maintenance costs, while concurrently improving overall asset utilization and management practices (Abdullahi et al., 2024; Moghadam et al., 2021; Yucesan & Viana, 2023). A summary of these diverse applications within predictive maintenance is provided in Table 4.2.

A key aspect of this predictive capability lies in early and accurate fault detection and diagnostics. DTs facilitate this by, for instance, enabling the creation of virtual sensor arrays that can augment existing physical sensor networks or even substitute for faulty or unavailable sensors. These systems often employ spatiotemporal models and data-driven estimators to intelligently identify sensor faults, verify the integrity of incoming data and reconstruct normal sensor behaviour, thereby enhancing the reliability and operational safety of wind turbines (Li & Shen, 2022; Wang et al., 2021). Furthermore, structural DTs are being developed to support tasks such as the dynamic identification of impacts on turbine blades. Such systems can achieve accurate diagnostics even when faced with limited real-world training data by leveraging techniques like transfer learning and adaptive neural networks, allowing for more efficient online deployment and use (Li et al., 2024).

Beyond mere detection, robust prognostics are essential for effective maintenance planning. DTs contribute significantly here by integrating physics-based understanding with data-driven approaches to quantify uncertainties associated with component degradation processes, such as the deterioration of grease in main bearings or the progression of fatigue cracks in structural elements. This hybrid modelling paradigm can provide reliable prognostic estimates of component health and predict fleet-wide unreliability trends, often reducing the dependence on extensive historical datasets that may not always be available, thus directly supporting more informed and proactive maintenance interventions (Moghadam et al., 2021; Yucesan & Viana, 2023).

Table 4.2 Summary of DT applications in wind turbine maintenance

Application area	DT role/methodology	AI/ML integration	Key benefits
Predictive maintenance	Real-time monitoring, lifetime prediction of components (Abdullahi et al., 2024; Moghadam et al., 2021; Yucesan & Viana, 2023)	Deep learning, predictive analytics (Abdullahi et al., 2024; Fahim et al., 2022)	Reduced downtime, optimised maintenance
Fault detection	Virtual sensor arrays, spatiotemporal models (Li & Shen, 2022; Wang et al., 2021)	Data-driven estimators, transfer learning (Li et al., 2024)	Early fault identification, improved safety
Diagnostics	Structural DTs, adaptive learning (Li et al., 2024; Wang et al., 2021)	Attention mechanisms, neural networks (Li et al., 2024)	Accurate diagnostics, efficient online use
Prognostics	Hybrid physics/data-driven models, uncertainty quantification (Moghadam et al., 2021; Yucesan & Viana, 2023)	ML, uncertainty modelling (Yucesan & Viana, 2023)	Reliable prognostics, proactive planning
Maintenance planning	Distributed DT frameworks, real-time feedback (Abdullahi et al., 2024; Fahim et al., 2022)	AI-driven optimisation (Chen et al., 2021; Fahim et al., 2022)	Intelligent scheduling, cost reduction

The power of DTs in predictive maintenance is further amplified when integrated with AI and ML algorithms, paving the way for intelligent, semi-autonomous systems for wind turbine operation and maintenance. These sophisticated frameworks can, for example, be used to train AI models specifically for predictive maintenance tasks and fault diagnosis, utilizing a rich combination of both simulated data generated by the DT and real-world operational data (Chen et al., 2021; Fahim et al., 2022). Such AI-enhanced DTs can then provide real-time feedback and drive the optimization of maintenance planning, often leveraging distributed architectures incorporating cloud and edge computing for scalable and responsive solutions (Abdullahi et al., 2024; Fahim et al., 2022). This synergy also redefines the role of human operators, supporting them with advanced decision-making tools and elevating their function to a supervisory capacity while AI algorithms handle much of the routine monitoring, diagnostics and initial planning, as detailed in Table 4.2.

Virtual Sensing and Load Reconstruction. As briefly alluded to in the preceding discussion on predictive maintenance, the concept of virtual sensing, facilitated by DT technology, represents a significant enhancement to data acquisition and system understanding, particularly for wind turbine structures. Virtual sensing effectively enables the estimation of system states, structural responses or environmental conditions at locations where physical sensors are either not installed are malfunctioning, or where direct instrumentation is impractical or prohibitively expensive. By adeptly integrating available sensor data from instrumented points with underlying physics-based models or data-driven emulators, DTs can infer or predict values at these unmeasured locations. This capability optimizes the placement and number of physical sensors required, potentially reducing instrumentation costs and complexity and additionally allows for a far more comprehensive and granular monitoring and assessment of the entire asset (Bado et al., 2022; Febrianto et al., 2022). The capacity for synthetic sensing, often powered by AI, further extends this by deriving richer environmental or operational information from even basic sensor inputs, thereby broadening the scope of what can be monitored and understood virtually (Minerva et al., 2022a). A summary of these concepts is presented in Table 4.3.

One of the most powerful applications of virtual sensing within the DT framework is in the reconstruction of loads experienced by wind turbine structures, especially under extreme or infrequently occurring conditions such as severe wind gusts, turbulence or significant wave impacts on offshore installations. By synthesizing the limited real-time data from existing sensors with robust physical models, the DT can estimate the true dynamic response and loading across the entire structure, including at critical but un-instrumented points, during these rare events (Bado et al., 2022; Febrianto et al., 2022). This ability to reconstruct detailed load histories is crucial for comprehensive long-term structural health monitoring. It enables more reliable what-if analyses, allowing for the assessment of structural integrity under various extreme loading scenarios and supporting the proactive management and design validation of wind turbine structures and other critical infrastructure components.

The insights derived from virtual sensing and load reconstruction feed directly and logically into enhancing predictive maintenance strategies, as discussed earlier. By providing a more continuous, comprehensive and spatially rich understanding of the asset's health and performance in real-time, these virtualized data streams become invaluable. The ability to predict system responses and identify subtle anomalies or trends across the structure, even in areas without direct measurement, can flag potential impending failures much earlier or with greater confidence. This, in turn, allows maintenance interventions to be scheduled based on the actual, evolving condition of the asset rather than on predetermined fixed intervals or limited sensor readings.

Table 4.3 Key aspects of virtual sensing and load reconstruction in DTs

Concept	Role in DTs	Key benefits
Virtual sensing	Estimates responses at unmeasured locations using models and sensor data (Bado et al., 2022; Febrianto et al., 2022; Minerva et al., 2022a)	Enhanced monitoring, optimized sensor placement, reduced instrumentation costs
Load reconstruction	Predicts structural loads under extreme conditions using synthesized data (Bado et al., 2022; Febrianto et al., 2022)	Reliable assessment, supports what-if analysis, improves safety and resilience
Predictive maintenance	Uses virtual sensing outputs to inform maintenance schedules (Bado et al., 2022; Febrianto et al., 2022)	Reduces downtime, extends asset life, enables condition-based maintenance

This data-driven approach, where the DT can continuously update its understanding and even compensate for predicted structural behaviours, optimizes maintenance activities, significantly reduces unplanned downtime, extends asset operational life and supports more informed, condition-based decision-making regarding maintenance actions and overall serviceability (Bado et al., 2022; Febrianto et al., 2022), as summarized in Table 4.3.

Lifecycle Management. The utility of DTs extends far beyond specific tasks like predictive maintenance or virtual sensing; they offer a holistic framework for managing wind turbine structures throughout their entire lifecycle, from initial conception (design and planning) to eventual decommissioning. This comprehensive approach facilitates optimized resource utilization and more informed decision-making at each stage, as summarized in Table 4.4.

During the design and planning phases, DTs serve as powerful virtual prototyping tools. By creating detailed digital representations of proposed wind turbine structures and their intended operational environments, engineers can simulate various design scenarios, assess potential performance characteristics and evaluate different structural configurations before any physical construction commences. This capability for upfront modelling and assessment supports more robust decision-making, aids in early risk identification and mitigation, and can lead to more optimized designs for both onshore and offshore wind turbines (Siddiqui et al., 2023; Stadtmann et al., 2023d).

Once a wind turbine is operational, its DT enables continuous, real-time monitoring and the integration of data from a network of sensors, as previously discussed. This allows for ongoing assessment of structural health, the early detection of faults

Table 4.4 Summary of DT contributions across lifecycle stages

Lifecycle stage	Key DT contributions	Resource optimization role
Design and planning	Virtual modelling, scenario simulation (Siddiqui et al., 2023; Stadtmann et al., 2023d)	Informed design, risk reduction
Operation and monitoring	Real-time health monitoring, fault detection (Sakr & Sadhu, 2024; Tao et al., 2018; Wang et al., 2021)	Minimized downtime, efficient maintenance
Maintenance and prognostics	Life prediction, prognostics and health management (PHM), data synthesis (Tao et al., 2018; Yucesan & Viana, 2023; Zhao et al., 2023a)	Optimized maintenance, cost reduction
Decommissioning	Lifecycle data management, planning (Siddiqui et al., 2023; Stadtmann et al., 2023d)	Sustainable resource recovery, compliance

or anomalies and the prediction of phenomena such as crack propagation or component fatigue. Such insights are fundamental to implementing condition-based maintenance strategies, which, by facilitating just-in-time interventions, help to minimize downtime and extend the operational service life of the asset (Sakr & Sadhu, 2024; Tao et al., 2018; Wang et al., 2021; Yucesan & Viana, 2023).

In the realm of maintenance and prognostics, DT-driven PHM systems leverage a combination of data-driven and physics-informed models to predict the RUL of critical components and optimize maintenance schedules. These systems can be particularly valuable in situations where historical data is scarce, as DTs can be used to generate synthetic datasets for model training or employ techniques like transfer learning to adapt models from similar components or conditions. This improves the accuracy of life predictions for crucial elements like main bearings, thereby supporting more efficient and cost-effective maintenance planning (Tao et al., 2018; Yucesan & Viana, 2023; Zhao et al., 2023a).

Even at the end-of-life stage, the DT continues to provide value. By maintaining a comprehensive and continuously updated digital record of the asset's history, including its design specifications, operational performance, maintenance interventions and material composition, the DT facilitates informed decision-making regarding decommissioning. This includes optimizing decommissioning planning, maximizing material and resource recovery and ensuring compliance with environmental regulations and sustainability objectives (Siddiqui et al., 2023; Stadtmann et al., 2023d).

The integrative frameworks upon which DTs are built are fundamental to this lifecycle management capability, which inherently promote resource optimization. By combining smart monitoring capabilities with big data analytics and advanced modelling techniques, these frameworks allow for the optimization of various resources. This includes maximizing power output through better operational strategies, minimizing maintenance costs through predictive interventions and improving energy forecasting accuracy. Predictive models embedded within the DT, for instance, can forecast wind speed and power generation more reliably, aiding in efficient energy management and the strategic allocation of operational resources (Fahim et al., 2022; Siddiqui et al., 2023; Stadtmann et al., 2023d). Furthermore, these frameworks often incorporate sophisticated uncertainty quantification methodologies, perhaps using surrogate models like Gaussian Processes. This enables robust reliability analysis and supports risk-informed decision-making even when faced with the variable environmental and operational conditions characteristic of wind energy (Jorgensen et al., 2023; Yucesan & Viana, 2023). To foster broader adoption and ensure consistency, efforts are also underway to develop structured, multi-level DT capability models (ranging from standalone systems to fully autonomous ones) which support scalable and standardized deployment across the wind energy sector. Such standardization facilitates consistent resource optimization practices and encourages industry-wide collaboration and learning (Siddiqui et al., 2023; Stadtmann et al., 2023b, d), as detailed in Table 4.4.

The practical benefits of DTs in fostering holistic lifecycle management are further underscored by several documented case examples across diverse wind turbine systems and operational contexts. These implementations illustrate how the integration of real-time data, advanced modelling and predictive analytics can yield tangible improvements in efficiency, reliability and cost-effectiveness throughout an asset's lifespan.

One illustrative case involves the application of a DT for comprehensive wind turbine asset management. This system effectively synthesizes various techniques, including system identification to characterize the turbine's dynamic behaviour, data-augmented modelling to refine predictions and rigorous verification and validation processes. By continuously updating the virtual model with real-time data streamed from the physical turbine, this DT enhances predictive maintenance capabilities and optimizes operational efficiency, thereby effectively managing the asset's lifecycle (Wagg et al., 2020).

The utility of DTs also extends to the manufacturing phase of wind turbine components. A notable example is the development of a DT-based virtual factory concept, which has been evaluated in scenarios involving both a wind blade manufacturing plant and a nacelle assembly line. This integrated simulation model supports the detailed modelling, simulation and evaluation of these complex manufacturing systems. According to industry experts, this approach facilitates better management of both product and system lifecycles by enabling manufacturers to adapt more effectively to dynamic production environments and to proactively improve manufacturing processes (Yildiz et al., 2021).

Focusing specifically on critical component health, a DT framework has been successfully applied for the predictive maintenance of wind turbine gearboxes. This system utilizes real-time SCADA data, incorporating parameters such as gearbox oil temperature, bearing temperature and rotor speed, to forecast the time to failure of these crucial components. The predictive module within the DT provides early warnings of impending issues, enabling more efficient operation and maintenance management, which in turn reduces costly downtime and extends the operational life of the gearboxes (Xu et al., 2025).

The unique challenges of floating offshore wind turbines have also spurred tailored DT solutions. One such development involves a multi-level DT (encompassing standalone, descriptive and prescriptive capabilities) for an operational FOWT. The standalone aspect of the DT aids significantly during the initial planning, design and construction phases. Subsequently, the descriptive and prescriptive functionalities support real-time operational monitoring and inform maintenance decisions throughout the turbine's service life. This comprehensive, multi-level approach demonstrably improves the lifecycle management of these complex offshore assets by enabling more robust planning and optimized maintenance strategies (Rasheed et al., 2024; Stadtmann et al., 2023d).

Beyond structural and major mechanical components, DTs are also being leveraged to enhance the reliability of crucial subsystems. For example, a novel control strategy, developed using DT models, has been implemented to mitigate thermal stress on IGBT modules within the power converters of permanent magnet synchronous generator (PMSG) wind turbines. By reducing the thermal cycling experienced by these sensitive electronic components, the DT-informed control strategy helps to decrease converter failures, thereby improving overall turbine reliability and lowering operation and maintenance costs. This case highlights how detailed modelling and control optimization, facilitated by DTs, can enhance subsystem resilience (Ahmedi et al., 2025).

Finally, to address the scale of entire wind energy installations, a distributed DT framework has been implemented for the predictive maintenance of multiple turbines within a wind farm. This architecture provides real-time condition monitoring and health management for key components across the entire fleet, improving overall asset utilization and significantly reducing instances of unplanned downtime. Such frameworks typically leverage industrial IoT data to create a connected ecosystem of DTs, thereby enhancing lifecycle management capabilities at the farm level (Abdullahi et al., 2024). These diverse examples affirm the versatile and impactful role of DTs in optimizing the management of wind energy assets across all lifecycle stages.

Aero- and Structural Control Integration. The sophisticated capabilities inherent in DT technology further extend to actively enhance the operational control of wind turbines, particularly in the critical domain of aero-structural integration. By providing a dynamic, high-fidelity virtual representation of the physical asset, DTs serve as powerful platforms for developing, testing and deploying advanced control strategies aimed at optimizing aerodynamic performance and managing structural responses under diverse and fluctuating conditions.

A primary application lies in enhancing dynamic control strategies for blade pitch and yaw to optimize aerodynamic efficiency and overall energy capture, especially within wind farm settings. DTs facilitate this by enabling real-time monitoring of individual turbine performance and simulating the complex wake interactions between turbines. This allows for the dynamic repositioning of turbines, primarily through yaw adjustments, to minimize detrimental wake effects on downstream units and thereby maximize the collective energy production of the entire farm (Kandemir et al., 2023; Siddiqui et al., 2023). Advanced control techniques, such as deliberate yaw misalignment or dynamic individual pitch control, can be developed and refined using DT simulations to actively manipulate wake characteristics, such as by enhancing wake mixing to increase wind speeds for adjacent turbines (Frederik et al., 2020; Munters & Meyers, 2018). For instance, the helix approach employs precisely timed individual pitch control to induce wake rotation, leading to faster wake recovery and improved energy capture for turbines operating in the wake (Frederik et al., 2020). Furthermore, the synergistic combination of yaw control with collective or individual pitch adjustments, optimized through DT modelling, has demonstrated effectiveness in tracking desired power reference signals for the entire wind farm, thereby optimizing overall performance while concurrently reducing the need for turbine derating (Starke et al., 2023).

In parallel with aerodynamic optimization, DTs are instrumental in developing and implementing more effective strategies for structural response management, crucial for ensuring the integrity and longevity of turbines, especially as they increase in size and are deployed in more challenging environments. Various structural control systems (broadly categorized as active, semi-active, passive and hybrid (Elgammal &

Ali, 2024)) can be designed, tested and tuned with greater efficacy using DTs. For example, advanced control methods like H-infinity control can be simulated and validated within a DT environment to demonstrate their effectiveness in reducing tower loading under both steady and turbulent wind conditions (Menezes & Araújo, 2023). As turbines become larger and more flexible, the dynamic loads they experience increase, making active and semi-active control techniques particularly important (Xie & Aly, 2020). Conventional pitch control and generator torque control in variable-speed turbines are primary methods for load reduction and power regulation (Bossanyi, 2003), and their performance can be further refined through DT-based optimization. Moreover, independent blade pitch control offers additional potential for load alleviation. Semi-active control systems, such as those employing magnetorheological dampers, can be modelled within a DT to evaluate their ability to limit base bending moments without detrimentally increasing top displacement (Caterino & Spizzuoco, 2017).

The enhancement of these control strategies by DT technology is multifaceted. DTs provide a virtual environment for extensive simulation and optimization of active, semi-active and passive control strategies across a wide spectrum of operating scenarios and environmental conditions, far exceeding what is feasible with physical testing alone (Bado et al., 2022; Rasheed et al., 2024), as summarized in Table 4.5. For active and semi-active systems, the DT can utilize its real-time state estimation capabilities as direct input for controllers, or it can be used offline to tune control parameters for optimal performance. In an online capacity, the DT can recommend or even automatically implement real-time control actions (such as precise adjustments to blade pitch, yaw angle or the settings of damping devices) to actively mitigate vibrations, reduce transient loads and enhance overall turbine stability (Bado et al., 2022; Rasheed et al., 2024). Recent investigations into floating offshore wind turbines have substantiated this approach, demonstrating that DT-enabled adaptive control strategies, such as dynamically adjusting rated rotor speeds, can significantly mitigate corrosion-fatigue in critical ring-flange connections and extend their operational lifespan (Zhang et al., 2025). For passive control systems, which are typically fixed at the design stage, DTs offer significant advantages during the initial design and optimization phase. They allow engineers to simulate the long-term structural response to various passive control mechanisms (e.g. tuned mass dampers), ensuring robust performance and durability throughout the turbine's intended lifecycle (Bado et al., 2022; Stadtmann et al., 2023d). Furthermore, specialized DT formulations, sometimes referred to as a 'Risk Twin', can integrate structural reliability analysis and Bayesian inference directly into the control loop. This allows for the visualization of evolving risk profiles and informs control decisions in real-time, enabling adaptive control strategies that are continuously refined based on updated Reliability Indices and operational feedback, thereby improving safety margins and potentially extending the operational life of the asset (Bado et al., 2022; Wang & Wang, 2025). This integrated approach, leveraging the DT's predictive power and comprehensive system understanding, marks a significant step towards more intelligent and adaptive control of wind energy structures.

Table 4.5 Comparison of control strategies enhanced by DTs

Control strategy	Enhancement via DT	Example functions enabled
Active	Real-time feedback and optimization	Adaptive blade pitch control, active damping, load mitigation (Bado et al., 2022; Rasheed et al., 2024)
Semi-active	Predictive and adaptive adjustment	Tuneable dampers, smart actuators, real-time parameter tuning (Bado et al., 2022; Rasheed et al., 2024)
Passive	Improved design and simulation	Optimal placement of passive devices, lifecycle performance prediction (Bado et al., 2022; Stadtmann et al., 2023d)

4.4 Challenges, Solutions and Future Directions

While the preceding discussions have highlighted the substantial advancements and diverse applications of DTs in enhancing the operation, maintenance and control of wind energy structures, the widespread adoption and full realization of their potential are contingent upon addressing a series of inherent complexities and practical hurdles. The sophisticated capabilities of these digital representations, from predictive maintenance to integrated aero-structural control, are indeed transformative; however, their development and deployment are not without significant challenges. These range from accurately modelling and adapting to complex and unpredictable environmental dynamics to managing the considerable data and computational demands that underpin their functionality. Furthermore, issues surrounding industry-wide standardization, seamless integration with existing operational workflows and human expertise and the overarching need for robust cybersecurity measures must be navigated. Successfully overcoming these obstacles is crucial, as is the continued exploration of emerging technological innovations that promise to further extend the capabilities and impact of DTs in the wind energy sector.

Complex Environmental Factors. One of the foremost challenges in developing effective DTs for wind turbine structures lies in accurately capturing and adapting to complex environmental factors, which are inherently dynamic and site-specific. Offshore wind turbines, for example, are particularly susceptible to the harsh marine environment, facing persistent threats from corrosion and accelerated fatigue damage to their support structures and other components (Leng et al., 2023; Okenyi et al., 2024). The accuracy and reliability of physical sensors, such as anemometers for wind speed measurement, can also be compromised by these environmental factors, potentially leading to failures or erroneous data that negatively impact turbine performance and the fidelity of the DT (Ibrahim et al., 2023). The complex interplay of wind and wave dynamics presents a particular challenge for offshore installations, including floating systems. While DTs offer the potential to update structural reliability assessments dynamically, quantify uncertainties in fatigue damage accumulation due to these combined loads and thereby optimize maintenance decisions (Augustyn et al., 2021), the process is not straightforward. For instance, efforts to update wave loading parameters within a DT based on new information can sometimes paradoxically reduce the perceived structural reliability if this updating introduces increased uncertainty into virtual sensing estimations (Augustyn et al., 2021). Nevertheless, techniques such as reduced-order modelling are being explored to create computationally efficient yet accurate DTs capable of predicting structural responses to combined wind–wave loading in near real-time (Zhao et al., 2023b). Furthermore, as noted previously, DTs themselves can act as virtual sensors, for instance, by estimating wind speed based on generated power, thus offering a way to overcome some limitations or failures of physical sensors (Ibrahim et al., 2023).

Addressing these site-specific challenges effectively necessitates robust strategies for real-time model updates and ensuring overall model robustness in the face of fluctuating conditions and imperfect data. The goal is to maintain a high-fidelity virtual representation that accurately reflects the physical asset's current state despite environmental variabilities and potential sensor degradation. Recent research has focused on developing such adaptive capabilities. Bayesian updating frameworks, employing sophisticated statistical methods like Transitional Markov Chain Monte Carlo, have shown promise in creating high-fidelity virtual models that are resilient to measurement noise and can assimilate new data to refine their accuracy over time (Xu et al., 2023). Hybrid modelling approaches, which synergistically combine the strengths of physics-based understanding with data-driven learning, are also pivotal. When coupled with advanced data fusion algorithms, these hybrid models can enable the development of interpretable, real-time predictive DTs capable of adapting to changing conditions (Kandemir et al., 2024). An example of such a concept involves using OpenFAST (National Renewable Energy Laboratory (NREL) 2025) linearization in conjunction with Kalman filtering techniques to estimate critical parameters like wind speed, operational loads and accumulated fatigue in real-time, achieving accuracy levels reportedly around 10% (Branlard et al., 2020). To specifically address identification uncertainty arising from varying operational and environmental conditions, strategies involving an intermediate model for environmental condition estimation, followed by Bayesian system identification for updating the structural model parameters, have been proposed (Zhu et al., 2020). These advanced approaches typically leverage a combination of SCADA measurements, dedicated IoT sensor data and sophisticated computational techniques to continuously enhance the accuracy, responsiveness and robustness of DTs, making them more reliable tools for managing wind turbines operating in complex and often unpredictable environments. However, challenges related to data integration from diverse sources, the need for efficient communication networks and the development of universally accepted standards also permeate broader aspects of DT implementation.

Data and Computational Constraints. The efficacy and scalability of DTs for wind energy structures are also profoundly influenced by data and computational constraints. The very foundation of a DT (i.e. its ability to accurately mirror a physical asset in real-time) necessitates the acquisition, processing and integration of vast quantities of diverse data, alongside substantial computational resources for simulation, analysis and optimization (Rasheed et al., 2020; Van Den Brand et al., 2021; Möhring et al., 2022).

A primary obstacle lies in multi-modal data integration and processing. As previously implied, DTs for wind turbines must assimilate data from a multitude of sources, including operational SCADA systems, dedicated condition monitoring sensors (vibration, acoustic, thermal), meteorological data, maintenance records, design specifications and potentially external market information. This data arrives in various formats (structured and unstructured), at different frequencies, and with varying levels of quality and completeness, creating significant integration challenges (Hildebrandt et al., 2024). The specific data requirements are demanding (as illustrated in Fig. 4.3): real-time or near real-time data streams are often essential for timely decision-making and control, leading to high-volume data generation, especially from numerous sensors across large wind farms (Correia et al., 2023; Hildebrandt et al., 2024; Omrany et al., 2023). This continuous influx necessitates robust data management architectures.

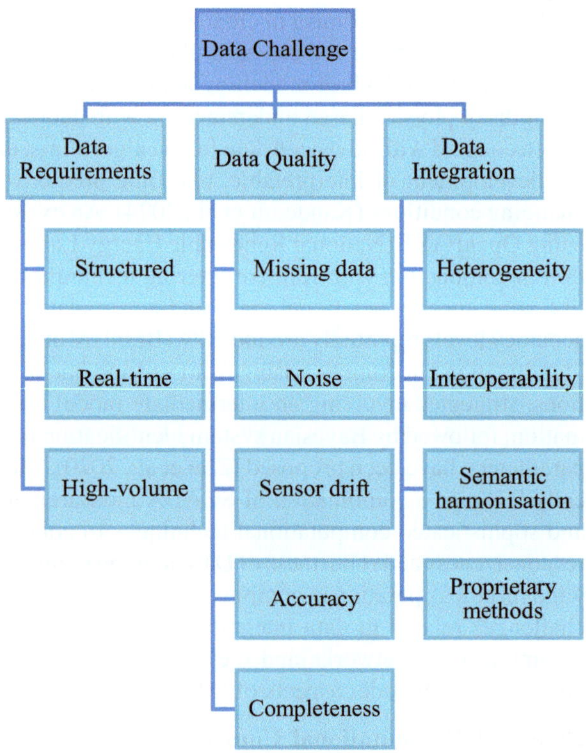

Fig. 4.3 Key data challenges in DT implementation

Compounding these integration challenges are pervasive data quality issues (Fig. 4.3). Missing data, due to sensor malfunctions, communication dropouts or intermittent data logging, can create gaps that hinder accurate modelling and analysis (Mohandes et al., 2024; Weil et al., 2023). Sensor data is frequently contaminated by noise, requiring sophisticated filtering and validation techniques to ensure its utility (Weil et al., 2023; Yang et al., 2024). Furthermore, sensor drift (i.e., the gradual loss of sensor accuracy over time due to environmental exposure or ageing) can introduce subtle but significant errors into the DT unless it is identified and corrected (Weil et al., 2023). Ensuring overall data accuracy and completeness remains a persistent concern, as flawed input data will inevitably lead to unreliable DT outputs and potentially misguided operational decisions (Mohandes et al., 2024; Omrany et al., 2023; Yang et al., 2024). The integration process itself (Fig. 4.3) is further complicated by the heterogeneity of data sources and formats, a lack of universally adopted interoperability standards and protocols, difficulties in achieving semantic harmonization (ensuring consistent meaning of data across different systems) and the prevalence of proprietary integration methods that can limit scalability and adaptability (Arin et al., 2023; Correia et al., 2023; Hildebrandt et al., 2024; Omrany et al., 2023; Weil et al., 2023).

The computational demands associated with running sophisticated DTs, particularly those incorporating high-fidelity multi-physics solvers or complex AI algorithms for real-time prediction and optimization, can also be substantial (Rasheed et al., 2020). Simulating the detailed aero-structural dynamics of a single turbine, let alone an entire wind farm, can strain current computational capabilities, especially when real-time or faster-than-real-time performance is required. One promising avenue for mitigating these computational burdens is the efficient use of surrogate models, also known as reduced-order models or metamodels (Bárkányi et al., 2021; Hartmann et al., 2018). These are essentially simplified mathematical approximations of more complex, computationally expensive simulation models. By training surrogate models on data generated from high-fidelity simulations or experimental tests, it is possible to achieve significantly faster prediction times while still retaining a reasonable level of accuracy for many applications, thus making real-time DT implementation more feasible.

Finally, as DTs become increasingly interconnected and reliant on continuous data streams between physical assets and digital platforms, cybersecurity considerations become paramount. The integration of physical and virtual systems inherently expands the potential attack surface, making wind turbine DTs and the operational technology they connect to targets for cyber threats (Siddique et al., 2023). Key concerns include ensuring secure data transmission channels, implementing robust access control mechanisms to prevent unauthorized interference, employing strong encryption techniques for data at rest and in transit and developing effective threat detection and response capabilities (Siddique et al., 2023). The convergence of these systems within IIoT environments, which often rely on high-fidelity models accurately reflecting real-time operations, further underscores the need for stringent security measures (Soudan, 2024). Addressing emerging threats and vulnerabilities necessitates the development and implementation of comprehensive cybersecurity frameworks that encompass data protection, secure communication protocols and resilience against attacks, starting from the earliest stages of DT design and development (Naveen et al., 2025). Ensuring the integrity, confidentiality and availability of both the data and the DT itself is critical for maintaining operational safety, protecting intellectual property and ensuring the overall resilience and competitiveness of wind energy operations.

Standardization and Industry Adoption. Following the discussion on data and computational constraints, further critical considerations for the successful and widespread implementation of DTs in the wind energy sector revolve around standardization, practical integration, scalability and industry adoption. The transition from isolated DT deployments to ubiquitous, interoperable systems requires addressing several interconnected challenges that span technical, organizational and collaborative domains.

A significant impediment, particularly evident in complex, multi-stakeholder environments like the architecture, engineering, construction and facility management (AEC-FM) industry, which shares parallels with large-scale energy projects, is the need for unified DT standards across different turbine types and project phases (Pinon Fischer et al., 2022). The current lack of standardized methodologies, data formats and communication protocols often leads to ad-hoc DT development approaches, resulting in limited interoperability between systems from different vendors or across various stages of a wind farm's lifecycle and generally inconsistent implementations (Nour El-Din et al., 2022; Pinon Fischer et al., 2022; Stadtmann et al., 2023b). This absence of a formalized standard hinders the broader adoption and inhibits the realization of the full potential of DTs, particularly in improving operational efficiency and reducing costs across the entire wind energy sector (Siddiqui et al., 2023; Tuton et al., 2024). To address this, researchers and industry bodies are exploring various avenues. One suggestion is to leverage and adapt existing standards from related fields, such as building information modelling (BIM) standards like ISO 19650, to promote interoperability in the data exchange and management aspects of DTs (Fonseca & Gaspar, 2021; Nour El-Din et al., 2022). Concurrently, the development of open standards specifically for DT data and model exchange is actively being advocated (Fonseca & Gaspar, 2021). Initiatives like the NorthWind project underscore the importance of developing hierarchical assessments of DT capabilities, which would greatly benefit from standardized definitions and classifications to ensure clarity and foster consistent, effective implementations across the diverse range of wind turbine designs and operational contexts, especially in offshore environments where varying turbine sizes and installation complexities exacerbate integration challenges (Ambrozaite et al., 2024; Rasheed et al., 2024). Establishing such unified standards is seen as crucial for streamlining operations, enhancing collaboration among stakeholders and ultimately accelerating the growth and maturity of the wind energy sector (Issa et al., 2023; Tuton et al., 2024).

Beyond standardization, the practical integration of DTs with existing operational systems and human operators/technicians is vital for their successful uptake. DTs cannot operate in a vacuum; they must seamlessly connect with legacy control systems, enterprise resource planning (ERP) software, maintenance management systems and, crucially, provide intuitive and actionable information to human personnel. In manufacturing, for instance, DTs equipped with virtual reality (VR) interfaces have shown potential for tasks like robot control, though considerations such as operator stress levels need to be managed (Kuts et al., 2022). More broadly in smart manufacturing, DTs facilitate enhanced human–machine interaction, supporting roles from real-time monitoring to complex decision-making (Wilhelm et al., 2021). For DTs to be truly effective, particularly in applications like predictive maintenance in the railway sector or quality management in manufacturing, integration with systems like manufacturing execution systems (MES) is considered essential. MES-integrated DTs can provide an intelligence layer that supports decision-making for tasks such as error state management or optimizing disassembly processes based on quality assessments (Dirnfeld et al., 2024; Negri et al., 2020). The goal is to empower human operators with enhanced insights and tools, rather than replacing them, fostering a collaborative intelligence environment.

Another significant challenge lies in scaling DT solutions from a single turbine to an entire wind farm or even across multiple wind farms. Obstacles include the previously mentioned lack of standardized definitions and descriptions, persistent data-related issues (compounded at scale), increasing model complexity when considering farm-level interactions like wake effects and general industrial acceptance hurdles (Ebrahimi, 2019; Stadtmann et al., 2023b). Strategies to overcome these scaling challenges often involve the development of hybrid modelling approaches that balance fidelity with computational feasibility, the robust integration of diverse data sources from numerous assets and the strategic leveraging of real-time systems incorporating edge and cloud computing architectures to handle distributed data processing and analytics (Kandemir et al., 2024). Predictive DTs for offshore wind farms, for example, are being developed on platforms that specifically incorporate advanced visualization tools, robust data communication protocols suitable for remote environments and sophisticated predictive algorithms designed to enhance farm-level operation and maintenance strategies (Haghshenas et al., 2023).

Eventually, cross-industry collaborations can play a pivotal role in accelerating the development and adoption of DT technology in the wind energy sector, as summarized in Table 4.6. While direct, large-scale collaborations between entirely disparate industries (e.g. wind energy and health care) on specific DT projects may still be emergent, the transfer of knowledge, technologies and best practices is increasingly common. Research consortia, such as the NorthWind project in Norway, bring together academic institutions, industry stakeholders and government agencies, fostering a multi-sectoral approach to advancing DT applications within wind energy (Rasheed et al., 2024). Technology integration is also evident, for instance, in the adoption of immersive visualization technologies like virtual reality, more commonly associated with the gaming and entertainment industries, for enhancing the monitoring and management of floating offshore wind farms (Stadtmann et al., 2023a). Furthermore, the wind energy sector benefits significantly from advancements in foundational technologies driven by other industries, such as the industrial IoT and cloud computing, which underpin distributed DT frameworks for predictive maintenance (Abdullahi et al., 2024; Kou et al., 2024). This synergistic adoption of methods and tools for data visualization, predictive analytics or IoT integration from other sectors indicates a valuable form of indirect cross-industry learning and collaboration (Zami et al., 2024).

Technological Innovations and Trends. The ongoing evolution of DT technology is marked by a confluence of technological innovations and emerging trends that promise to further expand its capabilities and address existing limitations, particularly within the demanding context of wind energy systems, as illustrated in Fig. 4.4. These advancements span computational paradigms, modelling techniques and control philosophies, all contributing to more intelligent, adaptive and integrated DT solutions.

A prominent trend is the increasing integration of advanced AI, particularly generative AI, with DT frameworks. While still in its relatively early stages of application, generative AI offers significant potential to enhance DTs by, for instance, creating synthetic data for training predictive models, simulating complex scenarios with greater realism or even automatically generating or optimizing DT model components (Li et al., 2025; Muhammad et al., 2024). In fields like telecommunications, generative AI is being explored to improve network DTs for predictive maintenance and real-time decision-making (Muhammad et al., 2024), and similar benefits are anticipated in other complex systems. The ability of generative AI to learn from vast datasets and produce novel, realistic outputs could revolutionize how DTs are built, validated and utilized, particularly in customizing models for specific needs in areas like climate science and sustainability, which have direct relevance to renewable energy (Bordukova et al., 2024; Emmert-Streib, 2023).

The computational architecture supporting DTs is also evolving, with a notable shift towards distributed models incorporating edge computing alongside traditional cloud-based solutions. Cloud computing has been instrumental in enabling the real-time data processing, analytics and visualization capabilities of modern DTs, facilitating applications from personal health management to structural health monitoring (Al-Ali et al., 2020; Dang et al., 2022; Liu et al., 2019). However, for applications

Table 4.6 Types of cross-industry collaboration in DTs

Collaboration type	Example in wind energy sector	Nature of cross-industry interaction
Research consortia	NorthWind project (Rasheed et al., 2024)	Academia, industry and government collaboration
Technology integration	VR for floating wind farms (Stadtmann et al., 2023a)	Adoption of VR/gaming tech in energy sector
Industrial IoT and cloud computing	Predictive maintenance frameworks (Abdullahi et al., 2024; Kou et al., 2024)	Leveraging information technology/tech advances for energy applications
Knowledge transfer/best practices	Data visualization and analytics (Zami et al., 2024)	Borrowing methods from other industries

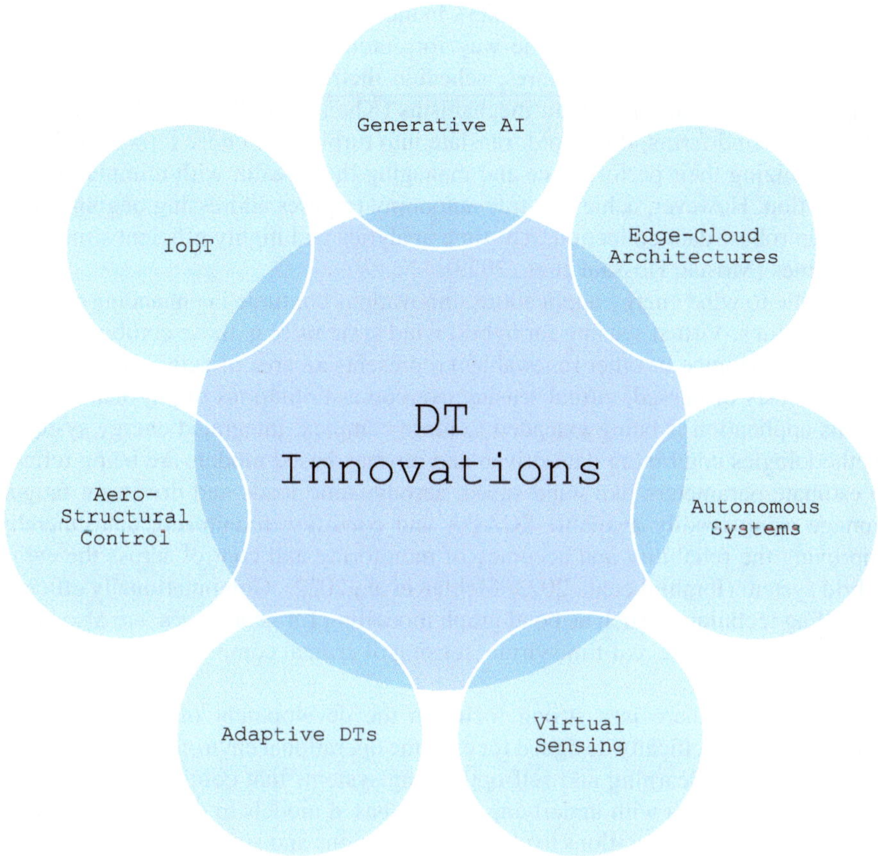

Fig. 4.4 Key technological innovations and emerging trends shaping the future of DT technology

requiring very low latency and rapid local responses, such as real-time control or immediate fault detection in wind turbines, edge computing offers advantages by processing data closer to the physical asset. Consequently, hybrid edge-cloud architectures are emerging as a common trend, aiming to balance the benefits of localized processing with the extensive storage and computational power of the cloud (Khan et al., 2022). This distributed approach is also foundational to the concept of an Internet of Digital Twins (IoDT), an emerging framework designed to enable seamless interconnection, real-time data exchange and cooperation between multiple, potentially heterogeneous, physical and virtual entities (Wang et al., 2023).

The drive towards more autonomous systems is another significant trend influencing DT development. DTs are increasingly seen as crucial enablers for autonomous operations, providing the real-time simulation, analysis and optimization capabilities necessary for physical systems to make independent decisions and take action (Mostaq Hossain et al., 2023). In the context of maintenance, the integration of AI with DTs is paving the way for autonomous maintenance paradigms, where systems can predict failures, schedule their own maintenance and potentially even initiate in-built repair mechanisms (Khan et al., 2020; Mahmoud et al., 2022). For wind farms, this could translate into turbines or entire farms capable of self-optimizing their performance and managing their health with minimal human intervention. However, achieving true autonomy requires addressing ongoing challenges in robust data collection, real-time analytics and highly efficient simulation capabilities (Mostaq Hossain et al., 2023).

Specific to wind energy applications, innovations continue in enhancing core DT functionalities. Virtual sensing for hybrid wind systems (e.g. those combining wind with energy storage or other renewables) represents an area of active development. As previously discussed, virtual sensing overcomes limitations of physical sensors, and its application is being extended to more complex, integrated energy systems. Methodologies combining data-driven and physics-based models are being refined to estimate parameters like wind speed, aerodynamic loads and drivetrain fatigue damage using readily available SCADA and condition monitoring data, thereby improving the reliability and accuracy of monitoring and control across the entire hybrid system (Ibrahim et al., 2023; Mehlan et al., 2022). Computationally efficient modelling techniques, such as bond graph modelling for gear stages, are also being employed to facilitate real-time virtual sensing of critical component loads (Mehlan et al., 2022).

Furthermore, there is a strong focus on the development of adaptive digital twins (ADTs) specifically designed for extreme operational environments. ADTs are conceived as self-learning and self-optimizing systems that continuously integrate real-time sensor data with underlying physics-based models to dynamically update and adapt their representations to changing conditions and accumulating knowledge (Miao et al., 2022; Walmsley et al., 2024). For instance, novel frameworks incorporating physics-informed ML have recently demonstrated the capability to enable real-time environmental adaptation and support stochastic opportunistic maintenance strategies (Ali et al., 2025). This adaptability is crucial for wind turbines operating in harsh offshore environments or facing unpredictable weather events. Successful applications in manufacturing systems, where ADTs enable online optimisation and real-time look-ahead simulations even under constant process changes, highlight their potential (Ogunsakin et al., 2023). The demonstrable improvements in accuracy achieved by integrating sensor data with physics-based models in ADTs, even in complex real-world building applications, underscore their value for enhancing decision-making in both design and operational phases (Miao et al., 2022; Xie et al., 2022).

Additionally, significant research effort is being directed towards advancing integrated control algorithms for enhanced aero-structural synergy within DT systems. As explored earlier, DTs provide an ideal platform for developing and testing sophisticated control strategies. Current innovations focus on leveraging the DT's real-time predictive capabilities to achieve more dynamic and holistic control. This includes using parallel computation within the DT framework to accelerate real-time optimization algorithms for closed-loop control systems, thereby improving execution time and convergence (Viola & Chen, 2023). The aim is to create control systems that can simultaneously optimize aerodynamic energy capture while actively managing structural loads and fatigue, leading to improved performance, increased reliability and extended operational life for wind turbines. These ongoing technological innovations collectively point towards a future where DTs become even more integral, intelligent and autonomous components of wind energy systems.

4.5 Digital Twin Framework for Corrosion-Fatigue Monitoring in Offshore Wind Turbine Towers

Corrosion-Fatigue Challenges and Framework Objectives. Following the discussion on broader technological innovations, the focus now shifts to a specific application illustrating the practical utility of DTs: the development of a framework for monitoring and managing corrosion-fatigue in offshore wind turbine towers. This particular degradation mechanism presents a substantial threat to the structural integrity and operational lifespan of these assets, especially in aggressive marine environments.

Corrosion-fatigue arises from the detrimental synergistic interaction between cyclic mechanical loading, induced primarily by wind and wave action and the corrosive effects of the marine surroundings. This combination can significantly accelerate material degradation and crack propagation, particularly in critical locations such as the tower base (Fathyunes & Mohtadi-Bonab, 2023). For instance, studies considering the harsh conditions of environments have indicated that corrosion-induced thinning alone can lead to a notable reduction in the fatigue life of steel towers, by approximately 8.9% (McAuliffe et al., 2024). Furthermore, the effects of pitting corrosion, when coupled with fatigue, may compromise the safety of support structures well before their intended 20-year design lifespan is reached, thereby underscoring the urgent need for proactive and informed maintenance strategies (Shittu et al., 2020). Recent comprehensive reviews have further underscored that accurately capturing these synergistic degradation mechanisms within a DT environment is essential for transitioning from static design rules to dynamic structural prognosis (Ali et al., 2026). The tower base region, especially when situated within the splash zone of a monopile foundation, is exceptionally vulnerable due to its constant exposure to seawater and the often amplified stress ranges experienced at this location, making it a critical area for diligent monitoring and assessment (Fathyunes & Mohtadi-Bonab, 2023; Zhang et al., 2024b).

To address these pressing challenges, a conceptual DT framework designed specifically to monitor the progression of corrosion-fatigue at the tower base of offshore wind turbines is introduced. The primary objective of this framework is to enable a shift towards condition-based maintenance, thereby facilitating interventions that can extend the structural life of the tower and minimize operational downtime. The proposed framework achieves this by integrating real-time or periodically updated sensor data, which include measurements of corrosion rate and stress cycle characteristics, with appropriate mathematical models. Specifically, it envisions the use of a linear corrosion model to estimate material loss over time, combined with a modified form of Paris' Law (Paris & Erdogan, 1963) to predict crack growth, explicitly accounting for the synergistic effects of corrosion and fatigue. By processing these inputs, the DT framework aims to provide dynamic predictions of crack propagation rates and estimate the RUL of the monitored section. These outputs are intended to inform timely and targeted maintenance decisions, thereby addressing the critical need for reliable and accessible tools to manage the accelerated degradation commonly observed in offshore wind turbine structures. This illustrative example also serves to provide a practical understanding of how DT principles can be applied to solve specific engineering challenges within the wind energy domain.

Digital Twin Architecture. To effectively monitor and manage the corrosion-fatigue challenges previously outlined, the conceptual DT is architected around a framework designed to systematically integrate real-time data with predictive mathematical models. This integration facilitates the estimation of crack growth and supports the optimization of maintenance schedules. For clarity and pedagogical focus, this framework is organized into three distinct, interconnected layers: a physical data collection layer, a virtual modelling layer and a decision-making layer. These layers monitor the integrity of wind turbine structural components particularly under the harsh conditions typical of offshore environments. This three-layer architecture represents a simplification compared to more complex four- or five-layer frameworks often encountered in advanced DT implementations (Adebiyi et al., 2024; Malakuti et al., 2019), which might include additional dedicated layers for tasks such as extensive data preprocessing or advanced analytics. The streamlined structure adopted here is deliberately chosen to grasp the core functionality and workflow of a DT without the added complexity of full-scale advanced systems, while still reflecting essential practical principles. The overall architecture, illustrating the flow of data from sensors through the models to inform maintenance decisions, is depicted in Fig. 4.5.

The physical data collection layer forms the interface with the actual wind turbine structure. It involves the deployment of sensors directly on the components to capture real-time, or periodically updated, environmental and structural response data. For instance, electrochemical probes could be utilized to measure the corrosion rate, providing a quantitative indicator of the marine environment's aggressiveness. Simultaneously, strain gauges would record the cyclic stress ranges experienced by the component under consideration due to the combined effects of wind and wave loading. These sensors provide critical input parameters for the subsequent modelling layer.

Data from the physical layer feeds into the virtual modelling layer, where mathematical models are employed to simulate the corrosion-fatigue degradation process. This layer typically incorporates a linear corrosion model to calculate the reduction in material thickness over time due to corrosive action. Importantly, it also includes a

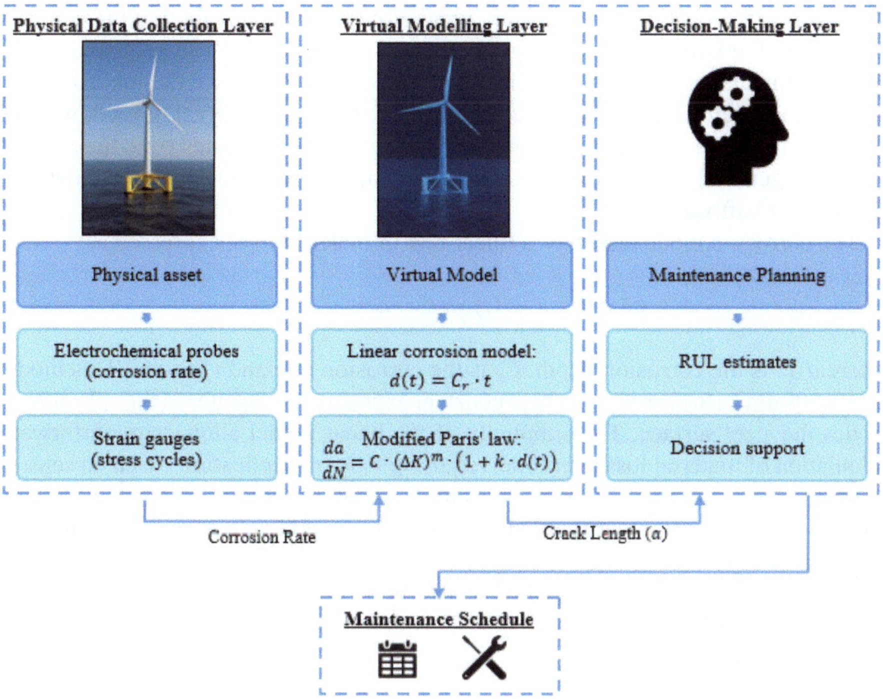

Fig. 4.5 Three-layer DT architecture for corrosion-fatigue monitoring

crack growth model, such as a modified form of Paris' Law (Paris & Erdogan, 1963). This modification specifically aims to incorporate a synergy factor to account for the accelerated crack propagation that occurs under the combined influence of corrosion and cyclic loading, a hallmark of corrosion-fatigue.

Finally, the decision-making layer integrates the processed sensor data from the physical layer with the predictive outputs from the virtual modelling layer. Its primary function is to estimate the current crack length and predict RUL of the monitored component. This layer incorporates predefined criteria, triggering a maintenance intervention when the estimated crack length exceeds a critical threshold.

Mathematical Modelling for Corrosion-Fatigue. The virtual modelling layer of the DT framework employs specific mathematical models to simulate the processes of material degradation within the marine splash zone. These models are formulated to capture the distinct effects of uniform corrosion and fatigue crack growth, as well as their critical synergistic interaction. By utilizing input data derived from sensors, these models predict the progression of cracks and thereby inform maintenance decisions. The modelling approach is intentionally kept relatively simple yet physically realistic, enabling a clear understanding of the underlying processes that drive corrosion-fatigue in offshore wind turbine towers and facilitating comprehension, potentially even allowing for manual calculations. The requisite mathematical understanding primarily involves basic algebra for modelling corrosion progression, elementary differential equations for describing crack growth and the concept of numerical integration for performing iterative updates over time, all of which are generally accessible with standard computational tools like a calculator or spreadsheet software.

The corrosion model assumes a linear rate of material loss, expressed as:

$$d(t) = C_r \cdot t, \tag{4.1}$$

where $d(t)$ is the corrosion depth, C_r is the corrosion rate and t is time. This model represents uniform corrosion, where constant exposure to seawater in the splash zone erodes the steel surface. The simplicity of the linear model allows straightforward calculation of material loss over time, requiring only multiplication based on sensor-measured corrosion rates. This provides the foundation for assessing how corrosion impacts the structural integrity of wind turbine structural components and influences fatigue crack growth.

Fatigue crack growth is modelled using Paris' Law (Paris & Erdogan, 1963), a standard approach in fracture mechanics, given by:

$$\frac{da}{dN} = C \cdot (\Delta K)^m, \tag{4.2}$$

where a is the crack length, N is the number of load cycles, C and m are material constants and ΔK is the stress intensity factor range, calculated as:

$$\Delta K = Y \cdot \Delta\sigma \cdot \sqrt{\pi a}, \tag{4.3}$$

where Y is a geometry factor and $\Delta\sigma$ is the cyclic stress range.

To capture the corrosion-fatigue synergy, a modified Paris' Law is used:

$$\frac{da}{dN} = C \cdot (\Delta K)^m \cdot (1 + k \cdot d(t)), \tag{4.4}$$

where k is a synergy factor that accounts for corrosion's acceleration of crack growth. This equation integrates the corrosion depth $d(t)$ from the linear model (Eq. (4.1)), amplifying the fatigue crack growth rate to reflect the combined degradation mechanism critical for offshore wind turbine structural components.

Decision-Making for Maintenance Interventions. The DT's decision-making module uses the crack length (a), calculated iteratively from the corrosion-fatigue models, to assess structural health and initiate maintenance interventions. The crack length, updated periodically using the modified Paris' Law (Eq. (4.4)), is compared to a predefined critical threshold (a_c). This threshold serves as a conservative limit to prevent structural failure due to corrosion-fatigue in the marine splash zone. When

$a \geq a_c$, the DT generates a maintenance alert, ensuring timely action to maintain the structural integrity of the wind turbine components.

Maintenance interventions are triggered by the DT's alert, recommending immediate inspection and repair. The decision-making module also estimates the RUL, defined as the time from the current year (t_0) until the crack length reaches a_c. RUL is forecasted by iterating the modified Paris' Law (Eq. (4.4)): for each year n, the corrosion depth is computed (Eq. (4.1)). Afterwards, the crack growth per cycle is calculated (Eq. (4.4)). Subsequently, the annual crack growth (Δa) becomes:

$$\Delta a = \frac{da}{dN} \cdot N. \tag{4.5}$$

The crack length is updated as $a_{n+1} = a_n + \Delta a$, continuing until $a_{n+1} \geq a_c$. The RUL is approximated using linear interpolation:

$$\text{RUL} = (t_n - t_0) + \frac{a_c - a_n}{a_{n+1} - a_n}(t_{n+1} - t_n), \tag{4.6}$$

where t_n is the last year with $a < a_c$, and the fractional term estimates the time within the year when a reaches a_c. This enables operators to schedule maintenance proactively, thus minimizing downtime and extending the tower's service life.

4.6 Case Study

This case study demonstrates the DT's potential to predict corrosion-fatigue crack growth at a floating offshore wind turbine's tower base, a critical structural component (McAuliffe et al., 2024), and inform maintenance strategies to ensure operational reliability. The tower base, situated in the splash zone of a monopile foundation, is modelled under North Sea conditions, where harsh marine exposure and cyclic wind and wave loads accelerate degradation. Using sensor data and a corrosion-fatigue framework, the DT simulates the progression of a single surface crack, enabling accurate predictions of crack growth and timely maintenance interventions to prevent structural failure.

Assumptions and Inputs

The case study focuses on the DTU 10 MW Reference Wind Turbine's tower base (Bak et al., 2013), as depicted in Fig. 4.6. The physical asset is defined as a cylindrical S355 steel (European Committee for Standardization (CEN), 2005) tower base with a diameter of 6 m and a wall thickness of 35 mm. This tower base is mounted on a monopile foundation which has a diameter of 7.5 m and a wall thickness of 85 mm.

Environmental conditions are modelled for the North Sea, characterized by an average salinity of 35 ppt and a water temperature of 10 °C. These conditions drive a variable corrosion rate (C_r) averaging 0.115 mm/year. This annual average is broken

Fig. 4.6 Illustration of the DTU 10-MW Reference Wind Turbine

down to 0.02875 mm/quarter to reflect seasonal variations, with 0.025 mm/quarter assumed for calm seasons and 0.0375 mm/quarter for stormy seasons. To assess the impact of environmental variability, additional scenarios are considered: mild conditions ($C_r = 0.08$ mm/year, 8 °C, 30 ppt) and severe conditions ($C_r = 0.15$ mm/year, 12 °C, 40 ppt).

Loading conditions assume a baseline cyclic stress range ($\Delta\sigma$) of 100 MPa. This stress range is comprised of approximately 70% wind-induced loads and 30% wave-induced loads. The structure is subjected to 1 million load cycles per year, which is equivalent to 250,000 cycles per quarter. Sensitivity analyses are planned to explore variations in $\Delta\sigma$ from 80 to 120. For probabilistic modelling, $\Delta\sigma$ is assumed to follow a normal distribution, $\Delta\sigma \sim N(100, 10)$ MPa.

Material properties for S355 steel (European Committee for Standardization (CEN), 2005) include a Young's modulus (E) of 210 GPa and Paris' Law constants $C = 3 \times 10^{-11}$ and $m = 3$. A geometry factor (Y) of 1.12 is used, appropriate for a semi-elliptical surface crack. The initial crack length (a_0) is varied across different scenarios to represent different manufacturing qualities or detection capabilities: $a_0 = 0.05$ mm for high-quality manufacturing, $a_0 = 0.1$ mm as the base case (assumed to be detectable via ultrasonic testing), and $a_0 = 0.2$ mm representing a potential larger initial defect. The critical crack length (a_c), signalling failure, is defined as 3.5 mm, which is 10% of the tower base wall thickness. A synergy factor (k), capturing the interactive effects of corrosion and fatigue, is analysed at three levels: 0.3 (low interaction), 0.5 (base case) and 0.7 (high interaction). Probabilistic modelling assumes k varies uniformly within this range [0.3, 0.7]. For probabilistic RUL estimation, the corrosion rate is modelled as $C_r \sim N(0.115, 0.01)$ mm/year and the initial crack length as $a_0 \sim N(0.1, 0.02)$ mm.

Several assumptions simplify the analysis for demonstration purposes while aiming to maintain realism and replicability. The variable corrosion rate averages seasonal effects but simplifies the more complex reality of storm intensity spikes for manual calculation tractability. Loading is assumed to be uniform, with a constant baseline $\Delta\sigma$ and cycle count, thereby neglecting variable amplitude loading effects often introduced by storms; however, sensitivity analyses explore some stress variations. The S355 steel (European Committee for Standardization (CEN), 2005) is considered homogeneous, ignoring potential weld imperfections or localized material variations. The crack is modelled as a single semi-elliptical surface crack to streamline the application of the geometry factor. A quarterly time step (equivalent to 3 months) is employed for the simulation, which improves resolution compared to an annual step while remaining computationally feasible; linear corrosion and crack growth are assumed within each quarter. Sensor data inputs to the DT, such as corrosion rates from electrochemical probes and stress cycles from strain gauges, are assumed to be accurate, error-free and averaged monthly for aggregation into quarterly inputs.

Maintenance scenarios are defined to evaluate different intervention strategies. These include the application of a protective coating in Year 5, which is assumed to reduce the corrosion rate (C_r) to 0.05 mm/year for a duration of 3 years, at an estimated cost of $30,000 and 1 day of downtime. A minor repair scenario is modelled in year 7, involving a 20% reduction in the existing crack length, costing an estimated $40,000 and 1.5 days of downtime. Another scenario considers the implications of deferring maintenance until Year 9.5, by which point failure may have already occurred or be imminent. Inspection frequencies are also varied: annual, biennial and a base case inspection at Year 9. These inspections are modelled to detect cracks at different stages of growth and have an associated cost of $20,000 per inspection. For comparative purposes, a simplified model using the basic Paris' Law without accounting for the corrosion-fatigue synergy (effectively $k = 0$, as per Eq. (4.4)) serves as a benchmark.

Results and Discussion. The DT analysis provides a detailed and multifaceted understanding of single surface crack progression due to corrosion-fatigue, further elucidated through the examination of multiple scenarios and the adoption of a quarterly time resolution for increased granularity.

In the foundational base case scenario, which assumes no active intervention prior to failure, the simulation commences with an initial crack length of 0.1 mm at Year 0 (designated as Quarter 0). As detailed in Table 4.7 and visually represented by the crack growth trajectory in Fig. 4.7, the crack achieves several key milestones: it measures 0.13 mm by the end of Year 1 (Quarter 4), advances to 0.44 mm by Year 2 (Quarter 8), reaches 0.92 mm by Year 4 (Quarter 16), grows to 1.92 mm by Year 6 (Quarter 24) and further extends to 3.01 mm by Year 8 (Quarter 32). The simulation indicates continued growth, with the crack attaining a length of 3.40 mm by Year 9 (Quarter 36). Subsequently, the crack length surpasses the predefined critical threshold of 3.5 mm within Quarter 37, corresponding to Year 9.25, at which point it is calculated to be 3.52 mm. This critical juncture defines the base case RUL as 9.25 years. If no intervention occurs, the simulation projects that the crack to reach 3.66 mm by the end of Year 10 (Quarter 40). A defining characteristic of this progression is the evident acceleration in crack growth, a hallmark of the synergistic effects of corrosion and fatigue. This acceleration is starkly highlighted by comparing quarterly growth rates: the average rate across the first year (Quarters 1 through 4) is a modest 0.0075 mm/quarter. In stark contrast, the growth observed in the single quarter leading directly to the defined failure point (from 3.40 mm at Quarter 36 to 3.52 mm at Quarter 37) is 0.12 mm. This represents a substantial 1500% increase in the quarterly growth rate when comparing the initial operational period to the period immediately preceding failure, vividly underscoring the escalating nature of damage accumulation.

The dynamic evolution of the actual RUL, re-estimated throughout the asset's simulated operational life, is illustrated in Fig. 4.8. This RUL metric declines from a theoretical 10-year design life at Year 0 (an assumption based on a hypothetical linear degradation or a different failure point) to zero years by Year 9.25. This decline is markedly nonlinear, with the rate of RUL consumption increasing as the crack propagates and the structural integrity diminishes. For instance, between Year 1 (RUL of 8.25 years) and Year 2 (RUL of 7.24 years), the actual RUL diminishes by

Table 4.7 Quarterly crack length and corrosion depth for the base case

Quarter	Year	Crack length (mm)	Corrosion depth (mm)
0	0.00	0.10	0.00
4	1.00	0.13	0.11
8	2.00	0.44	0.23
12	3.00	0.67	0.34
16	4.00	0.92	0.46
20	5.00	1.21	0.57
24	6.00	1.92	0.69
28	7.00	2.46	0.80
32	8.00	3.01	0.91
36	9.00	3.40	1.03
40	10.00	3.66	1.12

Fig. 4.7 Crack length progression over 10 years for the base case

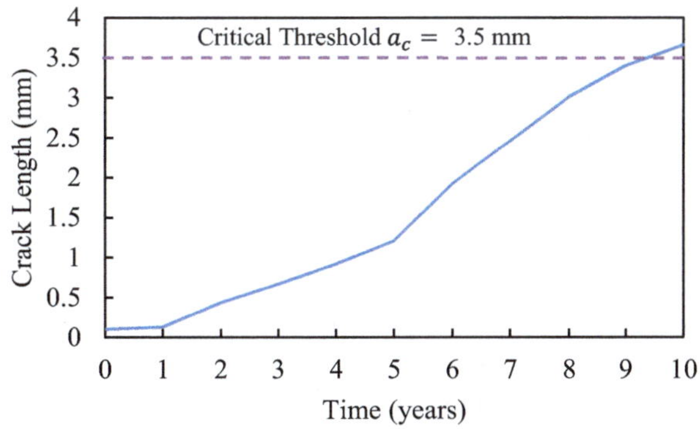

1.01 years. Later in the simulated life, between Year 7 (RUL of 2.13 years) and Year 8 (RUL of 1.08 years), the RUL decreases by 1.05 years, reflecting the heightened severity of the damage and the consequently faster depletion of the remaining operational lifespan. The benefits of proactive intervention are clearly demonstrated when considering a hypothetical maintenance action in Year 9, costing an assumed $50,000 and resulting in 2 days of operational downtime (Table 4.8). When compared against a reactive maintenance scenario, which would necessarily occur after failure (i.e. post-9.25 years) and is estimated to cost $500,000 and involve 20 days of downtime, the proactive approach yields considerable direct cost savings of $450,000 and an 18-day reduction in downtime. This illustrates a ninefold return on the maintenance investment in terms of direct costs alone, without factoring in potential secondary losses stemming from an unexpected structural failure.

Further insights into RUL extension and associated costs are provided by examining several distinct maintenance scenarios and the implications of model simplification, with key outcomes summarized in Table 4.9. The base case, as established, results in an RUL of 9.25 years, assuming no active intervention before failure. Implementing a coating strategy in Year 5, involving the application of a protective coating at an assumed cost of $30,000 and 1 day of downtime, significantly extends the RUL

Fig. 4.8 Actual RUL estimation over time for the base case

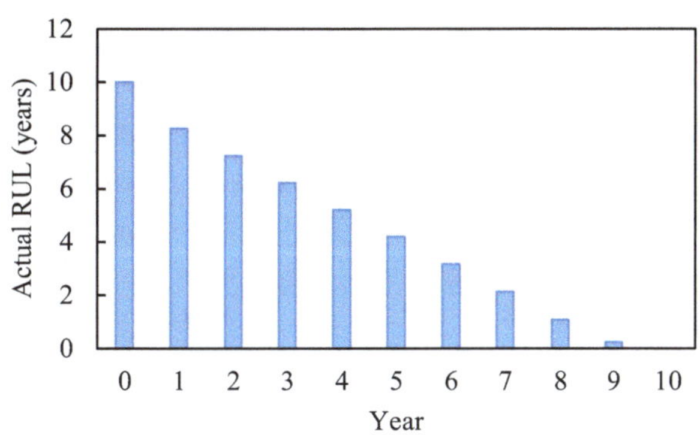

Table 4.8 Cost–benefit analysis of maintenance strategies for the base case

Strategy	Cost ($)	Downtime (days)	Savings ($)	Downtime reduction (days)
Proactive (Year 9)	50,000	2	450,000	18
Reactive (post-9.25 years)	500,000	20	–	–

to 12.0 years. This represents an increase of 2.75 years, or approximately 29.7%, over the base case, thereby highlighting the substantial effectiveness of early-stage corrosion mitigation measures in prolonging asset life. Alternatively, undertaking a minor repair in Year 7, which reduces the existing crack length by 20% (from an estimated 2.46 mm to approximately 1.97 mm, at an assumed cost of $40,000 and 1.5 days of downtime), extends the RUL to 9.8 years. This provides a more modest 0.55-year (or 5.9%) extension compared to the base case but still offers a tangible benefit, particularly if earlier coating interventions were not performed or are less effective at that advanced stage of degradation. The scenario labelled unmitigated deterioration (failure at 9.1 years) appears to represent a situation where the degradation process proceeds at a slightly faster rate than in the base case, leading to failure before a hypothetical planned intervention at year 9.5. This scenario underscores the inherent risk associated with fixed maintenance schedules if the actual rate of deterioration outpaces projections, potentially resulting in a shorter functional life than anticipated. Finally, employing a simplified model that entirely neglects the synergy factor (i.e. $k = 0$) predicts an RUL of 10.5 years. This is 1.25 years or 13.5%, longer than the DT's more comprehensive base case RUL of 9.25 years. This discrepancy clearly demonstrates that failing to account for the crucial interaction between corrosion and fatigue can lead to an overly optimistic and potentially unsafe overestimation of the structure's operational lifespan.

The prevailing environmental conditions exert a significant influence on corrosion rates and, consequently, on the RUL, as detailed in Table 4.10 and Fig. 4.9. Under simulated mild conditions (assuming a corrosion rate, $C_r = 0.08$ mm/year), the RUL is calculated to be 9.7 years, with a corresponding crack length of 2.70 mm at Year 8. This represents a 0.45-year (or 4.9%) increase in RUL compared to the typical scenario. The typical conditions scenario ($C_r = 0.115$ mm/year) serves as the base case, yielding an RUL of 9.25 years and a crack length of 3.01 mm at Year

Table 4.9 RUL and cost–benefit for different maintenance scenarios

Scenario/model	RUL (years)	Maintenance action	Cost ($)	Downtime (days)
Base case	9.25	None	–	–
Coating (Year 5)	12.0	Apply coating	30,000	1
Minor repair (Year 7)	9.8	20% crack reduction	40,000	1.5
Delay to year 9.5	9.1	None until failure	–	–
Simplified model	10.5	None (no synergy)	–	–

8. Conversely, under severe conditions (C_r = 0.15 mm/year), the RUL is reduced to 8.5 years, with the crack length reaching 3.30 mm by Year 8. This constitutes a 0.75-year (or 8.1%) decrease in RUL compared to the typical conditions. The total variation in RUL across these defined environmental scenarios amounts to 1.2 years (ranging from 8.5 to 9.7 years), a span that underscores the critical importance of incorporating accurate, site-specific environmental data and robust environmental modelling for achieving reliable life predictions.

The sensitivity of the model's predictions to variations in key input parameters was assessed to understand the robustness of the RUL estimations. Variations in the initial crack size (a_0), with data intended for Fig. 4.10, showed that an a_0 of 0.05 mm (indicative of high-quality manufacturing) resulted in an RUL of 9.5 years (crack at year 8 = 2.85 mm). The base case a_0 of 0.1 mm yielded an RUL of 9.25 years (crack at year 8 = 3.01 mm), while a larger initial defect of $a_0 = 0.2$ mm resulted in an RUL of 8.9 years (crack at year 8 = 3.15 mm). Halving the initial crack size from the base case thus increased the RUL by 0.25 years (2.7%), whereas doubling it decreased the RUL by 0.35 years (3.8%), demonstrating a moderate sensitivity to initial defect sizes and emphasizing the value of stringent manufacturing quality control and accurate initial non-destructive inspections.

Sensitivity to the synergy factor (k), shown in Fig. 4.11, indicated that for k = 0.3 (representing low corrosion-fatigue interaction), the RUL was 9.4 years (crack at Year 8 = 2.95 mm). The base case (k = 0.5) yielded an RUL of 9.25 years (crack at Year 8 = 3.01 mm) and for k = 0.7 (high interaction), the RUL was 9.0 years (crack at year 8 = 3.10 mm). Varying the synergy factor from 0.3 to 0.7 produced

Table 4.10 RUL and crack growth for different environmental scenarios

Scenario	Corrosion rate (mm/year)	Temperature (°C)	Salinity (ppt)	RUL (years)	Crack length at Year 8 (mm)
Mild	0.08	8	30	9.7	2.70
Typical	0.115	10	35	9.25	3.01
Severe	0.15	12	40	8.5	3.30

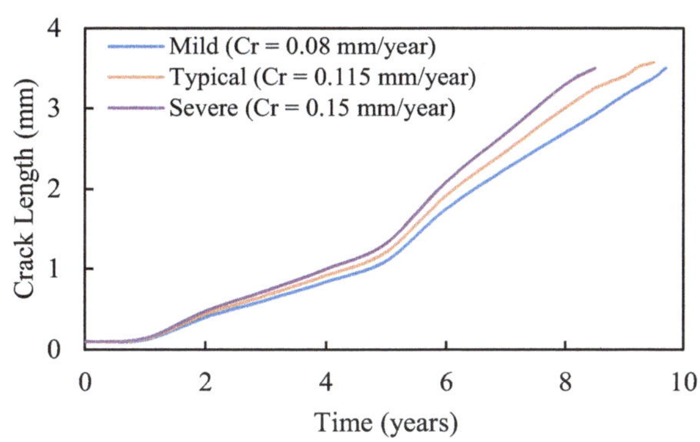

Fig. 4.9 Crack length progression under different environmental scenarios

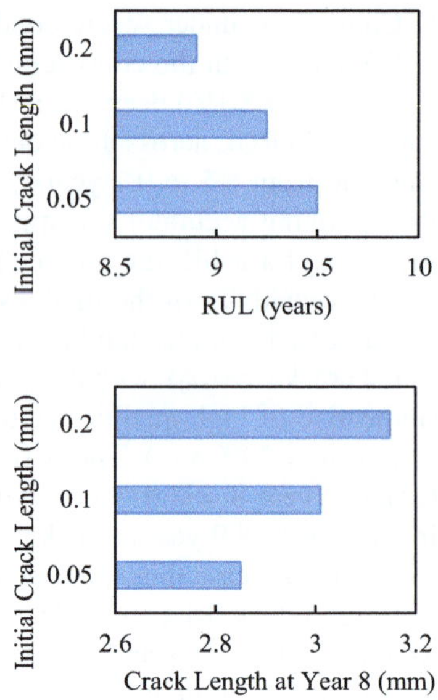

Fig. 4.10 Effect of initial crack length on RUL and reference crack length at Year 8

an RUL range of 0.4 years, suggesting that while this factor is an important model component, its precise value within this tested range exerts a less pronounced effect on the overall RUL compared to parameters like environmental corrosivity.

The specific corrosion rate (C_r) sensitivity analysis, with data for Fig. 4.12, showed the RUL decreasing from 9.7 years at $C_r = 0.08$ mm/year to 9.3 years at $C_r = 0.1$ mm/year and further to 8.8 years at $C_r = 0.12$ mm/year. Relative to the RUL at $C_r = 0.08$ mm/year (9.7 years), a 25% increase in corrosion rate (to 0.1 mm/year) led to a

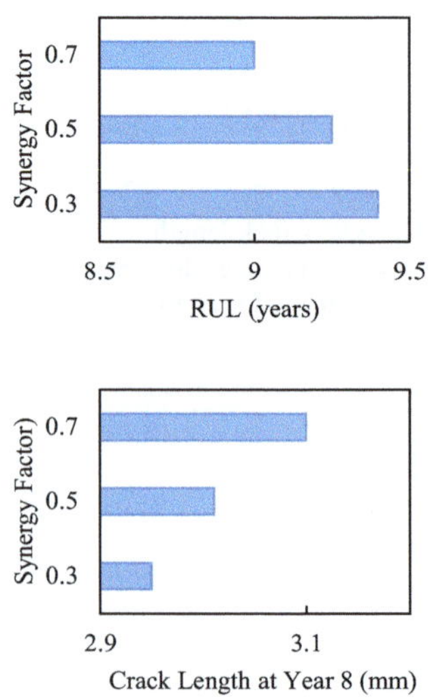

Fig. 4.11 Effect of synergy factor on RUL and reference crack length at Year 8

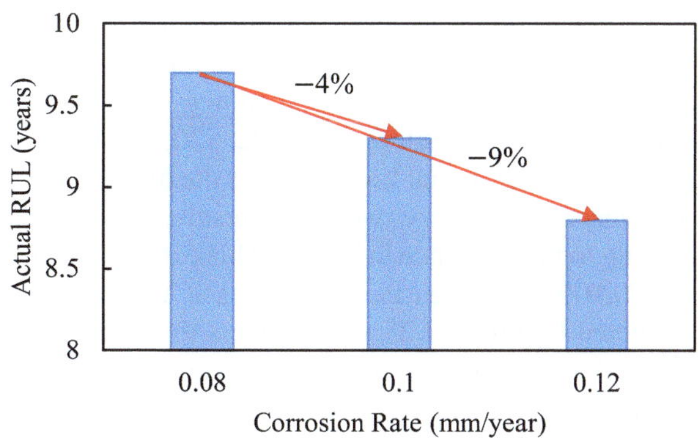

Fig. 4.12 Sensitivity analysis of RUL for the base case

4% RUL reduction, while a 50% increase in corrosion rate (from 0.08 to 0.12 mm/year) resulted in a cumulative 9% RUL reduction, reinforcing the strong, nonlinear influence of corrosion rate on structural longevity.

Acknowledging the inherent uncertainties associated with various input parameters, a probabilistic approach was also employed, as summarized in Table 4.11. This involved simultaneously varying the corrosion rate (C_r) assumed to follow a normal distribution $N(0.115, 0.01)$, the stress range ($\Delta\sigma$) as $N(100, 10)$ and the initial crack size (a_0) as $N(0.1, 0.02)$. The mean RUL derived from these probabilistic simulations was 9.25 years, aligning precisely with the deterministic base case RUL.

Adjusting the parameters by one standard deviation (SD) below their means ($C_r =$ 0.105 mm/year, $\Delta\sigma = 90$ MPa, $a_0 = 0.08$ mm) yielded an RUL of 9.80 years, which is 0.55 years (or 5.9%) longer than the mean life. Conversely, shifting parameters one SD above the mean ($C_r = 0.125$$ mm/year, $\Delta\sigma = 110$ MPa, $a_0 = 0.12$ mm) resulted in an RUL of 8.50 years, representing a reduction of 0.75 years (or 8.1%) from the mean life.

This probabilistic assessment therefore provides an RUL range from 8.50 to 9.80 years, with a spread of 1.3 years. This range quantifies the combined impact of these critical uncertainties and offers a more realistic envelope for life expectancy estimations.

The frequency of inspections directly influences when maintenance interventions are triggered, thereby affecting the achieved RUL and total incurred costs, as outlined in Table 4.12. The costs presented are interpreted as total strategy costs, encompassing inspection (s) and the subsequent major repair (assumed to be the $50,000 proactive

Table 4.11 Probabilistic RUL estimates with parameter uncertainty

Run	Corrosion rate (mm/year)	Cyclic stress (MPa)	Initial crack (mm)	RUL (years)
Mean	0.115	100	0.10	9.25
+ 1 SD	0.125	110	0.12	8.50
−1 SD	0.105	90	0.08	9.80

repair unless failure occurs). An annual inspection strategy is shown to lead to crack detection at 0.44 mm in Year 2, prompting maintenance in Year 3 and extending the RUL to 10.0 years at a total strategy cost of $100,000 (potentially comprising multiple inspections at, for example, $20,000 each, plus the repair cost). This yields the longest operational life but incurs the highest cost. A biennial inspection strategy, with inspections every two years, results in crack detection at 0.92 mm in Year 4, leading to maintenance in Year 5. This extends the RUL to 9.7 years with a total cost of $80,000. The base case (inspection at Year 9), where an inspection just before the predicted failure at 9.25 years detects the crack at 3.40 mm, results in an effective RUL of 9.25 years (as failure is averted by proactive repair) at a cost of $50,000. Comparing these, annual inspection provides an additional 0.75 years of RUL over the base strategy for an extra $50,000, while biennial inspection offers 0.45 years more RUL for an additional $30,000. The optimal strategy depends on operational risk tolerance, the economic valuation of extended life and confidence in crack growth prediction.

The DT analysis, as presented, offers robust insights into corrosion-fatigue crack propagation and the efficacy of various management strategies. However, certain limitations largely arising from the simplifying assumptions made for this illustrative case study, warrant acknowledgement. The use of an average corrosion rate does not capture the acute impact of short-duration storm events, and the quarterly time step might obscure very short-term fluctuations in loading or environmental severity. The assumption of a uniform baseline stress simplifies the complex stochastic nature of real-world offshore loading. Furthermore, the model's focus on a single, pre-existing crack with homogeneous material properties does not account for multiple crack initiations or the influence of features like weldments. Associated costs for maintenance and inspections are treated as fixed values, which may vary in practice. These simplifications, typical for broadly applicable case studies, highlight avenues for refinement in asset-specific DT development, such as incorporating more sophisticated stochastic environmental models, variable amplitude loading spectra, multi-crack analyses and dynamic costing. Despite these limitations, the current analysis effectively demonstrates the DT's capacity to quantify risks, compare intervention strategies and ultimately support more informed decision-making for structural integrity management.

Table 4.12 RUL and cost–benefit for different inspection frequencies

Frequency	Detection point (mm, Year)	Maintenance year	RUL (years)	Cost ($)	Downtime (days)
Annual	0.44, Year 2	3	10.0	100,000	5
Biennial	0.92, Year 4	5	9.7	80,000	4
Base (Year 9)	3.48, Year 9	9	9.25	50,000	2

4.7 Conclusion

This chapter examined the transformative role of DTs and data-driven methods in optimizing wind energy O&M, progressing beyond a review of the technology's potential to address lifecycle challenges through the development of a specialized framework for managing corrosion-fatigue in offshore towers. The application of this framework to the critical intersection of deterioration mechanisms and structural performance yielded several key insights regarding reliability and cost-efficiency:

1. The evolution of DT technology towards complex engineering applications is driven by the convergence of IoT sensing, AI/ML analytics and cloud/edge computing. This trajectory highlights DTs as essential tools for integrated, data-driven wind energy management.
2. Widespread deployment faces significant obstacles, including modelling complex site-specific dynamics (e.g. wind–wave interactions), managing heterogeneous data integration and addressing computational costs. Overcoming these requires industry-wide standardization and robust cybersecurity protocols to ensure interoperability and safety.
3. The DT framework quantitatively validates proactive maintenance. A $50,000 intervention in Year 9 (2 days downtime) averted a projected $500,000 post-failure reactive scenario (20 days downtime). This yielded $450,000 in direct savings and an 18-day downtime reduction, representing a ninefold return on investment.
4. Early maintenance interventions, guided by the DT, significantly extend operational lifespan. Applying a protective coating in Year 5 ($30,000) increased RUL by 2.75 years (29.7% over the 9.25-year base case). In contrast, a later minor repair in year 7 ($40,000) yielded only a 0.55-year (5.9%) extension, demonstrating the increased efficacy of early-stage mitigation.
5. Accurate modelling of environmental conditions and corrosion-fatigue synergy is critical. Varying corrosion rates (0.08–0.15 mm/year) caused a 1.2-year RUL variation (9.7 to 8.5 years). Furthermore, neglecting the synergy factor (k = 0) overestimated RUL by 1.25 years (13.5%), highlighting the safety risks of simplified operational assumptions.
6. Probabilistic analysis accounting for input uncertainties (corrosion rate, stress range, initial crack size) revealed an RUL range of 8.50 to 9.80 years (a 1.3-year spread) around the deterministic mean. This quantifies operational risk more realistically than deterministic models, demonstrating the necessity of probabilistic approaches for decision-making.

Future research should prioritize enhanced predictive fidelity and operational autonomy. Critical frontiers include leveraging generative AI, including large language models (LLMs), and adaptive architectures to improve degradation modelling in stochastic environments, while scaling aero-structural control to farm-level optimization. Furthermore, addressing identified modelling limitations (i.e. multi-crack analysis, dynamic costing and stochastic loading) is essential. Through standardised practices and industry-wide collaboration, these advancements will position DTs as a fundamental component of intelligent wind energy management.

Acknowledgements This work was supported by the National Key Research and Development Program of China (2024YFF0505400); National Natural Science Foundation of China (52208182, U24A20177); Sichuan Science and Technology Program (2025ZNSFSC1316); Qingdao Science and Technology Department Program (2511gjgg50hy); Fundamental Research Funds for the Central Universities (YJ202404) and China Scholarship Council (2024GSP006483, 2024GSP006413).

The last author gratefully acknowledges the Swiss National Science Foundation (SNSF) Scientific Exchanges Programme and RESSLAB, under the direction of Professor Dr. D. Lignos at the École Polytechnique Fédérale de Lausanne (EPFL), for hosting the author during the preparation of this chapter. Their support is sincerely appreciated. Any opinions, findings and conclusions expressed in this material are those of the authors and do not necessarily reflect the views of the sponsors.

References

Abayadeera, M. R., & Ganegoda, G. U. (2024). Digital twin technology: a comprehensive review. *International Journal of Innovative Science and Research Technology (IJISRT)*, 640–661. https://doi.org/10.38124/ijisrt/IJISRT24JUN425

Abdulhayan, S., & Ali Abd, S. (2024). Big data and digital twins. *Handbook of industrial and business applications with digital twins* (1st ed., pp. 154–183). CRC Press.

Abdullahi, I., Longo, S., & Samie, M. (2024). Towards a distributed digital twin framework for predictive maintenance in industrial internet of things (IIoT). *Sensors, 24*, 2663. https://doi.org/10.3390/s24082663

Adebiyi, T. A., Ajenifuja, N. A., & Zhang, R. (2024). Digital twins and civil engineering phases: Reorienting adoption strategies. *Journal of Computing and Information Science in Engineering, 24*, Article 100801. https://doi.org/10.1115/1.4066181

Agho, M. O., Eyo-Udo, N. L., & Onukwulu, E. C., et al. (2025). Digital twin technology for real-time monitoring of energy supply chains. *IJRIAS, IX*, 564–592. https://doi.org/10.51584/IJRIAS.2024.912049

Ahmedi, A., Barnes, M., Ng, C., et al. (2025). Mitigating converter thermal stress in PMSG wind turbines using enhanced control strategy and reduced order modeling. *IEEE Transactions on Energy Conversion, 40*, 478–489. https://doi.org/10.1109/TEC.2024.3449110

Al-Ali, A. R., Gupta, R., Zaman Batool, T., et al. (2020). Digital twin conceptual model within the context of internet of things. *Future Internet, 12*, 163. https://doi.org/10.3390/fi12100163

Ali, Y., Elgammal, A., Heng, J., et al. (2026). Insights into corrosion-fatigue deterioration in offshore wind turbine structures through digital twin applications. *Ocean Engineering, 343*, Article 123514. https://doi.org/10.1016/j.oceaneng.2025.123514

Ali, Y., Dai, K., & Elgammal, A., et al. (2025). AI-driven digital twin for corrosion-fatigue management and opportunistic maintenance in offshore wind turbines. *Wind and Structures, 41*, 393–411. https://doi.org/10.12989/was.2025.41.5.393

Ambarita, E. E., Karlsen, A., Osen, O., & Hasan, A. (2023). Towards fully autonomous floating offshore wind farm operation & maintenance. *Energy Reports, 9*, 103–108. https://doi.org/10.1016/j.egyr.2023.09.148

Ambarita, E. E., Karlsen, A., Scibilia, F., & Hasan, A. (2024). Industrial digital twins in offshore wind farms. *Energy Inform, 7*, 5. https://doi.org/10.1186/s42162-024-00306-6

Ambrozaite, O., Torabi, C., & Van Damme, N., et al. (2024). Standardization of turbine design and installation vessels to accelerate the offshore wind industry in the United States. *JSPG, 25*. https://doi.org/10.38126/JSPG250101

Arin, I. A., Meyliana, W. H. L. H. S., & Murad, D. F. (2023). A systematic literature review of recent trends and challenges in digital twin implementation. *2023 10th International Conference on ICT for Smart Society (ICISS)* (pp. 1–10). IEEE.

Augustyn, D., Ulriksen, M. D., & Sørensen, J. D. (2021). Reliability updating of offshore wind substructures by use of digital twin information. *Energies, 14*, 5859. https://doi.org/10.3390/en14185859

Bado, M. F., Tonelli, D., Poli, F., et al. (2022). Digital twin for civil engineering systems: An exploratory review for distributed sensing updating. *Sensors, 22*, 3168. https://doi.org/10.3390/s22093168

Bak, C., Zahle, F., & Bitsche, R., et al. (2013). *The DTU 10-MW reference wind turbine*. DTU Wind Energy, Roskilde, Denmark

Bárkányi, Á., Chován, T., Németh, S., & Abonyi, J. (2021). Modelling for digital twins—potential role of surrogate models. *Processes, 9*, 476. https://doi.org/10.3390/pr9030476

Blakseth, S. S., Rasheed, A., Kvamsdal, T., & San, O. (2022). Combining physics-based and data-driven techniques for reliable hybrid analysis and modeling using the corrective source term approach. *Applied Soft Computing, 128*, Article 109533. https://doi.org/10.1016/j.asoc.2022.109533

Bordukova, M., Makarov, N., Rodriguez-Esteban, R., et al. (2024). Generative artificial intelligence empowers digital twins in drug discovery and clinical trials. *Expert Opinion on Drug Discovery, 19*, 33–42. https://doi.org/10.1080/17460441.2023.2273839

Bossanyi, E. A. (2003). Wind turbine control for load reduction. *Wind Energy, 6*, 229–244. https://doi.org/10.1002/we.95

Branlard, E., Jonkman, J., Dana, S., & Doubrawa, P. (2020). A digital twin based on Open-FAST linearizations for real-time load and fatigue estimation of land-based turbines. *Journal of Physics: Conference Series, 1618*, Article 022030. https://doi.org/10.1088/1742-6596/1618/2/022030

Branlard, E., Jonkman, J., Brown, C., & Zhang, J. (2024). A digital twin solution for floating offshore wind turbines validated using a full-scale prototype. *Wind Energ Sci, 9*, 1–24. https://doi.org/10.5194/wes-9-1-2024

Caterino, N., & Spizzuoco, M. (2017). Structural control of a wind turbine accounting for second order effects. *IOP Conference Series: Materials Science and Engineering, 245*, Article 022040. https://doi.org/10.1088/1757-899X/245/2/022040

Chang, M., Hwang, S., & Bae, S. (2024). Physics-based and data-driven digital twin technology: Focusing on battery energy storage systems, electrolyzers, hydrogen fueling facilities, and electric vehicle charging facilities. *JIEIE, 38*, 452–457. https://doi.org/10.5207/JIEIE.2024.38.6.452

Chen, X., Eder, M. A., Shihavuddin, A., & Zheng, D. (2021). A human-cyber-physical system toward intelligent wind turbine operation and maintenance. *Sustainability, 13*, 561. https://doi.org/10.3390/su13020561

Chen, J., Meng, C., Gao, Y., & Liu, Y. (2022). Multi-fidelity neural optimization machine for Digital Twins. *Structural and Multidisciplinary Optimization, 65*, 340. https://doi.org/10.1007/s00158-022-03443-2

Chen, B.-Q., Liu, K., Yu, T., & Li, R. (2024). Enhancing reliability in floating offshore wind turbines through digital twin technology: A comprehensive review. *Energies, 17*, 1964. https://doi.org/10.3390/en17081964

Cohen, J., & Huan, X. (2024). Uncertainty-aware explainable AI as a foundational paradigm for digital twins. *Frontiers of Mechanical Engineering, 9*, 1329146. https://doi.org/10.3389/fmech.2023.1329146

Correia, J. B., Abel, M., & Becker, K. (2023). Data management in digital twins: A systematic literature review. *Knowledge and Information Systems, 65*, 3165–3196. https://doi.org/10.1007/s10115-023-01870-1

Costa, Á. M., Orosa, J. A., Vergara, D., & Fernández-Arias, P. (2021). New tendencies in wind energy operation and maintenance. *Applied Sciences, 11*, 1386. https://doi.org/10.3390/app11041386

Dang, H., Tatipamula, M., & Nguyen, H. X. (2022). Cloud-based digital twinning for structural health monitoring using deep learning. *18*, 3820–3830. https://doi.org/10.1109/TII.2021.3115119

Daraki, M.-S., Mina, B., Almasoudi, M., et al. (2024). Applications and challenges of digital twins of floating wind turbines. *SimTerm proceedings 2024-zbornik radova* (pp. 53–60). University of Niš.

De Angelis, M., Gray, A., Ferson, S., & Patelli, E. (2023). Robust online updating of a digital twin with imprecise probability. *Mechanical Systems and Signal Processing, 186*, Article 109877. https://doi.org/10.1016/j.ymssp.2022.109877

Dirnfeld, R., De Donato, L., Somma, A., et al. (2024). Integrating AI and DTs: Challenges and opportunities in railway maintenance application and beyond. *Simulation, 100*, 903–917. https://doi.org/10.1177/00375497241229756

Donnelly, O., Anderson, F., & Carroll, J. (2024). Operation and maintenance cost comparison between 15 MW direct-drive and medium-speed offshore wind turbines. *Wind Energy Science, 9*, 1345–1362. https://doi.org/10.5194/wes-9-1345-2024

Ebrahimi, A. (2019). Challenges of developing a digital twin model of renewable energy generators. *2019 IEEE 28th international symposium on industrial electronics (ISIE)* (pp. 1059–1066). IEEE.

Elgammal, A., & Ali, Y. (2024). A novel hysteretic restoring force model for shear link dampers: A machine learning approach. *Structures, 70*, Article 107848. https://doi.org/10.1016/j.istruc.2024.107848

Emmert-Streib, F. (2023). What is the role of AI for digital twins? *AI 4*, 721–728. https://doi.org/10.3390/ai4030038

Errandonea, I., Beltrán, S., & Arrizabalaga, S. (2020). Digital twin for maintenance: A literature review. *Computers in Industry, 123*, Article 103316. https://doi.org/10.1016/j.compind.2020.103316

European Committee for Standardization (CEN). (2005). Eurocode 3: Design of steel structures-Part 1–1: General rules and rules for buildings

Fahim, M., Sharma, V., Cao, T.-V., et al. (2022). Machine learning-based digital twin for predictive modeling in wind turbines. *IEEE Access, 10*, 14184–14194. https://doi.org/10.1109/ACCESS.2022.3147602

Fakhraian, E., Marata, F., Dezobry, V., et al. (2024). A robust digital twin framework for coaxial helicopters: Development, validation, and practical implications. *2024 IEEE 4th International conference on digital twins and parallel intelligence (DTPI)* (pp. 1–7). IEEE.

Fathyunes, L., & Mohtadi-Bonab, M. A. (2023). A review on the corrosion and fatigue failure of gas turbines. *Metals, 13*, 701. https://doi.org/10.3390/met13040701

Febrianto, E., Butler, L., Girolami, M., & Cirak, F. (2022). Digital twinning of self-sensing structures using the statistical finite element method. *DCE, 3*, Article e31. https://doi.org/10.1017/dce.2022.28

Fei, T., Weiran, L., & Meng, Z. (2019). Digital twin five-dimensional model and ten domain applications. *Computer Integrated Manufacturing Systems, 25*, 1–18.

Ferrigno, E., & Barsola, G. A. (2023). 3D real time digital twin. In: SPE Latin American and Caribbean petroleum engineering conference. SPE, Port of Spain, Trinidad and Tobago, p. D021S010R006

Fonseca, Í. A., & Gaspar, H. M. (2021). Challenges when creating a cohesive digital twin ship: A data modelling perspective. *Ship Technology Research, 68*, 70–83. https://doi.org/10.1080/09377255.2020.1815140

Frederik, J. A., Doekemeijer, B. M., Mulders, S. P., & Van Wingerden, J. (2020). The helix approach: Using dynamic individual pitch control to enhance wake mixing in wind farms. *Wind Energy, 23*, 1739–1751. https://doi.org/10.1002/we.2513

Fu, Y., Xu, W. X., & Liu, L. J. (2015). Study of operation and maintenance tactics for offshore wind turbines. *Journal of, 31*, 219–222.

Githens, G. (2007). Product lifecycle management: Driving the next generation of lean thinking by Michael Grieves. *Journal of Product Innovation Management, 24*, 278–280. https://doi.org/10.1111/j.1540-5885.2007.00250_2.x

Glaessgen, E., & Stargel, D. (2012). The digital twin paradigm for future NASA and U.S. Air Force Vehicles. In: 53rd AIAA/ASME/ASCE/AHS/ASC Structures, Structural Dynamics and Materials Conference
20th AIAA/ASME/AHS Adaptive Structures Conference
14th AIAA. American Institute of Aeronautics and Astronautics, Honolulu, Hawaii

Global Wind Energy Council. (2025). *Global wind report 2025*. Belgium.

Grieves, M. W. (2005). Product lifecycle management: The new paradigm for enterprises. *IJPD, 2*, 71. https://doi.org/10.1504/IJPD.2005.006669

Grieves, M. (2011). *Virtually perfect: Driving innovative and lean products through product lifecycle management*. Space Coast Press.

Grieves, M. W. (2014). *Digital twin: Manufacturing excellence through virtual factory replication*. USA

Haces-Fernandez, F., Cruz-Mendoza, M., & Li, H. (2022). Onshore wind farm development: Technologies and layouts. *Energies, 15*, 2381. https://doi.org/10.3390/en15072381

Haghshenas, A., Hasan, A., Osen, O., & Mikalsen, E. T. (2023). Predictive digital twin for offshore wind farms. *Energy Inform, 6*, 1. https://doi.org/10.1186/s42162-023-00257-4

Hartmann, D., Herz, M., & Wever, U. (2018). Model order reduction a key technology for digital twins. In W. Keiper, A. Milde, & S. Volkwein (Eds.), *Reduced-order modeling (ROM) for simulation and optimization* (pp. 167–179). Springer International Publishing.

Hassan, Q., Viktor, P., & Al-Musawi T. J., et al. (2024). The renewable energy role in the global energy transformations. *Renewable Energy Focus, 48*, 100545. https://doi.org/10.1016/j.ref.2024.100545

Hayman, B., Wedel-Heinen, J., & Brøndsted, P. (2008). Materials challenges in present and future wind energy. *MRS Bulletin, 33*, 343–353. https://doi.org/10.1557/mrs2008.70

Hildebrandt, G., Dittler, D., Habiger, P., et al. (2024). Data integration for digital twins in industrial automation: A systematic literature review. *IEEE Access, 12*, 139129–139153. https://doi.org/10.1109/ACCESS.2024.3465632

Hong, S., McMorland, J., Zhang, H., et al. (2024). Floating offshore wind farm installation, challenges and opportunities: A comprehensive survey. *Ocean Engineering, 304*, Article 117793. https://doi.org/10.1016/j.oceaneng.2024.117793

Hua, E. Y., Lazarova-Molnar, S., & Francis, D. P. (2022). Validation of digital twins: Challenges and opportunities. *2022 Winter simulation conference (WSC)* (pp. 2900–2911). IEEE.

Ibrahim, M., Rassõlkin, A., Vaimann, T., et al. (2023). Digital twin as a virtual sensor for wind turbine applications. *Energies, 16*, 6246. https://doi.org/10.3390/en16176246

Issa, R., Hamad, M. S., & Abdel-Geliel, M. (2023). Digital twin of wind turbine based on Microsoft® Azure IoT platform. In: 2023 IEEE conference on power electronics and renewable energy (CPERE). IEEE, Luxor, Egypt, pp. 1–8

Jia, Y., & Jiang, X. (2023). Research on the method of digital twin operation and maintenance platform for intelligent early warning of wind turbine tower. PHMAP_CONF 4. https://doi.org/10.36001/phmap.2023.v4i1.3666

Jorgensen, J., Hodkiewicz, M., Cripps, E., & Hassan, G. M. (2023). Requirements for the application of the Digital Twin Paradigm to offshore wind turbine structures for uncertain fatigue analysis. *Computers in Industry, 145*, Article 103806. https://doi.org/10.1016/j.compind.2022.103806

Kaewbumrung, M., Plengsa-Ard, C., Pansang, S., & Palasai, W. (2024). Preventive maintenance of horizontal wind turbines via computational fluid dynamics-driven wall shear stress evaluation. *Results in Engineering, 22*, Article 102383. https://doi.org/10.1016/j.rineng.2024.102383

Kale, C. P., & Gawali, M. B. (2023). Survey on energy generation prediction for offshore wind turbines using digital twins. *2023 4th International conference on computation, automation and knowledge management (ICCAKM)* (pp. 1–6). IEEE.

Kandemir, E., Liu, J., & Hasan, A. (2023). Digital twin-driven dynamic repositioning of floating offshore wind farms. *Energy Reports, 9*, 208–214. https://doi.org/10.1016/j.egyr.2023.08.091

Kandemir, E., Hasan, A., Kvamsdal, T., & Abdel-Afou Alaliyat, S. (2024). Predictive digital twin for wind energy systems: A literature review. *Energy Inform, 7*, 68. https://doi.org/10.1186/s42162-024-00373-9

Kapteyn, M. G., Knezevic, D. J., Huynh, D. B. P., et al. (2022). Data-driven physics-based digital twins via a library of component-based reduced-order models. *Numerical Methods in Engineering, 123*, 2986–3003. https://doi.org/10.1002/nme.6423

Kessels, B. M., Fey, R. H. B., & Van De Wouw, N. (2025). Uncertainty quantification in real-time parameter updating for digital twins using Bayesian inverse mapping models. *Nonlinear Dynamics, 113*, 7613–7637. https://doi.org/10.1007/s11071-024-10608-9

Khan, S., Farnsworth, M., McWilliam, R., & Erkoyuncu, J. (2020). On the requirements of digital twin-driven autonomous maintenance. *Annual Reviews in Control, 50*, 13–28. https://doi.org/10.1016/j.arcontrol.2020.08.003

Khan, L. U., Saad, W., Niyato, D., et al. (2022). Digital-twin-enabled 6G: Vision, architectural trends, and future directions. *IEEE Communications Magazine, 60*, 74–80. https://doi.org/10.1109/MCOM.001.21143

Kochunas, B., & Huan, X. (2021). Digital twin concepts with uncertainty for nuclear power applications. *Energies, 14*, 4235. https://doi.org/10.3390/en14144235

Kou, G., Dinçer, H., Yüksel, S., & Deveci, M. (2024). Synergistic integration of digital twins and sustainable industrial internet of things for new generation energy investments. *Journal of Advanced Research, 66*, 39–46. https://doi.org/10.1016/j.jare.2023.11.023

Kuts, V., Marvel, J. A., Aksu, M., et al. (2022). Digital twin as industrial robots manipulation validation tool. *Robotics, 11*, 113. https://doi.org/10.3390/robotics11050113

Lachvajderová, L., Trebuňa, M., & Kádárová, J. (2024). Unlocking industry potential: The evolution and impact of digital twins. *Acta Mechanica Slovaca, 28*, 46–51. https://doi.org/10.21496/ams.2024.009

Leng, J., Gardoni, P., Wang, M., et al. (2023). Condition-based structural health monitoring of offshore wind jacket structures: Opportunities, challenges, and perspectives. *Structural Health Monitoring, 22*, 3558–3575. https://doi.org/10.1177/14759217221148688

Li, Y., & Shen, X. (2022). A novel wind speed-sensing methodology for wind turbines based on digital twin technology. *IEEE Transactions on Instrumentation and Measurement, 71*, 1–13. https://doi.org/10.1109/TIM.2021.3139698

Li, D., Ho, S.-C.M., Song, G., et al. (2015). A review of damage detection methods for wind turbine blades. *Smart Materials and Structures, 24*, Article 033001. https://doi.org/10.1088/0964-1726/24/3/033001

Li, T., Luan, Y., Pang, Z., & Zhang, W. (2024). Structural digital twin modeling and adaptive pretrain-finetune learning for dynamic impact identification on wind turbine blades. *20*, 10292–10303. https://doi.org/10.1109/TII.2024.3393111

Li, T., Long, Q., Chai, H., et al. (2025). Generative AI empowered network digital twins: Architecture, technologies, and applications. *ACM Computing Surveys, 57*, 1–43. https://doi.org/10.1145/3711682

Liu, Y., Zhang, L., Yang, Y., et al. (2019). A novel cloud-based framework for the elderly healthcare services using digital twin. *IEEE Access, 7*, 49088–49101. https://doi.org/10.1109/ACCESS.2019.2909828

Liu, D., Wang, S., Shi, J., & Liu, D. (2024). Digital twin dynamic-polymorphic uncertainty surrogate model generation using a sparse polynomial chaos expansion with application in aviation hydraulic pump. *Chinese Journal of Aeronautics, 37*, 231–244. https://doi.org/10.1016/j.cja.2024.10.008

Luo, H., Wang, L., Sun, W., & Lu, C. (2023). Intelligent monitoring and maintenance of wind turbine blades driven by digital twin technology. *2023 3rd International conference on new energy and power engineering (ICNEPE)* (pp. 626–630). IEEE.

Mahmoud, E. G., Darwish, A., & Hassanien, A. E. (2022). The future of digital twins for autonomous systems: Analysis and opportunities. In A. E. Hassanien, A. Darwish, & V. Snasel (Eds.), *Digital twins for digital transformation: Innovation in industry* (pp. 187–200). Springer International Publishing.

Malakuti, S., Schmitt, J., Platenius-Mohr, M., et al. (2019). A four-layer architecture pattern for constructing and managing digital twins. In T. Bures, L. Duchien, & P. Inverardi (Eds.), *Software architecture* (pp. 231–246). Springer International Publishing.

Marykovskiy, Y., Clark, T., Deparday, J., et al. (2024). Architecting a digital twin for wind turbine rotor blade aerodynamic monitoring. *Frontiers in Energy Research, 12*, 1428387. https://doi.org/10.3389/fenrg.2024.1428387

Mazumder, A., Banerjee, P. S., & Karmakar, A., et al. (2023). Digital twin for industry 5.0: A vision, taxonomy, and future directions. In S. Bhattacharyya, J. S. Banerjee, D. De, M. Mahmud (Eds.), *Intelligent human centered computing*. Springer Nature Singapore, Singapore, pp. 246–259

McAuliffe, J., Baisthakur, S., Broderick, B., & Fitzgerald, B. (2024). Corrosion fatigue analysis of NREL's 15-MW offshore wind turbine with time-varying stress concentration factors. *Journal of Physics: Conference Series, 2767*, Article 062023. https://doi.org/10.1088/1742-6596/2767/6/062023

Mehlan, F. C., Pedersen, E., & Nejad, A. R. (2022). Modelling of wind turbine gear stages for Digital Twin and real-time virtual sensing using bond graphs. *Journal of Physics: Conference Series, 2265*, Article 032065. https://doi.org/10.1088/1742-6596/2265/3/032065

Menezes, E. J. N., & Araújo, A. M. (2023). Wind turbine structural control using H-infinity methods. *Engineering Structures, 286*, Article 116095. https://doi.org/10.1016/j.engstruct.2023.116095

Miao, B. H., Dong, Y., & Wu, Z. Y., et al. (2022). Integration of physics-based building model and sensor data to develop an adaptive digital twin: poster abstract. In *Proceedings of the 9th ACM international conference on systems for energy-efficient buildings, cities, and transportation*. ACM, Boston Massachusetts, pp. 282–283

Minerva, R., Awan, F. M., & Crespi, N. (2022a). Exploiting digital twins as enablers for synthetic sensing. *IEEE Internet Computing, 26*, 61–67. https://doi.org/10.1109/MIC.2021.3051674

Minerva, R., Patrikakis, C. Z., & Margaria, T. (2022b). Digitally cloning the physical. *IT Prof, 24*, 18–19. https://doi.org/10.1109/MITP.2022.3224263

Moghadam, F. K., Rebouças, G. F. D. S., & Nejad, A. R. (2021). Digital twin modeling for predictive maintenance of gearboxes in floating offshore wind turbine drivetrains. *Forschung Im Ingenieurwesen, 85*, 273–286. https://doi.org/10.1007/s10010-021-00468-9

Moghadam, F. K., Keprate, A., & Gao, Z. (2024). Editorial: Online monitoring of wind power plants using digital twin models. *Front Energy Res, 12*, 1531689. https://doi.org/10.3389/fenrg.2024.1531689

Mohandes, S. R., Singh, A. K., Fazeli, A., et al. (2024). Determining the stationary digital twins implementation barriers for sustainable construction projects. *SASBE*. https://doi.org/10.1108/SASBE-11-2023-0344

Möhring, M., Keller, B., & Radowski, C. –F., et al. (2022). Empirical insights into the challenges of implementing digital twins. In A. Zimmermann, R. J. Howlett, & L. C. Jain LC (Eds.), *Human centred intelligent systems*. Springer Nature Singapore, Singapore, pp. 229–239

Mostaq Hossain, S. M., Kumar Saha, S., Banik, S., & Banik, T. (2023). A new era of mobility: Exploring digital twin applications in autonomous vehicular systems. *2023 IEEE world AI IoT congress (AIIoT)* (pp. 0493–0499). IEEE.

Muhammad, K., David, T., Nassisid, G., & Farus, T. (2024). *Integrating generative AI with network digital twins for enhanced network operations*

Munko, M. J., Vidmar, M., & Ó Brádaigh, C. M., et al. (2024). Do digital twins require physical simulations? A study of developing digital twins with varying reliance on physics-based models. In: 2024 IEEE international conference on engineering, technology, and innovation (ICE/ITMC). IEEE, Funchal, Portugal, pp. 1–8

Munters, W., & Meyers, J. (2018). Dynamic strategies for yaw and induction control of wind farms based on large-eddy simulation and optimization. *Energies, 11*, 177. https://doi.org/10.3390/en11010177

National Renewable Energy Laboratory (NREL). (2025). OpenFAST

Naveen, P., Maheswar, R., & Ragupathy, U. S. (2025). *Digital twins and cybersecurity: Safeguarding the future of connected systems* (1st ed.). Wiley.

Negri, E., Berardi, S., Fumagalli, L., & Macchi, M. (2020). MES-integrated digital twin frameworks. *Journal of Manufacturing Systems, 56*, 58–71. https://doi.org/10.1016/j.jmsy.2020.05.007

Nour El-Din, M., Pereira, P. F., Poças Martins, J., & Ramos, N. M. M. (2022). Digital twins for construction assets using BIM standard specifications. *Buildings, 12*, 2155. https://doi.org/10.3390/buildings12122155

Ogunsakin, R., Mehandjiev, N., & Marin, C. A. (2023). Towards adaptive digital twins architecture. *Computers in Industry, 149*, Article 103920. https://doi.org/10.1016/j.compind.2023.103920

Okenyi, V., Bodaghi, M., Mansfield, N., et al. (2024). A review of challenges and framework development for corrosion fatigue life assessment of monopile-supported horizontal-axis offshore wind turbines. *Ships and Offshore Structures, 19*, 1–15. https://doi.org/10.1080/17445302.2022.2140531

Omrany, H., Al-Obaidi, K. M., Husain, A., & Ghaffarianhoseini, A. (2023). Digital twins in the construction industry: A comprehensive review of current implementations, enabling technologies, and future directions. *Sustainability, 15*, 10908. https://doi.org/10.3390/su151410908

Pandit, A. (2025). Digital twins in construction: Creating real-time replicas of large-scale projects. *IJSREM, 09*, 1–7. https://doi.org/10.55041/IJSREM40755

Paris, P., & Erdogan, F. (1963). A critical analysis of crack propagation laws. *Journal of Basic Engineering, 85*, 528–533. https://doi.org/10.1115/1.3656900

Peng, H., Li, S., Shangguan, L., et al. (2023). Analysis of wind turbine equipment failure and intelligent operation and maintenance research. *Sustainability, 15*, 8333. https://doi.org/10.3390/su15108333

Pinon Fischer, O. J., Matlik, J. F., Schindel, W. D., et al. (2022). Digital twin: Reference model, realizations, and recommendations. *Insight, 25*, 50–55. https://doi.org/10.1002/inst.12373

Poddar, H., & VP, R. (2024). A look into the future: Advances in digital twin technology. In L. Ashok Kumar, & S. Angalaeswari, K. Mohana Sundaram, et al. (Eds.), *Advances in computational intelligence and robotics*. IGI Global, pp. 1–27

Rangel-Ramírez, J. G., & Sørensen, J. D. (2008). Optimal risk-based inspection planning for offshore wind turbines. *International Journal of Steel Structures, 8*, 295–303.

Rasheed, A., San, O., & Kvamsdal, T. (2020). Digital twin: Values, challenges and enablers from a modeling perspective. *IEEE Access, 8*, 21980–22012. https://doi.org/10.1109/ACCESS.2020.2970143

Rasheed, A., Stadtmann, F., Fonn, E., et al. (2024). Digital twin for wind energy: Latest updates from the North Wind project. In: *Volume 1: offshore technology*. American Society of Mechanical Engineers, Singapore, Singapore, p. V001T01A045

Raşit Cesur, M., Cesur, E., & Harun Çelik, A. (2024). Digital twins in industry 4.0 automation. In *Handbook of industrial and business applications with digital twins*, 1st edn. CRC Press, New York, pp. 203–220

Rosen, R., Von Wichert, G., Lo, G., & Bettenhausen, K. D. (2015). About the importance of autonomy and digital twins for the future of manufacturing. *IFAC-PapersOnLine, 48*, 567–572. https://doi.org/10.1016/j.ifacol.2015.06.141

Sakr, M., & Sadhu, A. (2024). Recent progress and future outlook of digital twins in structural health monitoring of civil infrastructure. *Smart Materials and Structures, 33*, Article 033001. https://doi.org/10.1088/1361-665X/ad2bd7

Salic, T., Charpentier, J. F., Benbouzid, M., & Le Boulluec, M. (2019). Control Strategies for floating offshore wind turbine: Challenges and trends. *Electronics, 8*, 1185. https://doi.org/10.3390/electronics8101185

Santos, F., Teixeira, Â. P., & Soares, C. G. (2015). Modelling and simulation of the operation and maintenance of offshore wind turbines. *Proceedings of the Institution of Mechanical Engineers, Part o: Journal of Risk and Reliability, 229*, 385–393. https://doi.org/10.1177/1748006X15589209

Saracco, R. (2019). Digital twins: Bridging physical space and cyberspace. *Computer, 52*, 58–64. https://doi.org/10.1109/MC.2019.2942803

Schaumann, P., Lochte-Holtgreven, S., & Steppeler, S. (2011). Special fatigue aspects in support structures of offshore wind turbines. *Materialwissenschaft Werkst, 42*, 1075–1081. https://doi.org/10.1002/mawe.201100913

Segovia, M., & Garcia-Alfaro, J. (2022). Design, modeling and implementation of digital twins. *Sensors, 22*, 5396. https://doi.org/10.3390/s22145396

Semeraro, C., Lezoche, M., Panetto, H., & Dassisti, M. (2021). Digital twin paradigm: A systematic literature review. *Computers in Industry, 130*, Article 103469. https://doi.org/10.1016/j.compind.2021.103469

Shafto, M., Conroy, M., & Doyle, R., et al. (2012). NASA technology roadmap: Modeling, simulation, information technology & processing roadmap technology area. NASA (National Aeronautics and Space Administration), USA

Shittu, A. A., Mehmanparast, A., Shafiee, M., et al. (2020). Structural reliability assessment of offshore wind turbine support structures subjected to pitting corrosion-fatigue: A damage tolerance modelling approach. *Wind Energy, 23*, 2004–2026. https://doi.org/10.1002/we.2542

Siddique, S., Haque, M. A., Rifat, R. H., et al. (2023). Cyber security issues in the industrial applications of digital twins. *2023 IEEE symposium series on computational intelligence (SSCI)* (pp. 873–878). IEEE.

Siddiqui, M. S., Keprate, A., Yang, L., & Malmedal, T. (2023). Towards an integrative framework for digital twins in wind power. *2023 IEEE international conference on industrial engineering and engineering management (IEEM)* (pp. 0264–0268). IEEE.

Singh, U., Rizwan, M., Malik, H., & García Márquez, F. P. (2022). Wind energy scenario, success and initiatives towards renewable energy in India—a review. *Energies, 15*, 2291. https://doi.org/10.3390/en15062291

Soudan, B. (2024). Cybersecurity of digital twins in industrial IoT environments. *2024 Advances in science and engineering technology international conferences (ASET)* (pp. 1–6). IEEE.

Stadtmann, F., Mahalingam, H. P., & Rasheed, A. (2023a). Data integration framework for virtual reality enabled digital twins. *2023 IEEE 9th world forum on internet of things (WF-IoT)* (pp. 1–6). IEEE.

Stadtmann, F., Rasheed, A., Kvamsdal, T., et al. (2023b). Digital twins in wind energy: Emerging technologies and industry-informed future directions. *IEEE Access, 11*, 110762–110795. https://doi.org/10.1109/ACCESS.2023.3321320

Stadtmann, F., Rasheed, A., & Rasmussen, T. (2023c). Standalone, descriptive, and predictive digital twin of an onshore wind farm in complex terrain. *Journal of Physics: Conference Series, 2626*, Article 012030. https://doi.org/10.1088/1742-6596/2626/1/012030

Stadtmann, F., Wassertheurer, H. A. G., & Rasheed, A. (2023d). Demonstration of a standalone, descriptive, and predictive digital twin of a floating offshore wind turbine. In *Volume 8: Ocean renewable energy*. American Society of Mechanical Engineers, Melbourne, Australia, p. V008T09A039

Starke, G. M., Meneveau, C., King, J., & Gayme, D. F. (2023). Yaw-augmented control for wind farm power tracking. *2023 American control conference (ACC)* (pp. 184–191). IEEE.

Summerfield-Ryan, O., & Park, S. (2023). The power of wind: The global wind energy industry's successes and failures. *Ecological Economics, 210*, Article 107841. https://doi.org/10.1016/j.ecolecon.2023.107841

Sung, C., & Tuo, R. (2024). A review on computer model calibration. *Wires Computational Stats, 16*, Article e1645. https://doi.org/10.1002/wics.1645

Tao, F., Zhang, M., Liu, Y., & Nee, A. Y. C. (2018). Digital twin driven prognostics and health management for complex equipment. *CIRP Annals, 67*, 169–172. https://doi.org/10.1016/j.cirp.2018.04.055

Thelen, A., Zhang, X., & Fink, O., et al. (2023). A comprehensive review of digital twin—part 2: Roles of uncertainty quantification and optimization, a battery digital twin, and perspectives. *Structural and Multidisciplinary Optimization, 66*:1. https://doi.org/10.1007/s00158-022-03410-x

Tuegel, E. J., Ingraffea, A. R., Eason, T. G., & Spottswood, S. M. (2011). Reengineering aircraft structural life prediction using a digital twin. *International Journal of Aerospace Engineering, 2011*, 1–14. https://doi.org/10.1155/2011/154798

Tuton, E., Ma, X., & Dethlefs, N. (2024). Redefining digital twins–a wind energy operations and maintenance perspective. *Journal of Physics: Conference Series, 2767*, Article 052001. https://doi.org/10.1088/1742-6596/2767/5/052001

Van Den Brand, M., Cleophas, L., Gunasekaran, R., et al. (2021). Models meet data: challenges to create virtual entities for digital twins. *2021 ACM/IEEE international conference on model driven engineering languages and systems companion (MODELS-C)* (pp. 225–228). IEEE.

Van Dinter, R., Tekinerdogan, B., & Catal, C. (2022). Predictive maintenance using digital twins: A systematic literature review. *Information and Software Technology, 151*, Article 107008. https://doi.org/10.1016/j.infsof.2022.107008

Vanderhorn, E., & Valluri, S. (2024). Guidance on the verification and validation of digital twins. In: Day 1 Tue, February 20, 2024. SNAME, Houston, Texas, USA, p D011S004R001

Viola, J., & Chen, Y. (2023). Parallel enabled and stability-aware self optimizing control with globalized constrained nelder-mead optimization algorithm. *IEEE Journal of Radio Frequency Identification, 7*, 178–181. https://doi.org/10.1109/JRFID.2023.3264218

Wagg, D. J., Worden, K., Barthorpe, R. J., & Gardner, P. (2020). Digital twins: State-of-the-art and future directions for modeling and simulation in engineering dynamics applications. *ASCE-ASME Journal of Risk and Uncertainty in Engineering Systems, Part b: Mechanical Engineering, 6*, Article 030901. https://doi.org/10.1115/1.4046739

Walker, J., Coraddu, A., Collu, M., & Oneto, L. (2022). Digital twins of the mooring line tension for floating offshore wind turbines to improve monitoring, lifespan, and safety. *8*, 1–16. https://doi.org/10.1007/s40722-021-00213-y

Walmsley, T. G., Patros, P., Yu, W., et al. (2024). Adaptive digital twins for energy-intensive industries and their local communities. *Digital Chemical Engineering, 10*, Article 100139. https://doi.org/10.1016/j.dche.2024.100139

Wang, Z., & Wang, Z. (2025). Risk twin: Real-time risk visualization and control for structural systems. *68*, 85. https://doi.org/10.1007/s00158-025-04014-x

Wang, M., Wang, C., Hnydiuk-Stefan, A., et al. (2021). Recent progress on reliability analysis of offshore wind turbine support structures considering digital twin solutions. *Ocean Engineering, 232*, Article 109168. https://doi.org/10.1016/j.oceaneng.2021.109168

Wang, Y., Su, Z., Guo, S., et al. (2023). A survey on digital twins: architecture, enabling technologies, security and privacy, and future prospects. *IEEE Internet of Things Journal, 10*, 14965–14987. https://doi.org/10.1109/JIOT.2023.3263909

Weil, C., Bibri, S. E., Longchamp, R., et al. (2023). Urban digital twin challenges: A systematic review and perspectives for sustainable smart cities. *Sustainable Cities and Society, 99*, Article 104862. https://doi.org/10.1016/j.scs.2023.104862

Wilhelm, J., Petzoldt, C., Beinke, T., & Freitag, M. (2021). Review of digital twin-based interaction in smart manufacturing: Enabling cyber-physical systems for human-machine interaction.

International Journal of Computer Integrated Manufacturing, 34, 1031–1048. https://doi.org/10.1080/0951192X.2021.1963482

Xie, F., & Aly, A.-M. (2020). Structural control and vibration issues in wind turbines: A review. *Engineering Structures, 210*, Article 110087. https://doi.org/10.1016/j.engstruct.2019.110087

Xie, G., Yang, K., Xu, C., et al. (2022). Digital twinning based adaptive development environment for automotive cyber-physical systems. *18*, 1387–1396. https://doi.org/10.1109/TII.2021.3064364

Xu, P., Chen, J., Li, J., et al. (2023). Using Bayesian updating for monopile offshore wind turbines monitoring. *Ocean Engineering, 280*, Article 114801. https://doi.org/10.1016/j.oceaneng.2023.114801

Xu, T., Zhang, X., Sun, W., & Wang, B. (2025). Intelligent operation and maintenance of wind turbines gearboxes via digital twin and multi-source data fusion. *Sensors, 25*, 1972. https://doi.org/10.3390/s25071972

Yang, W., Tavner, P. J., Crabtree, C. J., et al. (2014). Wind turbine condition monitoring: Technical and commercial challenges. *Wind Energy, 17*, 673–693. https://doi.org/10.1002/we.1508

Yang, Z., Tang, C., Zhang, T., et al. (2024). Digital twins in construction: Architecture, applications, trends and challenges. *Buildings, 14*, 2616. https://doi.org/10.3390/buildings14092616

Yildiz, E., Møller, C., & Bilberg, A. (2021). Demonstration and evaluation of a digital twin-based virtual factory. *International Journal of Advanced Manufacturing Technology, 114*, 185–203. https://doi.org/10.1007/s00170-021-06825-w

Yu, J., Song, Y., Tang, D., & Dai, J. (2021). A Digital Twin approach based on nonparametric Bayesian network for complex system health monitoring. *Journal of Manufacturing Systems, 58*, 293–304. https://doi.org/10.1016/j.jmsy.2020.07.005

Yucesan, Y. A., & Viana, F. A. C. (2023). Physics-informed digital twin for wind turbine main bearing fatigue: Quantifying uncertainty in grease degradation. *Applied Soft Computing, 149*, Article 110921. https://doi.org/10.1016/j.asoc.2023.110921

Zami, M. B. A., Shaon, S., Quy, V. K., & Nguyen, D. C. (2024). *Digital twin in industries: A comprehensive survey*

Zhang, J., Heng, J., Dong, Y., et al. (2024b). Coupling multi-physics models to corrosion fatigue prognosis of high-strength bolts in floating offshore wind turbine towers. *Engineering Structures, 301*, Article 117309. https://doi.org/10.1016/j.engstruct.2023.117309

Zhang, J., Heng, J., Dong, Y., et al. (2025). A digital twinning framework for intelligent operation and maintenance of floating offshore wind turbines under corrosion-fatigue. *IOP Conference Series: Earth and Environmental Science, 1552*, Article 012023. https://doi.org/10.1088/1755-1315/1552/1/012023

Zhang, E., Shen, F., Liu, S., et al. (2024a). Offshore wind power digital twin modeling system for intelligent operation and maintenance applications. *E3S Web of Conferences, 546*, 02010. https://doi.org/10.1051/e3sconf/202454602010

Zhao, W., Zhang, C., Wang, J., et al. (2023a). Research on main bearing life prediction of direct-drive wind turbine based on digital twin technology. *Measurement Science & Technology, 34*, Article 025013. https://doi.org/10.1088/1361-6501/ac99f4

Zhao, X., Dao, M. H., & Le, Q. T. (2023b). Digital twining of an offshore wind turbine on a monopile using reduced-order modelling approach. *Renewable Energy, 206*, 531–551. https://doi.org/10.1016/j.renene.2023.02.067

Zhao, X., Quang Tuyen, L., & My Ha, D. (2022). Digital twining of horizontal axis wind turbine with reduced-order modelling approach. In Volume 7: CFD and FSI. American Society of Mechanical Engineers, Hamburg, Germany, p. V007T08A003

Zhu, Y.-C., Wagg, D., Cross, E., & Barthorpe, R. (2020). Real-time digital twin updating strategy based on structural health monitoring systems. In Z. Mao (Ed.), *Model validation and uncertainty quantification* (Vol. 3, pp. 55–64). Springer International Publishing.

Chapter 5
Evolution and State-of-the-Art of Intelligent Vibration Control for Wind Turbine Structures

Kaosshan Dai, Qinlin Cai, Ahmed Elgammal, Tong Wang, and Chenchen Deng

Abstract Modern Wind Turbines (WTs), characterized by increasingly flexible structures, encounter significant vibration challenges due to complex aero-hydro-servo-elastic couplings. Although passive strategies remain widely used in the industry, their performance is constrained by issues such as limited bandwidth and frequency detuning. This chapter provides a critical review of the evolution of structural control, tracing the trajectory from traditional model-based control toward intelligent data-driven systems for WTs. It synthesizes state-of-the-art technologies across passive, semi-active, and active control systems, highlighting the shift toward high-authority intelligent control. Special emphasis is placed on the adaptive intelligence layer. The review concludes that the future of vibration mitigation lies in floating absorber-based hydrodynamic control, hydrodynamic-mechanical-aerodynamic integrated control, self-powered vibration control, and physics-informed artificial intelligence-based vibration control.

K. Dai · Q. Cai (✉) · A. Elgammal · T. Wang · C. Deng
Department of Civil Engineering, Sichuan University, Chengdu, China
e-mail: qinlin.cai@scu.edu.cn

K. Dai
e-mail: kdai@scu.edu.cn

A. Elgammal
e-mail: ahmed_elgammal@stu.scu.edu.cn

T. Wang
e-mail: wangtong2@stu.scu.edu.cn

C. Deng
e-mail: dengchenchen@stu.scu.edu.cn

K. Dai
State Key Laboratory of Intelligent Construction and Healthy Operation and Maintenance of Deep Underground Engineering, Sichuan University, Chengdu, China

C. Baniotopoulos and E. Marino (eds.), *Advances in Wind Energy in the Era of Artificial Intelligence*, CISM International Centre for Mechanical Sciences 625,
https://doi.org/10.1007/978-3-032-28815-8_5

5.1 Introduction

Why Do We Need Vibration Control for Wind Turbines? The global energy transition and rising demand for renewable energy have positioned wind power as a central element in the reconfiguration of energy systems worldwide, owing to its low-carbon characteristics and resource sustainability (GWEC, 2025). The development of Wind Turbines (WTs) has progressed steadily toward large-scale installations with higher unit capacities, as illustrated in Fig. 5.1 (GWEC, 2025). Such WT configurations generally exhibit evident flexibility and low inherent damping, rendering them highly sensitive to wind-induced loads. Meanwhile, WTs are increasingly deployed at deep-water offshore sites (McMorland et al., 2022) and in seismically active regions (Wang et al., 2025b), particularly in recent years, when wind energy development in plain areas has tended toward saturation. As a result, these WTs inevitably experience more severe multiphysics loading conditions at these deployment locations. To address the vibration-related issues of WTs and ensure structural safety during their service lives, a series of load-reduction and vibration-mitigation strategies has been developed for the WTs, which are generally categorized as pitch control and vibration control (Elkodama et al., 2023; Xie & Aly, 2020; Zuo et al., 2020). Pitch control adjusts the blade pitch angle to regulate the wind load, thereby mitigating structural vibrations. However, its operating mechanism is often coupled with the objective of maintaining/optimizing generated power, potentially compromising power performance. This aspect is not the focus of the present chapter.

How About Applying Civil-Building Vibration Control for Wind Turbines? Structural vibration control is not a novel concept in civil and mechanical engineering (Housner et al., 1997). The typical vibration control device, the Tuned Mass Damper (TMD), was first introduced in the 1920s (Ormondroyd and Den Hartog, 1928). Over the past century, structural vibration control strategies, with various

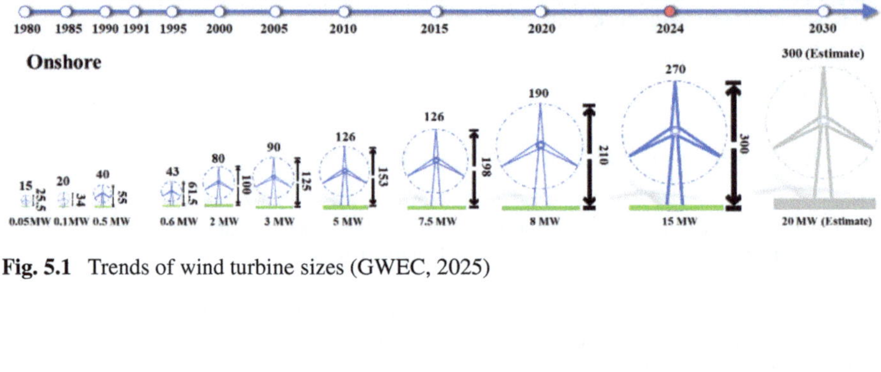

Fig. 5.1 Trends of wind turbine sizes (GWEC, 2025)

control configurations, have been developed, accompanied by numerous studies on theoretical modeling (Zhu et al., 2012), numerical simulations (Cai et al., 2025c), laboratory experiments (Shen et al., 2021), and field tests/applications (Gao et al., 2024). However, it was not until 1996 that these strategies were applied to WTs (Enevoldsen & Mørk, 1996). Prior to this, vibration control research had primarily focused on other structural configurations, including, but not limited to, civil buildings (Fujino & Sun, 1993), bridges (Shelley et al., 1995), and vehicle suspensions (Oueslati & Sankar, 1994). These strategies cannot be directly applied to WTs, whose operational mechanisms are considerably more complex. Compared to civil buildings, the following points illustrate the differences between WTs and civil buildings:

(1) WTs are characterized by a heavy hub–nacelle–blade assembly located at the tower top, which can account for over half of the total structural mass (Jonkman et al., 2009). This configuration yields a highly top-heavy structural system, fundamentally different from conventional civil buildings. Consequently, under seismic or extreme dynamic loading, the resulting tower-top inertial forces can be considerable.
(2) WTs operate with rotating blades that generate aerodynamic damping (Jahani et al., 2022), a phenomenon far exceeding the complexity of conventional civil structures, where such mechanisms are typically absent. Moreover, offshore WTs are subjected to hydrodynamic damping arising from wave-structure interactions (Jahani et al., 2022), a factor that is generally not considered in civil-building design.
(3) WTs are highly functional structures, with limited available space within the tower and nacelle (Cai et al., 2025b). This imposes significant constraints on the installation and implementation of vibration control devices. In contrast, space limitations are usually not a critical concern in conventional civil buildings.
(4) As WT development progresses toward deep-sea and seismic-prone regions (McMorland et al., 2022; Wang et al., 2025b), turbines are increasingly exposed to severe and highly coupled loading conditions, including wind, wave, current, and seismic excitations. Such multi-hazard loading scenarios are generally beyond the scope of traditional civil structural design.

For these reasons, vibration control strategies developed initially for civil buildings are not directly transferable to WTs.

General Development of Wind Turbine Vibration Control. Since the first proposal of TMD for WTs, its application in WTs has progressed rapidly. Figure 5.2 shows the annual number of publications retrieved from *Scopus* using relevant keywords. The results reveal an exponential increase in related publications over time, reflecting growing interest in vibration-control strategies for WTs. In recent years, several review papers have been published from different perspectives, providing comprehensive overviews and valuable guidance for research on WT vibration control. For example, Xie and Aly (2020) reviewed vibration issues in WTs, including tower, nacelle, blade, and drivetrain vibrations, as well as relevant operational noises,

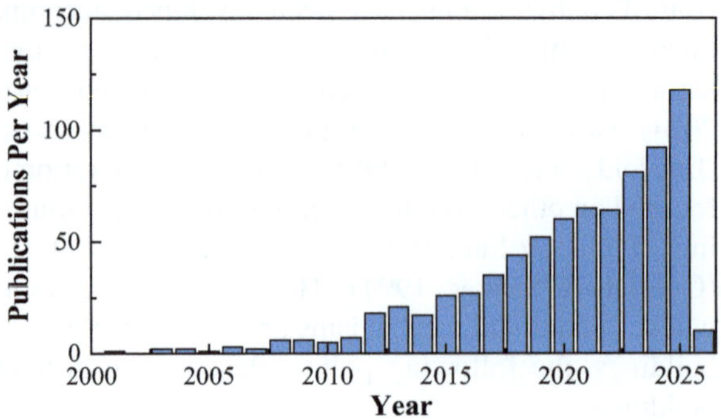

Fig. 5.2 Publications per year on WT vibration control (searched in *Scopus* by keywords, "wind turbine" and "vibration control" or "vibration reduction" or "vibration mitigation")

and presented corresponding vibration control strategies. Focusing on component-specific approaches, Zuo et al. (2020) conducted a review of control strategies applied to the WT tower and blades. Targeting the aeroelasticity, hydroelasticity, and vibrations of the offshore WTs, Jahani et al. (2022) provided a state-of-the-art review of relevant in-situ and laboratory experiments on offshore WT dynamics.

According to Fig. 5.2, a substantial number of studies have emerged since 2020, accounting for nearly half of the total publications on WT vibration control. These studies most often identify four differences between WT and civil structures and have consequently led to the development of novel vibration-control configurations and strategies. Accordingly, this chapter aims to provide a comprehensive review of recent vibration control strategies for WTs, with particular emphasis on intelligent control approaches, following the conventional classification framework, namely passive, semi-active, and active control strategies. It should be noted that the effectiveness of both semi-active and active control strategies relies heavily on control algorithms; they, in a broader sense, can thus be regarded as forms of intelligent vibration control. By critically assessing the trade-offs among control capacity, energy consumption, and computational complexity, this work provides a roadmap for the future of intelligent vibration mitigation in the era of large-scale aero-hydro-servo-elastic systems.

The remainder of the chapter is organized to mirror the evolutionary trajectory. Section 5.2 establishes the general theoretical fundamentals. The subsequent discussion systematically evaluates the control mechanisms, progressing from passive ones (Sect. 5.3) to the high-capacity semi-active and active systems (Sect. 5.4). The energy-harvesting and self-powered vibration control in WTs are addressed in a dedicated subsection. Section 5.5 reviews the intelligent algorithms that govern the control systems in WTs, tracing the shift from classical model-based control to data-driven and artificial intelligence-based control. Finally, Sect. 5.6 synthesizes these developments to identify the potential technological barriers and to outline a strategic vision for future WT vibration control.

5.2 Theoretical Fundamentals

Structural Dynamics. The dynamic response of WTs is governed by complex aero-servo-elastic couplings involving interactions among aerodynamic forces, structural dynamics, and control system responses (also potential hydrodynamic forces in offshore WTs) (Li et al., 2019; Tang et al., 2021). Note that the relevant representations, such as aerodynamic damping (Liu et al., 2017; Zhu et al., 2019), structural coupling (Yang & He, 2020; Zhang & Høeg, 2020), and/or hydrodynamic damping (Chen & Duffour, 2018), mainly arise from these coupling effects. Accurate modeling of the combined coupling mechanisms is therefore essential for reliably predicting WT vibration behavior and fatigue life.

Capturing these complex dynamics poses a computational challenge. High-fidelity finite element models established using commercial software can provide detailed representations of turbine dynamics but are often too slow for the fast computations required in real-time control loops (Latif et al., 2025). In contrast, mathematical parameter models simplify analyses by employing theoretical beam elements, typically formulated in lumped- or distributed-mass representations, with the rotor–nacelle-blades modeled as a lumped mass (Cai et al., 2025c; Li et al., 2024). While these models capture the overall inertial effects, they often neglect blade flexibility and centrifugal stiffening, potentially leading to inaccuracies in predicting structural responses and failure probabilities (Ali et al., 2020).

The above-mentioned modeling method focuses solely on the structure. However, the coupling between structural dynamics and external excitations must also be taken into account. Two main methodologies are commonly adopted. The first approach establishes the WT model by using the aforementioned commercial software or theoretical formulations, while the external loads are generated by separate aero-hydro-servo simulation tools, such as OpenFAST (Jonkman et al., 2010). This approach typically corresponds to a weakly coupled framework (Cai et al., 2025c; Li et al., 2024). The second method formulates the WT based on rigid-body dynamics and models the aerodynamic loads using the blade element momentum theory, representing a relatively strongly-coupled model (Sun, 2018a, 2018b). This simplified approach typically accounts for only the fundamental vibration modes of the tower and blades. Notably, OpenFAST allows the inclusion of multiple modes of the blades as well as the first two bending modes of the tower (Jonkman et al., 2009).

Regardless of which modeling method is used, the general governing equation of the WT with excitation can be expressed as:

$$\mathbf{M}\ddot{\mathbf{X}}(t) + \mathbf{C}\dot{\mathbf{X}}(t) + \mathbf{K}\mathbf{X}(t) = \mathbf{F}_{env}(t) + \mathbf{H}\mathbf{u}(t) \tag{5.1}$$

where $\mathbf{M}$, $\mathbf{C}$, and $\mathbf{K}$ represent the mass, damping, and stiffness matrices of the controlled WT, respectively; $\mathbf{X}(t)$ denotes the displacement vector; $\mathbf{F}_{env}(t)$ represents environmental excitation (e.g., wind and waves); and $\mathbf{u}(t)$ is the control force vector mapped by $\mathbf{H}$. The corresponding state-space representation can be given as

$$\dot{\mathbf{z}}(t) = \mathbf{A}\mathbf{z}(t) + \mathbf{B}\mathbf{u}(t) + \mathbf{E}\mathbf{d}(t) \tag{5.2}$$

$$\mathbf{y}(t) = \mathbf{C}_{out}\mathbf{z}(t) + \mathbf{D}\mathbf{u}(t) \tag{5.3}$$

where $\mathbf{z}(t)$ is the state vector (typically containing displacements and velocities), and $\dot{\mathbf{z}}(t)$ represents its time derivative. $\mathbf{A}$ and $\mathbf{B}$ define the system dynamics and control input influence, respectively, while $\mathbf{E}$ maps the disturbance vector $\mathbf{d}(t)$ (derived from $\mathbf{F}_{env}$) into the state space. The output vector $\mathbf{y}(t)$ represents the measured quantities available for the controller, related to the state by the output matrix $\mathbf{C}_{out}$ and the feedthrough matrix $\mathbf{D}$. This state-space framework facilitates the application of advanced control techniques, such as Linear Quadratic Regulator (LQR) controllers, thereby enabling efficient vibration suppression under varying environmental conditions.

Classification of Control Strategies. Structural vibration control strategies are generally categorized into three types: passive, active, and semi-active systems (Hosseini Lavassani et al., 2022; Ma et al., 2025). As illustrated in Fig. 5.3a, passive systems operate without external power input and rely solely on fixed-parameter mechanical components, such as dampers or absorbers, to mitigate structural vibrations. Active systems (Fig. 5.3b) necessitate an external power supply to drive actuators that generate control forces based on sensor feedback. Semi-active systems (Fig. 5.3c) represent an intermediate class, consuming only a small amount of energy, typically of the order of battery-level power requirements, to adaptively adjust damper properties, such as stiffness or damping, without directly injecting mechanical energy into the structure.

Passive control, although cost-effective and simple, offers control forces solely through structural deformation/vibrations, and its characteristics are inherently fixed and difficult to adapt to varying operational conditions (Xie & Aly, 2020). Moreover, passive control is typically optimized for a single target mode and may have limited multimodal vibration control capability, while WTs exhibit time-varying frequencies due to changes in rotor speed, soil-structure interaction, and material degradation (Lin et al., 2021; Zhang et al., 2015), under which the effectiveness of passive control may degrade significantly/be suboptimal.

Active control injects energy into the structure by using high-authority actuators, enabling effective vibration suppression over a broad frequency range (Soong & Spencer, 2002). However, the implementation of active control includes several limitations. First, a continuous and reliable power supply is required for sensors, actuators, and controllers. Second, the direct injection of external energy may potentially induce structural instability under unfavorable control actions. Finally, active control performance relies heavily on sophisticated control algorithms; the associated complexity, together with measurement noise and actuation errors, can compromise system robustness and reliability (Li et al., 2024; Qiu et al., 2009; Wang et al., 2018).

Semi-active control offers a compelling compromise by providing the adaptability of active control with the inherent stability of passive devices (Soong & Spencer, 2002). Such systems are generally considered stable because they do not inject

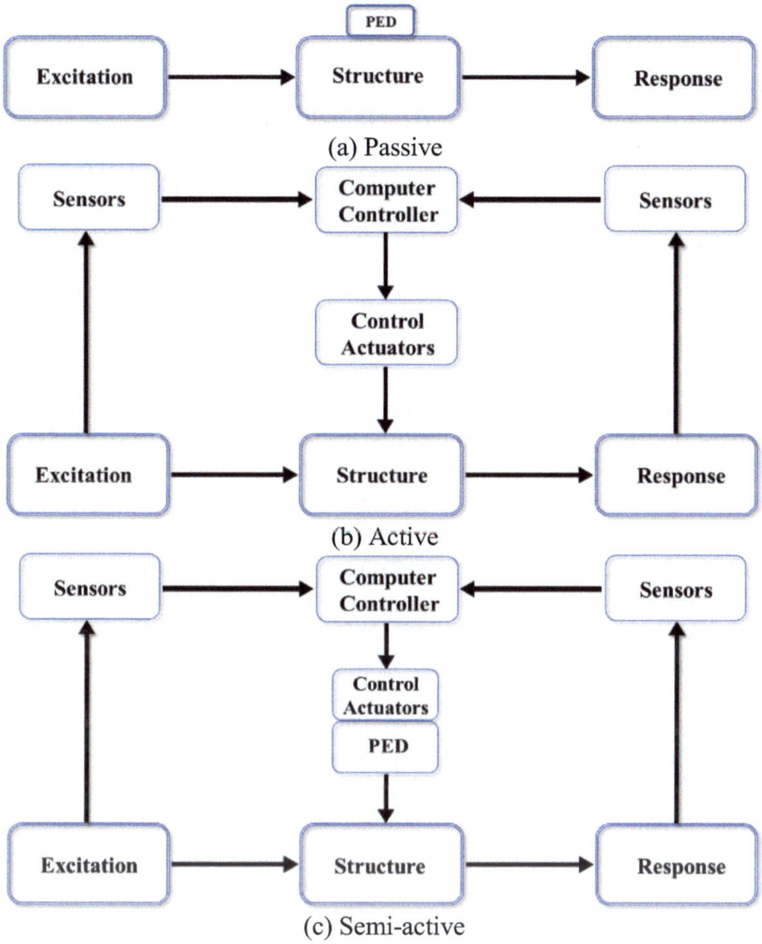

Fig. 5.3 Schematic of vibration control systems (Soong & Spencer, 2002)

mechanical energy directly but rather supply power for controllers and sensors only (Farahpour & Hejazi, 2023; Yang et al., 2002). These devices function as adjustable passive components; even if the control algorithm fails, they remain be power dissipative and retain certain control capacity. Although time delays and input saturation in the feedback loop still necessitate careful control law design, semi-active strategies effectively mitigate the destabilization risks inherent in fully active systems while maintaining lower power consumption. (Lu, 2004). Note that hybrid control (Soong & Spencer, 2002), which integrates two or more of the aforementioned control strategies, is sometimes classified as an additional vibration-control category.

Control Performance Metrics. The design of effective vibration control systems relies on defining objective functions that balance competing structural requirements. The displacement, acceleration, and fatigue loads are all reported as quantities targeted for control (Chen et al., 2021; ISO-20816–21, 2025; Yang & He, 2020). Generally, performance metrics quantify reductions in the Root-Mean-Square (RMS) of acceleration and/or the standard deviation of displacement at critical locations, particularly at the tower top (Chen et al., 2021; Hemmati et al., 2019a). Notably, in WTs, the displacement response generally includes a static component due to steady wind loading, whereas conventional vibration control strategies are effective only in mitigating the dynamic displacement.

Serviceability can be evaluated via fatigue-load indicators, such as by minimizing tower-base bending moments and blade-root moments (Yang & He, 2020). Nacelle acceleration serves as a critical performance metric because it directly influences the reliability of the gearbox and main shaft bearings (Ghane et al., 2018; ISO-20816–21, 2025). High accelerations at the nacelle indicate excessive dynamic loads that accelerate wear, leading to increased downtime and maintenance costs (Benbouzid et al., 2021). The aforementioned performance metrics are generally not in conflict; thus, both single- and multi-objective optimization approaches are applicable.

In addition to dynamic responses and fatigue loads, structural stability and power regulation are also key performance indices for WTs. Blade pitch control, although primarily intended for power optimization/maintaining, regulates aerodynamic loads that couple with structural vibrations. Individual pitch control strategies can suppress vibrations and mitigate unbalanced loads; however, aggressive implementation may degrade power production or amplify platform motions in offshore applications (Sarkar et al., 2021). To account for this coupling, performance indices typically incorporate penalties on pitch actuator usage, considering both the pitch angle magnitude and its rate of change (Ren et al., 2016). Such penalties help prevent excessive actuator wear while balancing power quality with fatigue load reduction. In this scene, advanced control frameworks can weigh actuator effort against power output stability to achieve an optimal trade-off. Nevertheless, it is worth noting that blade pitch control, as an inherently active control system for WTs, falls outside the conventional concept of structural vibration control and is therefore only briefly discussed here.

5.3 Passive Control

Passive structural control systems are widely explored in WTs to mitigate vibrations and enhance operational lifespan without requiring external power. Valued for their simplicity and unconditional stability compared to active counterparts, these systems leverage various damping mechanisms, such as eddy-current or electromagnetic damping, to dissipate energy (Cai et al., 2024; Sorge et al., 2024; Zhang et al., 2025a). Based on their physical mechanisms, these devices can be broadly categorized into two classes: mass-based absorbers and direct energy-dissipation elements.

Mass-based absorbers mitigate WT vibrations by transferring vibration energy from the WT to the auxiliary oscillating mass(es), which is subsequently dissipated via damping mechanisms (Han et al., 2025; Sun & Jahangiri, 2019). Direct energy-dissipation systems attenuate vibrations by dissipating energy without introducing additional oscillating mass (Dai et al., 2021; Tang et al., 2025). It should be noted, however, that this classification is not strictly exclusive, as hybrid configurations that integrate features of both categories are also possible.

Mass-Based Absorbers. Mass-based absorbers are widely recognized as effective passive control technologies in WTs. They operate by redistributing kinetic energy from the controlled structure to a secondary oscillating mass, which then dissipates it. These include, but are not limited to, TMDs, Tuned Liquid Dampers (TLDs), Tuned Liquid Column Dampers (TLCDs), Nonlinear Energy Sinks (NESs), particle dampers, and their derivatives, all of which generally incorporate the oscillating mass(es).

Solid Mass**:** Among these devices, TMDs and TLDs/TLCDs are relatively mature control configurations and have even been successfully implemented in real-world applications (Gao et al., 2024). Numerous studies have demonstrated the effectiveness of TMDs; Chen et al. (2021) reported reductions of 32% and 29% in the standard deviation of tower-top displacement and acceleration, respectively, for jacket-type offshore WTs equipped with TMDs. Gao et al. (2024) reported a 50% reduction in the amplitude of the tower-base side-to-side bending moment when applying TMD. However, the practical application of TMDs is constrained by a trade-off between mass ratio and stroke length. Increasing the auxiliary mass improves control capacity, but it is generally not suggested for top-heavy WTs (Chen et al., 2021). Reducing the mass necessitates large relative displacements (stroke), which often exceeds the nacelle spatial limits and potentially introduces the risk of damper saturation or mechanical failure (Jahangiri & Sun, 2020).

The Single-Pendulum TMD (SPTMD) is one of the representative TMD configurations that have been implemented in real-world WTs (Gao et al., 2024). Sun and Jahangiri (2018) investigated the bidirectional control performance of a three-dimensional (3D) SPTMD for a monopile WT, reporting superior reduction performance in tower-top RMS and peak displacements compared with two linear TMDs. Such 3D-SPTMDs and dual-TMDs were also explored for floating WTs, targeting platform roll and pitch motions and nacelle displacement (Luo et al., 2024). However, due to the low-frequency matching requirement for TMDs, SPTMDs typically necessitate a considerably long pendulum length. To satisfy the frequency optimization, a wave energy converter-inspired Double-Mass Pendulum Damper (DMPD) has been developed. This configuration employs two oscillating masses mounted on the same bar, as shown in Fig. 5.4, and exhibits a natural frequency independent of the pendulum length (Cai & Zhu, 2021; Cai et al., 2025a; 2025b). Figure 5.4 also shows the schematic of the SPTMD application. Numerical investigations indicated that the DMPD demanded significantly less installation space than the SPTMD, albeit with a slightly reduced control effectiveness (Cai et al., 2025b). Although not originally

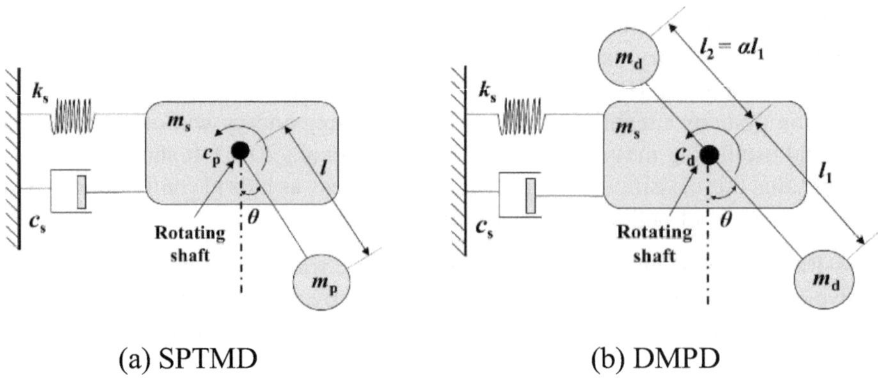

Fig. 5.4 Schematic of SPTMD and DMPD

intended for vibration control, DMPD nonlinear and energy-harvesting characteristics have been experimentally investigated, providing insights into its mechanical design (Cai & Zhu, 2024; Cai et al., 2025a). Parallel to these pendulum innovations, rolling-mass concepts, such as the ball vibration absorber, have been introduced to eliminate suspension arms entirely by employing curved tracks to generate restoring forces (Wang et al., 2025a). Notably, the track geometry governs the resulting control force characteristics and may introduce nonlinearities into the system.

The TMDs, including pendulum TMDs, are often optimized for a single controlled mode. Studies on Multiple TMDs (MTMDs) distributed along the tower height have reported enhanced control performance (Zuo et al., 2017). By strategically placing smaller TMD units at points of maximum modal amplitude, along the tower height, MTMDs can target both fundamental and higher-order modes simultaneously. This capability is crucial for multi-hazard scenarios involving wind, seismic, and wave excitations, where the WTs may undergo multi-mode vibrations (McNamara et al., 2024; Zuo et al., 2017). Numerical studies demonstrate that optimally distributed TMDs can achieve tower-top RMS fore-aft and side-to-side displacements of 10% and 42%, respectively, with a total mass ratio of 10% (McNamara et al., 2024).

Although TMDs or MTMDs have demonstrated varying degrees of success in previous studies, their performance is highly sensitive to frequency detuning. Such detuning commonly arises in WTs due to structural degradation or changing operational environment (Dai et al., 2021; Wang et al., 2020). To enhance the robustness of control performance, research has pivoted toward NESs that utilize targeted energy transfer to channel energy from the primary structure irreversibly. Unlike linear TMDs, which are effective only near a tuned frequency, NES devices employ polynomial restoring forces to engage with the primary structure across a broad spectrum of excitation frequencies (Lu et al., 2025; Zhang et al., 2024b). Innovations in restoring force profiles have further refined this concept. Track NES introduces geometric nonlinearity by constraining an auxiliary mass to move along a polynomial-shaped curved path, thereby eliminating the need for complex nonlinear springs, as shown in Fig. 5.5a. Zuo and his colleagues (Zuo & Zhu, 2022; Zuo et al., 2023) investigated

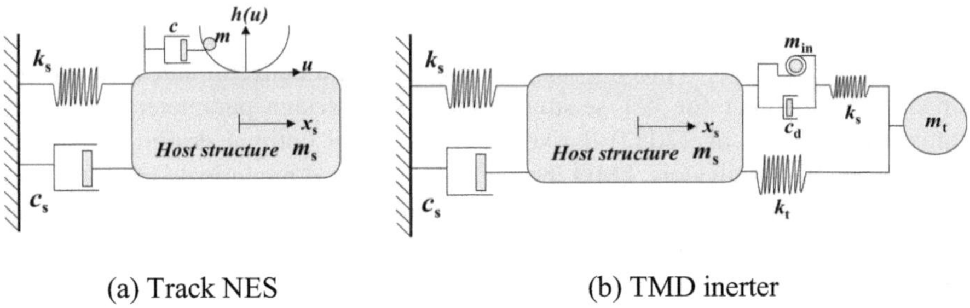

Fig. 5.5 Schematic of the track NES and TMD inerter

two different track profiles in their works: one employs positive quadratic and quartic components, yielding a monostable track NES, whereas the other adopts a negative quadratic term together with a positive quartic term, leading to a bistable track NES. In WTs, the track NESs exhibit superior robustness against detuning; however, their vibration control performance remains inferior to that of optimal TMDs (Zuo & Zhu, 2022; Zuo et al., 2023). They subsequently incorporated an electromagnetic damping mechanism into the NES, constructing an energy-harvesting track NES for WT vibration control (Cai et al., 2024). Note that, by using a circular track profile, the track NES likely functioned as an SPTMD. Building upon this geometric nonlinearity concept and extending it to multi-axis vibration control, a 3D nonlinear TMD was developed. By employing a spatial arrangement of springs, nonlinear restoring forces were generated simultaneously in the pitch, roll, and heave directions, thereby achieving 3D vibration control. Its control performance was superior to that of linear TMDs (Jahangiri & Sun, 2022).

Another solution for enhancing the control performance robustness is the use of particle TMDs, which consist of a tuned container filled with rolling balls; as the primary structure vibrates, vibration energy is dissipated through the nonlinear momentum transfer, inelastic collisions, and friction between the particles and the container walls (Lu et al., 2023). Unlike traditional linear TMDs, particle TMDs feature an inherently nonlinear damping mechanism that enhances performance during large-displacement events, such as earthquakes, when conventional systems may become saturated (Dong et al., 2026). Experimental validations on scaled WT towers demonstrated that particle TMDs exhibited a wider effective frequency bandwidth and greater resilience against detuning compared to linear counterparts (Lu et al., 2023).

Inerter, a two-terminal device, is a novel structural component defined in early 2000s. It generates an output force proportional to the relative acceleration between its terminals, effectively producing an apparent inertial mass that can far exceed its physical weight. Different mechanisms, such as ball screws, rack-and-pinion mechanisms, hydraulic motors, and electromechanical couplings, can be used to construct an inerter (Li & Zhu, 2018; Smith & Wang, 2003; Wang & Su, 2008; Wang et al., 2011).

The inerter has been explored in various configurations to enhance the control performance. Zhang et al. (2019) developed a tuned parallel inerter mass system, as shown in Fig. 5.5b, replacing the damping unit of a conventional TMD with a tuned inerter damper for WT seismic control, with design parameters optimized numerically. Dai et al. (2022) derived the theoretical optimal design parameters for an integrated equivalent TMD inerter and performed numerical simulations on WTs with this inerter configuration. Their results demonstrated that adopting an inerter can, to some extent, address the tower-top mass constraints and reduce the P-Delta instability risks associated with adding large conventional absorbers to slender towers. Shi and Zhu (2025) developed a geometrical nonlinear TMD inerter for WTs under parametric uncertainties, suggesting the need for an optimization approach to address high-dimensional uncertainties. Furthermore, other advanced hybrid designs, such as inerter-based NESs (Zeng et al., 2025), which integrate inertial amplification with nonlinear stiffness to broaden the effective working frequency range, have also been reported. In addition to the physical mass and the inerter, the feasibility of hydrodynamically induced added mass for vibration control has also been explored (Wang et al., 2024).

Note that mass-based absorbers might face certain challenges when applying for WTs. In blade applications, the fundamental challenge lies in centrifugal stiffening, which introduces nonlinearities in the blade's effective natural frequency due to rotation. An absorber tuned to the stationary frequency may therefore become suboptimal during operation (Holierhoek, 2023). In the context of floating WTs, integrating passive control may pose a stability risk due to strong coupling between tower and platform dynamics. Yao et al. (2024) reported that a nacelle-based TMD tuned to the tower's first bending mode can reduce base moments, but inadvertently amplify low-frequency platform pitch motions. To address this trade-off, distributed configurations employing separate absorbers for the nacelle (tower modes) and platform (pitch modes) are recommended to maintain overall system stability (Li et al., 2022; Luo et al., 2024).

***Liquid Mass*:** Using liquid as the mass element for the mass-based damper forms a TLD or TLCD, as shown in Fig. 5.6, another widely used passive control strategy in WTs (Bhattacharyya et al., 2017; Zhang, 2022). Generally, TLDs or TLCDs dissipate vibrational energy by mobilizing fluid momentum, with control forces generated by mechanisms such as boundary-layer friction and orifice head loss. A defining advantage of this class of absorbers in WTs is the potential for dual-use integration; unlike the solid mass dampers, liquid systems can utilize existing water reserves required for fire protection or ballast. This integration minimizes the structural weight penalties while providing a cost-effective structural control solution.

Energy dissipation of TLDs occurs principally through the nonlinear sloshing of the free surface and the impact of the fluid against the container walls. TLDs act as effective countermeasures for relatively high frequency vibrations, such as the edgewise modes of WT blades, where the first sloshing mode was tuned to match the structural natural frequency (Zhang et al., 2015). However, the stochastic nature of wave breaking introduces significant modeling complexities. In shallow-water

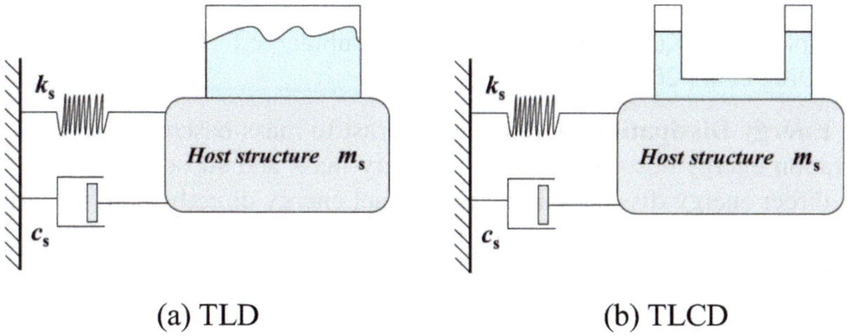

Fig. 5.6 Schematic of TLD and TLCD

TLDs, the damping force is highly amplitude-dependent, resulting in nonlinearities that challenge accurate prediction within conventional linear control frameworks (Love et al., 2011).

To address the unpredictability inherent in free-surface sloshing, the TLCD confines the fluid within a geometric conduit, typically a U-shaped tube. This geometric configuration replaces stochastic sloshing with a deterministic oscillating column, with damping precisely regulated through orifice plates. The restoration force is provided by gravity acting on the displaced liquid columns, while the head loss coefficient of the orifice governs energy dissipation. Numerous studies have demonstrated the effectiveness of TLCDs for WT vibration control (Hemmati et al., 2019b; Zhang & Høeg, 2020). Hemmati et al. (2019a) reported that TLCDs can reduce WT fragility by 6% in normal operation and 12% in parked conditions, with a mass ratio of 2.5%. Later, they conducted a systematic comparison between TMD and TLCD across different operation modes, finding that TLCDs were more effective in parked mode, while TMDs may perform better during normal operation, indicating the merit of a combined TMD–TLCD strategy (Hemmati et al., 2019b).

Hochrainer and Ziegler (2006) first developed the concept of the tuned liquid column gas damper, which seals the vertical columns and pressurizes the gas above the fluid surface. This creates a gas-spring effect, allowing the frequency to be tuned via initial pressure adjustment rather than physical dimensioning (Bargi et al., 2016). Numerical applications in Jacket-type WTs demonstrated that it can achieve at least 10% reduction in the standard deviation of nacelle displacement and acceleration (Bargi et al., 2016). Considering the multi-directional vibrations of the WTs, Ding et al., (2022, 2023) proposed a toroidal TLCD, which utilizes a circular or ring-shaped container that wraps around the WT tower or nacelle internals, ensuring that the liquid column can oscillate in alignment with the structural bending mode regardless of the wind direction. Numerical results demonstrated its superior bidirectional control capabilities, particularly in the side-to-side direction, where aerodynamic damping is low. Other novel TLD/TLCD configurations included, but were not limited to, multi-layer TLDs for generating sufficient counter-moment (Ha & Cheong, 2016), tuned liquid multi-column dampers essentially utilizing the fluid inside the semisubmersible platform (Xue et al., 2022), and non-submerged TLCD for heave motion control (Ding et al., 2025).

Direct Energy Dissipation Devices. In contrast to mass-based absorbers that rely on vibration energy absorption by an auxiliary mass and subsequent energy dissipation, direct energy dissipation devices extract energy directly from the structural deformation/vibrations. While a wide range of such devices has been developed in civil engineering, only certain configurations have been successfully explored for WTs. A distinct evolution in this field is the removal of the mass element to realize lightweight design.

The simplest form of direct energy dissipation is the use of pure viscous dampers. Installing viscous dampers in the lattice sections of WTs has been reported to extend structural fatigue life by up to three times by mitigating stress cycles induced by lateral wind and wave loads (Emami et al., 2022). In addition, viscous dampers have also been integrated with brace mechanical systems in WTs, including blades and towers, where the braces are designed to amplify displacements and enhance overall damping performance. Zhang et al. (2025a) proposed a toggle-brace mechanical system, as shown in Fig. 5.7, incorporating a viscous damper to reduce both flapwise and edgewise vibrations of blades. Similarly, Zhao et al. (2019) proposed a scissor-jack braced viscous damper for WT tower vibration control, wherein the scissor-jack brace employs a toggle-linkage geometry to amplify the damper stroke while simultaneously reducing the required damping force.

Cable-based energy-dissipation systems represent another class of direct energy-dissipation devices. In the WT tower application, prestressed cables distributed along the tower height transmit bending deformations to viscous dampers near the base (Dai et al., 2021). This configuration effectively converts the elongation of the tower wall into axial motion at the cable base, enabling concentrated energy dissipation

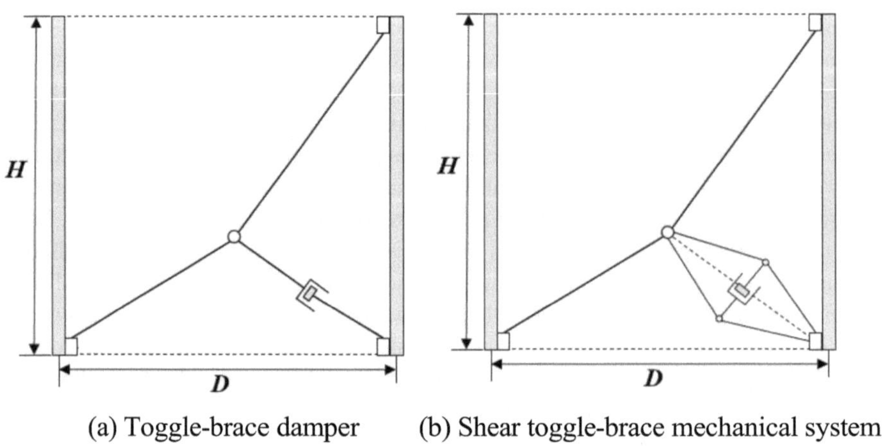

Fig. 5.7 Schematic of toggle-brace type damper (Zhang et al., 2025a)

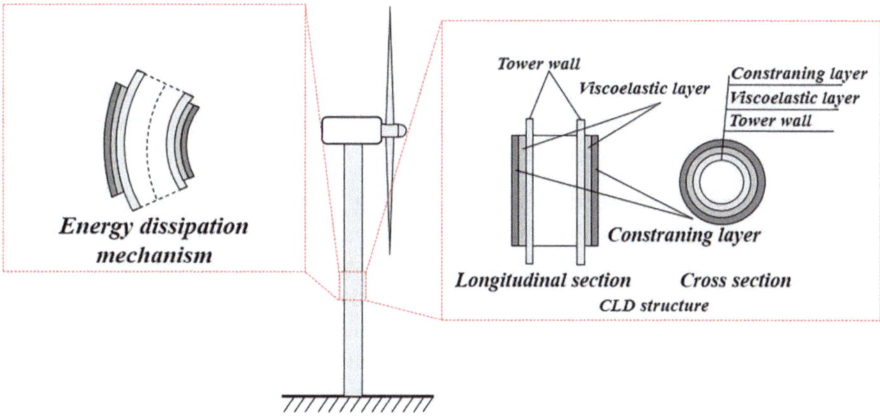

Fig. 5.8 Schematic of constrained-layer damping treatment for WTs (Yin et al., 2025)

without adding mass to the nacelle. Further refinement of this concept is observed in the tuned cable-inerter system, which integrates an inerter to amplify the energy flow through the mechanism (Dai et al., 2023). Similar configurations have also been explored for WT blade vibration control (Jiang et al., 2024b; Tang et al., 2025). Recently, constrained-layer damping treatments have emerged as another novel energy-dissipation approach for WTs, as shown in Fig. 5.8. This method involves sandwiching a viscoelastic material between the tower shell and a stiff constraining layer (Yin et al., 2025; Zhao et al., 2023). Studies demonstrated that, within appropriate temperature ranges, such constrained-layer damping treatments can increase the modal damping ratio of prototype towers by more than 1%, thereby effectively reducing the response under broadband excitation (Yin et al., 2025). In floating WT, the mooring system is a fundamental structural component. Yang et al. (2025) proposed a high-damping mooring system that incorporated a spring-damper assembly into the anchoring lines, achieving up to a 40% reduction in the standard deviation of platform translational displacements while simultaneously decreasing dynamic cable tension under parked condition. Recently, the concept of using floating absorbers (i.e., independent bodies coupled to the platform via dampers) for WT vibration control has been reported (Alotta et al., 2023). Floating absorbers may be classified as specialized mass-based absorbers tailored for offshore WTs.

5.4 Semi-Active and Active Control

Semi-active Control Systems. Passive structural control systems offer reliable baseline vibration control; however, their parameters are often fixed once designed for specific operating conditions. In practice, WTs experience considerable structural/excitation spectral variation due to factors such as fluctuating rotor speeds and shifts in operational states; thus, fixed-parameter passive control might exhibit reduced effectiveness or suboptimal performance. Semi-active structural control systems offer a solution to these limitations by incorporating controllable damping and/or stiffness elements, functioning as intelligent devices that respond effectively to variable operational demands to some extent.

***Variable-Damping Systems*:** Variable-damping semi-active structural control systems dynamically adjust the damping properties of the control device in response to external excitations. Unlike active control systems, these devices do not inject energy into the structure but rather modulate the damping parameters of the control device; thus, they prevent the system from inducing instability regardless of the control signal (Emami et al., 2023). Generally, there are two categories of variable-damping systems: discrete on–off switching between high and low damping states and continuously variable modulation in which the damping coefficient is adjusted within a specific range (Cai et al., 2025c; Liu et al., 2005).

The Magneto-Rheological (MR) damper is one of the representative technologies in variable-damping devices, offering a favorable balance of high force density, fast response time, and low power consumption (Spencer et al., 1997; Yanik, 2020). The mechanism employs a carrier fluid with micron-sized magnetizable particles that form chain-like structures under a magnetic field, inducing a rapid and reversible transition to a semi-solid state. This field-dependent rheology enables real-time, continuous tuning of the damping force by adjusting the coil current (Spencer et al., 1997). The integration of MR technologies into WTs has demonstrated effectiveness. Caterino (2015) experimentally investigated MR dampers installed at the WT tower base to reduce base bending stress. Results demonstrated that this control configuration reduced peak stress but tended to amplify tower-top displacement. Tian et al. (2025) explored integrating the MR damper within a pulley-cable mechanism to reduce the vibrations of a floating WT tower, reporting 60–70% and 30–40% reductions in tower-top maximum displacement and acceleration, respectively, under operational sea states. Similar explorations using the MR dampers as direct-energy-dissipation devices were also reported by Rezaee et al. (2018), as shown in Fig. 5.9. The MR dampers were also explored for incorporation with a mass-based absorber (Martynowicz & Santos, 2020; Park et al., 2019). Wang et al. (2026) studied an MR Elastomer (MRE) TMD, in which the stiffness and damping of the MRE device can be tuned by varying the applied current. It exhibited a 4.3% reduction of tower-top RMS displacement at the rated wind speed condition, with aerodynamic damping considered, but still outperformed the passive TMD. Another novel application of MR fluid is that Sarkar and Chakraborty (2018, 2019) employed MR fluid to replace water in TLCD, forming an MR-TLCD configuration that enabled the damper damping mechanism to respond quickly to changing conditions.

In addition to MR fluid, eddy-current and electromagnetic damping are two other representative mechanisms that can provide semi-active damping (Cai et al., 2025c; Jiang et al., 2024a). Jiang et al. (2024a) investigated the feasibility of semi-active eddy-current TMD for WTs. The numerical application showed that the maximum reduction in tower-top RMS acceleration was up to 52.66%, while those of its passive

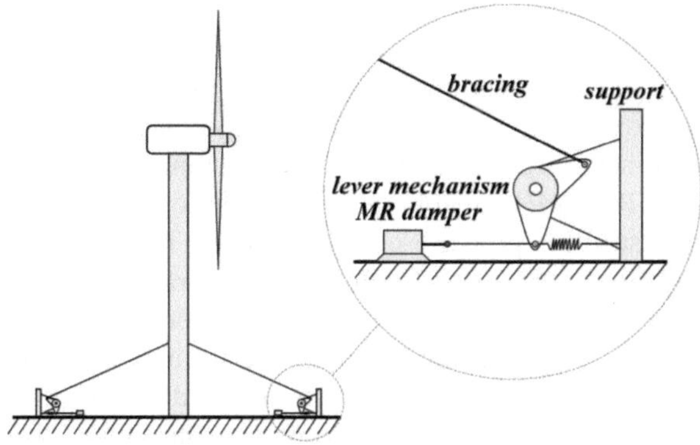

Fig. 5.9 Schematic of MR damper with a lever mechanism for WTs (Rezaee et al., 2018)

counterpart and TLCD were 39.53% and 28.07%, respectively, under shutdown condition. Cai et al. (2025c) replaced the TMD damping unit with semi-active electromagnetic damping; despite only a slight improvement in control performance over the passive counterpart, the electromagnetic damping mechanism can provide a self-powered capacity. Although eddy current and electromagnetic damping mechanisms utilize the principle of electromagnetic induction, their damping tuning strategies are different: the former realizes variable damping by adjusting the magnetic flux density or conductive area (Jiang et al., 2024a), while the latter is tuned by varying the equivalent resistance of the connected energy harvesting circuit/pure resistor (Cai et al., 2025c). The MR, eddy current, and electromagnetic dampers can be employed in passive control configurations as energy-dissipation elements or updation to realize a semi-active configuration, thereby improving control performance. Moreover, there are still other solutions to realize variable damping. For example, Dezvareh and Nazokkar (2025) and Nazokkar and Dezvareh (2022) proposed semi-active liquid-column gas dampers that exhibited variable hydraulic head loss and frequency for floating WT vibration. In controlled hydraulic head-loss mode, it exhibited a tens-of-percent reduction in the maximum and standard deviation of the platform pitch motions.

***Variable Stiffness Systems*:** Variable-stiffness tuning control, or its corresponding frequency tuning control, is also a form of semi-active control. The mechanism for achieving this adaptability involves the use of smart materials, most notably Shape-Memory Alloys (SMAs). SMAs possess the physical property of variable stiffness across different phases (Mani et al., 2020). In WT blades, embedding SMA wires into glass fiber-reinforced polymer composites has demonstrated tens of percent reduction in displacement amplitude compared to conventional blades by adaptively altering the blade's stiffness properties (Mani et al., 2020). Furthermore, SMA actuators enable shape-adaptive morphing of the trailing edge of the blade, reducing the flap-wise bending moment at the blade root (Karakalas et al., 2019). Notably, SMA can also be considered a damping material that enhances the damping performance of the attached blade, but its effectiveness is strain-dependent; the excitation level of the normal wind profile may be insufficient to activate it (Nicoletti & Liebich, 2018).

Employing MREs provides an alternative approach for realizing semi-active stiffness control (Wang et al., 2026). Unlike MR fluids that change viscosity, MREs change their shear modulus (stiffness) when exposed to a magnetic field. Leng and co-workers (Leng et al., 2021, 2023) conducted a series of works on MRE-based semi-active control. Leng et al. (2021) integrated the MRE with a TMD to develop an MRE-TMD capable of tracking the optimal tuning frequency through the variable stiffness property of MREs. MRE-TMD was reported to effectively suppress both first- and second-order modal vibrations of the WT, maintaining robustness even when the structural stiffness degrades due to damage. Later, they explored the multidirectional vibration control performance of WT using a 3D MRE-TMD design (Leng et al., 2023). Recently, the energy dissipation mechanism via pounding, which promotes the lightweight design, was introduced for MRE-TMD (Li et al., 2025). Targeting multi-hazard excitations, Sun (2018a) proposed vibration control of WT using a semi-active TMD with a short-time Fourier transformation-based control algorithm.

Active Structural Control. Active structural control employs actuators relying on external power source to provide control force (Zhang et al., 2020, 2022). While this capability enables active systems to achieve control performance beyond that of passive and semi-active approaches, it requires substantial, continuous power consumption and precise stability management (Hu & He, 2017).

Active Mass and Tension Systems**:** Active mass systems, commonly referred to as mass actuators, function by applying forces to a structure through reaction against an auxiliary mass (Lesieutre et al., 2003). Active TMDs (ATMDs) serve as one of the representative active mass systems for vibration reduction, reliability improvement, and fatigue life extension in WTs. In contrast to passive dampers, ATMDs employ feedback mechanisms for force adjustment. Brodersen et al. (2017) investigated the feasibility of the ATMD for WTs and validated its superiority over the passive TMD in both frequency and time domains. However, these performance gains were achieved at the expense of increased mass displacement. Such an ATMD was reported to reduce WT tower displacement, thereby enhancing reliability, as supported by the fragility curves analysis (Fitzgerald et al., 2018). Lackner and Rotea (2011) compared the fatigue life extension performance of TMD and ATMD. Optimized passive TMDs typically achieve fatigue load reductions of approximately 10%, whereas active configurations could exceed 30%. Beyond tower applications, the ATMDs were also adapted for simultaneous installation in the tower and blades for vibration control (Cong, 2019). Moreover, for floating WTs, Chen et al. (2025) enhanced vibration mitigation by combining a nacelle-mounted ATMD with an active inerter-based TMD installed on the platform. Note that the active inerter TMD in this context functions as an inerter variable system. Numerical results demonstrated that this combination was effective in reducing platform pitch motion, tower fore-aft motion, drive-train torsional motion, and blade flapwise low-frequency vibrations.

In addition to the ATMD configuration, other active mass-based controls have also been explored for WTs. Hu and He (2017) developed a stroke-limited hybrid mass damper, in which an active actuator was added on the opposite side of the oscillating mass (on the side opposite the stiffness and damping units), and stroke limits were introduced. Results demonstrated its effectiveness in reducing tower base moments and platform pitch while strictly adhering to actuator stroke constraints. Li et al. (2024) applied the active mass damper for WT vibration control. Instead of highlighting the active mass damper configuration, the novelty of their studies lies in the self-powered capability of the active configuration, enabled by a dedicated power-balance control logic. Figure 5.10 shows the schematic of ATMD and the active mass damper. Recently, Tang et al. (2024) proposed an integrated control strategy that combines pitch control with an active TMD inerter, using a unified control algorithm. The integrated control strategy showed enhanced vibration control and power generation compared to that of passive control and collective pitch control. Note that the active TLCD has also been reported. By actively modulating the internal pressure via gas valves or driving the fluid flow with pumps, these systems extended the operational bandwidth of traditional sloshing dampers (Hochrainer, 2015).

Active tendon systems differ fundamentally from mass-based systems, which rely on an auxiliary mass to generate control forces. Instead, active tendons directly modify the effective structural stiffness and damping by actively regulating the tension in pre-stressed cables, as shown in Fig. 5.11 (Ghaffarzadeh & Younespour, 2014; Nigdeli & Boduroğlu, 2013). Through dynamic tendon force adjustment, these systems can effectively suppress both translational and rotational responses under wind or seismic excitations (Nigdeli & Boduroğlu, 2013). A notable advantage of active tendon systems is their massless actuation, which enables high control forces without introducing heavy oscillatory components, thereby avoiding additional static gravitational loads on the structure. To date, active tendon systems have been primarily investigated for WT blade vibration control, typically installed longitudinally within the blade structure (Staino et al., 2012; Tao et al., 2018).

***Active Aerodynamic Systems*:** Active aerodynamic system embeds distributed sensors and actuators within the blade structure, enabling real-time modification of local aerodynamic characteristics to mitigate unsteady loads induced by turbulence and wind shear (Barlas & Van Kuik, 2010). In contrast to traditional pitch control, which rotates the entire blade assembly, active aerodynamic surfaces offer localized, rapid load alleviation. This technology has the potential to reduce fatigue damage

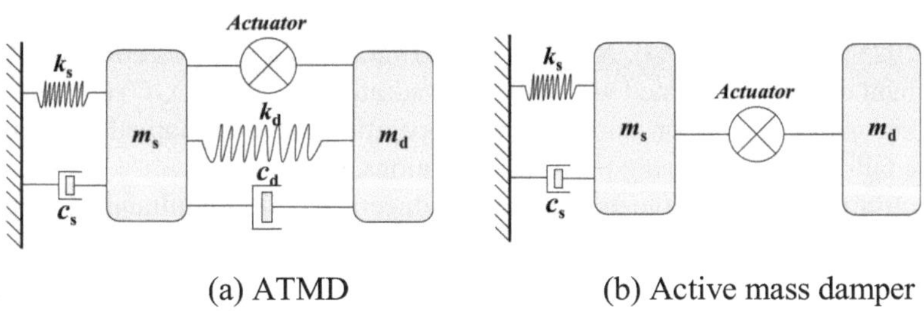

Fig. 5.10 Schematic of ATMD and active mass damper

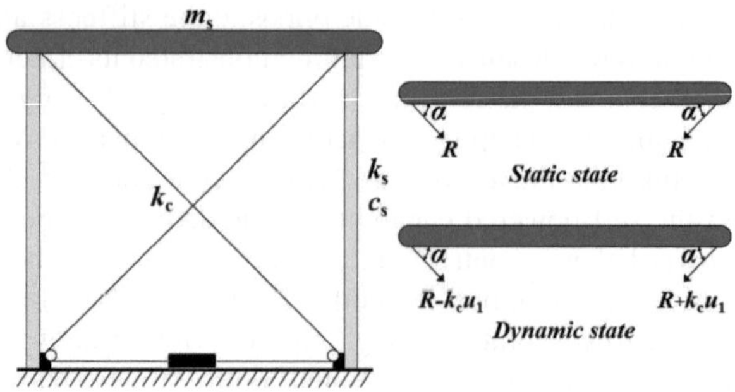

Fig. 5.11 Schematic of active tendon control (Nigdeli & Boduroğlu, 2013)

across blades, shafts, and towers and to increase power performance (Bernhammer et al., 2016; Nabhani et al., 2025).

The active trailing-edge flap is one of the mature applications of this technology. These components, either hinged or deformable, are typically installed along the outer length of a WT blade. Their primary function is to deflect in real time, allowing for immediate modulation of lift and drag forces (Bak et al., 2010; Pettas et al., 2016). By adjusting the flap angle in response to local aerodynamic conditions, active trailing-edge flaps effectively reduce the variability of blade-root bending moments. Full-scale field tests conducted on a Vestas V27 turbine have demonstrated their efficacy (Castaignet et al., 2014). When operated using model predictive control algorithms, the active trailing-edge flap achieved a 14% reduction in average flapwise fatigue loads and a 20% reduction in the amplitude of the 1P loading (Castaignet et al., 2014). Furthermore, by further integrating the active trailing-edge flaps with pitch control, lower damage equivalent loads by over 50% compared to using collective pitch control alone were reported (He et al., 2018). Beyond mechanical flap systems, active flow control approaches based on synthetic jets have also been explored (Nabhani et al., 2025).

Energy-Harvesting and Self-Powered Systems. The reliance of semi-active and active control systems on external power supplies introduces a critical vulnerability: During extreme events such as grid failures or seismic-induced blackouts, control capabilities may be compromised or entirely lost precisely when they are most needed. To address this, energy-harvesting and self-powered semi-active or active vibration control has emerged as a pivotal evolution in structural control (Cai et al., 2025c; Li et al., 2024). Xu et al. (2025) validated a self-powered electromagnetic shunt damper integrated with Linear Quadratic Gaussian (LQG) control, which maintains active regulation during normal operation and automatically reverts to a passive fail-safe mode during power interruptions.

Electromagnetic and piezoelectric transducers enable simultaneous vibration control and energy harvesting (Cai & Zhu, 2022; Huang & Yang, 2023). An electromagnetic transducer offers higher power density for large-scale applications. By replacing the conventional damping unit with an electromagnetic transducer, the control configuration becomes an energy-harvesting or self-powered counterpart.

Such a novel application traces its origins back to a US patent in the 1970s in automotive engineering and has since been extended to civil and other fields (Cai & Zhu, 2022). Jahangiri and Sun (2019) explored the feasibility of energy-harvesting bidirectional TMDs with an electromagnetic transducer as the damping unit for WTs and reported kilowatt power generation capacity, while the control performance remained promising. A similar observation was made by Elias and Beer (2024), who used multiple TMD-inerter with multiple electromagnetic transducers for WT vibration control. Note that they did not consider the energy harvesting circuit. With further incorporation of the electrical circuit, Cai et al. (2024) explored energy-harvesting bistable track NESs with a specifically designed circuit in a WT subjected to wind and wave loads. However, these studies are still limited to passive control. Cai et al. (2025c) proposed replacing the damping unit of a conventional TMD with semi-active electromagnetic damping, thereby constructing a self-powered semi-active TMD for WT vibration control. The vibration control and energy-harvesting performance of the novel device were evaluated under normal operation and emergency shutdown modes of the WT. Li et al. (2024) proposed the self-powered active mass damper for WT vibration control. The power-harvesting and dissipation modes were realized when the relative velocity of the electromagnetic transducer terminals generally was in phase with and opposite to the actuator control force, respectively. In addition to electromagnetic transducers, smart materials, such as dielectric elastomers, were explored for integration with TMDs in WTs to control vibrations and harvest energy simultaneously (Zhang et al., 2024a).

5.5 Intelligent Control Algorithms

The performance of the semi-active and active structural control systems described above depends on, to some extent, the effectiveness of the adopted control algorithms. Although high-authority actuators are capable of delivering the required control forces, advanced command strategies are essential to ensure structural stability and achieve the desired control performance. So far, many control algorithms have been proposed for structural vibration control. In this section, they are generally categorized into model-based and data-driven types.

Advanced Model-Based Control. The LQR remains one of the basic control algorithms. Its primary function is to minimize a cost function that weighs deviations in system states against the effort required for control. When combined with Kalman filters to form an LQG controller, this framework extends its capabilities to state estimation. The controller can then reconstruct structural responses that are not directly measured, even when sensor data are affected by noise (Rosół & Martynowicz, 2016). Experimental studies on a scaled flexible tower model have demonstrated its effectiveness in noise compensation, showing reliable vibration suppression. Recent advancements in this area have introduced wavelet analysis to update the controller's weighting matrices adaptively (Fitzgerald et al., 2018). It was reported that the wavelet controller for blade vibration control achieved a considerable reduction in out-of-plane responses compared to standard LQR or industry-standard proportional-integral controllers. The self-powered active mass damper in Li et al. (2024) also utilized an adaptive weight matrix to realize power balance. Although LQR and LQG controllers are mathematically optimal within their respective design frameworks, their reliance on linear system models is a notable weakness. WTs, in particular, exhibit significant nonlinear aerodynamic behavior during operation; linearized models at a specific operating point are generally employed. As a result, their performance can deteriorate or even become unstable when the system operates outside this linear region or when actual system parameters differ from those assumed in the model (Hur & Leithead, 2017). Instead of a Kalman filter, Hur and Leithead (2017) demonstrated that using a modified Luenberger observer, whose gain was chosen to minimize the effects of uncertainty, in the LQG algorithm can enhance robustness in control performance.

Also, to address robustness limitations, H∞ control synthesizes compensators that explicitly account for model uncertainties and external disturbances. In contrast to LQR methods, which focus on optimizing nominal performance, H∞ control aims to minimize the maximum gain from disturbance inputs to controlled outputs. This approach ensures system stability with respect to vibration or pitch control objectives when aerodynamic or structural parameters vary (Cong, 2019; Moradi & Vossoughi, 2015). In the field of vibration control, H∞ control has been effectively implemented for decentralized management of ATMDs. It addresses stroke constraints and actuator saturation while suppressing coupled vibrations between the tower and blades (Cong, 2019). However, the relatively conservative control gain might lead to slower control responses, and a trade-off exists between control performance and robustness. The selection of weighting functions in H∞ control design is, to some extent, experience-driven. In addition to the LQR, LQG, and H∞ control algorithms, Model Predictive Control (MPC) constitutes yet another important class within the family of model-based control algorithms. (Takács and Rohal'-Ilkiv, 2014). MPC is a receding horizon strategy in which, at each instant, the horizon shifts forward, and only the first control signal from the calculated sequence is applied (Camacho et al., 2026). Such an MPC algorithm has been applied to active tendons in WT blade flapwise vibration control (Staino and Basu, 2012).

Data-Driven and Neural Control. Model-based controllers generally rely on fixed rules established from pre-existing models. In this context, they may be susceptible to performance issues when operating conditions deviate from the original design assumptions. For instance, factors such as structural aging, ice accumulation on blades, or complex wake interactions can introduce discrepancies that model-based policies are not well equipped to handle. Accordingly, there is growing attention to learning-based architectures, in which the control policy is not explicitly programmed, but developed through data analysis and accumulated operational experience, enabling greater adaptability to changing real-world conditions.

Fuzzy logic control represents a significant departure from conventional mathematical modeling by substituting differential equations with linguistic rules designed to address uncertainty. This approach has proven especially effective for active vibration suppression, as it can approximate nonlinear sensor responses and account for actuator nonlinearities that compromise the performance of classical controllers (Qu et al., 2004). Chen et al. (2015) explored a semi-active controller that dynamically adjusts the voltage via fuzzy inference for the MR damper and reported success in reducing edgewise blade vibrations, even under severe wind conditions. In the area of pitch control, fuzzy Proportional–Integral–Derivative (PID) architectures have demonstrated greater adaptability than traditional PID controllers. By adjusting gains in real time, these systems were reported to achieve faster pitch-angle positioning and slightly higher power generation (Chavero-Navarrete et al., 2021).

Although fuzzy logic is effective at managing uncertainty, artificial neural networks offer a direct approach to modeling and controlling the complex, nonlinear aerodynamics of WTs. The frontier of this evolution is Deep Reinforcement Learning (DRL). In this approach, the controller, referred to as an agent, develops optimal strategies through trial-and-error interactions with the turbine environment. DRL addresses the temporal credit assignment problem by enabling the agent to take actions in the present, such as adjusting blade pitch, to mitigate anticipated load spikes several seconds later. In the context of floating WTs, DRL agents have achieved a 40% reduction in standard derivation of platform pitch displacement under extreme operating conditions (Zhang et al., 2022). The application of DRL extends beyond individual WTs. Multi-agent DRL frameworks have been used to coordinate pitch and yaw adjustments across entire wind farms. This coordination helped manage wake effects and enabled the simultaneous optimization of energy production and fatigue damage. This level of performance was beyond the reach of decentralized classical controllers (Padullaparthi et al., 2022). The adoption of learning-based control methods, however, presents several significant challenges; the foremost is the substantial computational cost of training (Huang & Zhao, 2025; Mei et al., 2025). Note that hybrid strategies that integrate RL-based control with a PID regulator and a learning observer have been reported to achieve promising learning convergence (Sierra-Garcia et al., 2022).

Digital Twin Integration. A digital twin, effectively bridging the gap between simulation and real-world operations (Leon-Medina et al., 2025), is likely to be integrated with the vibration control strategy. In vibration control, a digital twin is more than a 3D visualization. It serves as a bidirectional framework that continuously updates the structural state by integrating real-time sensor data with physics-based models (Ali et al., 2026; Yucesan & Viana, 2023). This approach enables virtual sensing, where mechanics-based model updating techniques, such as Bayesian inference, and reconstructing comprehensive vibration histories from limited accelerometer data. In accordance with that, digital twins can effectively compensate for missing or malfunctioning physical sensors. By quantifying uncertainties in these vibration estimates, the digital twin provides the feedback required for advanced controllers, particularly in challenging offshore environments where direct measurement is often not feasible (Moghadam & Nejad, 2022).

One of the most significant advancements in control engineering is the deployment of digital twins as training environments for artificial intelligence. Note that a digital twin allows for the exploration of advanced load-mitigation strategies that would be too risky to test on physical prototypes (Xia et al., 2021). In addition, multi-fidelity digital twin architectures support the training of multi-agent systems that manage fatigue across entire wind farms, balancing individual turbine loads with overall energy production (Padullaparthi et al., 2022). Moreover, digital twins enable health-aware control, a significant shift in which control objectives are dynamically adjusted based on the remaining useful life of the structure. By integrating real-time supervisory control and data acquisition system data with degradation models, the digital twin can predict specific failure modes in components, such as tower bases (Ali et al., 2025). This predictive capability can be incorporated directly into the control loop. For example, if a digital twin identifies accelerated fatigue in a blade root, the controller can automatically switch to a life-extension mode, reducing stress cycles and accepting a slight decrease in power output until maintenance is feasible (Sanchez et al., 2018). Recent research indicates that adaptive strategies based on digital twin-driven reliability indices can extend the operational lifespan of floating WTs by more than five years compared to standard control approaches (Zhang et al., 2025b). The integration of digital twins, therefore, transforms structural control systems from reactive damping mechanisms into proactive asset managers, optimizing the balance between energy production and structural integrity throughout the lifecycle.

5.6 Conclusions and Future Outlook

This chapter presents a review of the evolution of vibration control strategies for WTs. The discussion traces the progression from traditional passive control methods to intelligent vibration control systems. The synthesis of the literature establishes that passive systems continue to serve as the industry standard. Their appeal lies in their mechanical simplicity and inherent stability. However, these systems face inherent limitations, particularly in effective working bandwidth. As WTs increase in size, adopt multi-megawatt flexible designs, and move into floating/mountainous settings, passive control is insufficient for addressing the complex, nonlinear interactions between aerodynamic/hydrodynamic forces and structural dynamics. In response to these challenges, the field has shifted toward semi-active and active control systems.

However, the effectiveness of high-authority actuators/dampers depends entirely on the sophistication of the control logic. There has been a shift from model-based methods, such as LQR and $H\infty$ control, which often struggle with aerodynamic nonlinearity and uncertainty, toward adaptive and learning-based frameworks. The emergence of artificial intelligence, particularly DRL, has allowed controllers to develop optimal strategies through interaction rather than explicit programming. Additionally, the integration of digital twin technology has enhanced control by enabling virtual sensing and predictive health monitoring. This advancement enables the structural control system to transition from a reactive damping mechanism to a proactive asset manager, balancing power generation with the turbine's remaining useful life.

Looking forward, the following directions are desirable, with vibration control serving as an effective supplement to pitch control in WTs:

(1) Field tests of vibration control on WTs are still rare, and most rely on passive control methods, typically limited to TMD and TLD/TLCD. More field tests, especially of active vibration control strategies, are desirable after validations in theory, simulation, and lab tests. Note that in the current WT control framework, pitch and yaw controls are mostly established using active control mechanisms. The adoption of active vibration control for the WT tower or blades should not represent an insurmountable technical barrier.
(2) In the current WT control, most studies focus on mechanical control (i.e., conventional vibration control) or aerodynamic control (i.e., pitch control). In fact, hydrodynamic control using floating absorbers is a promising approach for floating WT vibration control. Note that in wind-wave combined structures, such as WT combined with wave energy converters, the wave energy converters should provide significant potential for simultaneous vibration control and wave energy harvesting. The dual-objective optimizations are desirable.
(3) In the current design, vibration control and pitch control are two separate control frameworks in WTs. An integrated control strategy that combines pitch control and active vibration control using a unified control algorithm provides enhanced vibration and power solutions for WTs. Furthermore, inspired by the above direction, the integrated aerodynamic-hydrodynamic-mechanical control shows potential for further enhancing performance in load reduction and vibration mitigation for floating WTs.
(4) Intelligent vibration control, including semi-active and active control, generally necessitates a power supply. A self-powered vibration control system should be developed to enable the intelligent vibration control modules to function as an independent control framework, remaining effective even when the WT is grid-disconnected or temporarily shut down.
(5) Intelligent vibration control using artificial intelligence algorithms should, to some extent, address the limitations of conventional semi-active and active control. Future research should emphasize physics-informed machine learning, ensuring that neural networks comply with fundamental dynamical laws and that artificial intelligence-driven control schemes remain physically consistent, stable, and safe for practical engineering applications.

Acknowledgements This work was supported by the National Key Research and Development Program of China (2024YFF0505400), the National Natural Science Foundation of China (52578596, U24A20177), and the China Scholarship Council (CSC) (2024GSP006413).

References

Ali, A., De Risi, R., & Sextos, A. (2020). Finite element modeling optimization of wind turbine blades from an earthquake engineering perspective. *Engineering Structures, 222*, Article 111105. https://doi.org/10.1016/j.engstruct.2020.111105

Ali, Y., Dai, K., Elgammal, A., Luo, Y., Heng, J., & Baniotopoulos, C. (2025). AI-driven digital twin for corrosion-fatigue management and opportunistic maintenance in offshore wind turbines. *Wind and Structures, 41*(5), 393.

Ali, Y., Elgammal, A., Heng, J., Dai, K., Saudi, G. N., & Salem, H. (2026). Insights into corrosion-fatigue deterioration in offshore wind turbine structures through digital twin applications. *Ocean Engineering, 343*, Article 123514. https://doi.org/10.1016/j.oceaneng.2025.123514

Alotta, G., Laface, V., Failla, G., Ruzzo, C., & Arena, F. (2023). A novel concept of floating absorber for motion mitigation in floating offshore wind turbines. *Engineering Structures, 294*, Article 116554. https://doi.org/10.1016/j.engstruct.2023.116554

Bak, C., Gaunaa, M., Andersen, P. B., Buhl, T., Hansen, P., & Clemmensen, K. (2010). Wind tunnel test on airfoil Risø-B1-18 with an active trailing edge flap. *Wind Energy, 13*(2–3), 207–219. https://doi.org/10.1002/we.369

Bargi, K., Dezvareh, R., & Mousavi, S. A. (2016). Contribution of tuned liquid column gas dampers to the performance of offshore wind turbines under wind, wave, and seismic excitations. *Earthquake Engineering and Engineering Vibration, 15*(3), 551–561. https://doi.org/10.1007/s11803-016-0343-z

Barlas, T. K., & Van Kuik, G. A. M. (2010). Review of state of the art in smart rotor control research for wind turbines. *Progress in Aerospace Sciences, 46*(1), 1–27. https://doi.org/10.1016/j.paerosci.2009.08.002

Benbouzid, M., Berghout, T., Sarma, N., Djurović, S., Wu, Y., & Ma, X. (2021). Intelligent Condition Monitoring of Wind Power Systems: State of the Art Review. *Energies, 14*(18), 5967. https://doi.org/10.3390/en14185967

Bernhammer, L. O., Van Kuik, G. A. M., & De Breuker, R. (2016). Fatigue and extreme load reduction of wind turbine components using smart rotors. *Journal of Wind Engineering and Industrial Aerodynamics, 154*, 84–95. https://doi.org/10.1016/j.jweia.2016.04.001

Bhattacharyya, S., Ghosh, A. (Dey), & Basu, B. (2017). Nonlinear modeling and validation of air spring effects in a sealed tuned liquid column damper for structural control. *Journal of Sound and Vibration, 410*, 269–286. https://doi.org/10.1016/j.jsv.2017.07.046

Brodersen, M. L., Bjørke, A.-S., & Høgsberg, J. (2017). Active tuned mass damper for damping of offshore wind turbine vibrations: Active tuned mass damper for damping of offshore wind turbine vibrations. *Wind Energy, 20*(5), 783–796. https://doi.org/10.1002/we.2063

Cai, Q., Dai, K., & Zhu, S. (2025a). Double-mass pendulum energy harvester with tunable ultra-low frequency: An experimental study. *Energy, 329*, Article 136738. https://doi.org/10.1016/j.energy.2025.136738

Cai, Q., Dai, K., & Zhu, S. (2025b). Wave energy converter-inspired ultra-low-frequency double-mass pendulum damper for vibration control of offshore wind turbines. *Ocean Engineering, 340*, Article 122213. https://doi.org/10.1016/j.oceaneng.2025.122213

Cai, Q., Hua, Y., Zhu, S., Zhang, X., & Zuo, H. (2024). Bistable energy-harvesting track nonlinear energy sink in offshore wind turbines. *Mechanical Systems and Signal Processing, 215*, Article 111407. https://doi.org/10.1016/j.ymssp.2024.111407

Cai, Q., Wang, J., Dai, K., Cheng, Y., Li, Y., Ma, W., & Sun, Z. (2025c). Vibration reduction of offshore wind turbines using self-powered-feasible semi-active tuned mass damper. *Ocean Engineering, 318*, Article 120182. https://doi.org/10.1016/j.oceaneng.2024.120182

Cai, Q., & Zhu, S. (2021). Applying double-mass pendulum oscillator with tunable ultra-low frequency in wave energy converters. *Applied Energy, 298*, Article 117228. https://doi.org/10.1016/j.apenergy.2021.117228

Cai, Q., & Zhu, S. (2022). The nexus between vibration-based energy harvesting and structural vibration control: A comprehensive review. *Renewable and Sustainable Energy Reviews, 155*, Article 111920. https://doi.org/10.1016/j.rser.2021.111920

Cai, Q., & Zhu, S. (2024). Nonlinear double-mass pendulum for vibration-based energy harvesting. *Nonlinear Dynamics, 112*(7), 5109–5128. https://doi.org/10.1007/s11071-023-09236-6

Camacho, E. F., Bordons, C., & Maestre, J. M. (2026). Introduction to model predictive control. In E. F. Camacho, C. Bordons, & J. M. Maestre (Eds), *Model Predictive Control* (pp. 1–23). Springer Nature Switzerland. https://doi.org/10.1007/978-3-031-87596-0_1

Castaignet, D., Barlas, T., Buhl, T., Poulsen, N. K., Wedel-Heinen, J. J., Olesen, N. A., Bak, C., & Kim, T. (2014). Full-scale test of trailing edge flaps on a Vestas V27 wind turbine: Active load reduction and system identification. *Wind Energy, 17*(4), 549–564. https://doi.org/10.1002/we.1589

Caterino, N. (2015). Semi-active control of a wind turbine via magnetorheological dampers. *Journal of Sound and Vibration, 345*, 1–17. https://doi.org/10.1016/j.jsv.2015.01.022

Chavero-Navarrete, E., Trejo-Perea, M., Jáuregui-Correa, J. C., Carrillo-Serrano, R. V., Ronquillo-Lomeli, G., & Ríos-Moreno, J. G. (2021). Pitch angle optimization for small wind turbines based on a hierarchical fuzzy-PID controller and anticipated wind speed measurement. *Applied Sciences, 11*(4), 1683. https://doi.org/10.3390/app11041683

Chen, C., & Duffour, P. (2018). Modelling damping sources in monopile-supported offshore wind turbines. *Wind Energy, 21*(11), 1121–1140. https://doi.org/10.1002/we.2218

Chen, D., Huang, S., Huang, C., Liu, R., & Ouyang, F. (2021). Passive control of jacket–type offshore wind turbine vibrations by single and multiple tuned mass dampers. *Marine Structures, 77*, Article 102938. https://doi.org/10.1016/j.marstruc.2021.102938

Chen, J., Yuan, C., Li, J., & Xu, Q. (2015). Semi-active fuzzy control of edgewise vibrations in wind turbine blades under extreme wind. *Journal of Wind Engineering and Industrial Aerodynamics, 147*, 251–261. https://doi.org/10.1016/j.jweia.2015.10.012

Chen, M., Li, S., Sun, Y., Zhao, Y., & Wang, R. (2025). MODWT-based multiscale wavelet LQR strategy and active-inerter tuned mass damper for structural vibration control of floating offshore wind turbines. *Ocean Engineering, 336*, Article 121678. https://doi.org/10.1016/j.oceaneng.2025.121678

Cong, C. (2019). Using active tuned mass dampers with constrained stroke to simultaneously control vibrations in wind turbine blades and tower. *Advances in Structural Engineering, 22*(7), 1544–1553. https://doi.org/10.1177/1369433218817892

Dai, K., Tang, J., & Zhang, S. (2022). An integrated equivalent tuned-mass-inerter vibration absorber and its optimal design. *Structural Control and Health Monitoring, 29*(11). https://doi.org/10.1002/stc.3089

Dai, K., Fang, C., Zhang, S., & Shi, Y. (2021). Conceptual design and numerical study on a cable-based energy dissipating system for the vibration reduction of tower-like structures. *Engineering Structures, 237*, Article 112034. https://doi.org/10.1016/j.engstruct.2021.112034

Dai, K., Jiang, Z., Fang, C., Li, P., & Zhang, S. (2023). A tuned cable-inerter system for vibration reduction of towers. *International Journal of Mechanical Sciences, 248*, Article 108199. https://doi.org/10.1016/j.ijmecsci.2023.108199

Dezvareh, R., & Nazokkar, A. (2025). Enhancing dynamic performance of OC4-DeepCwind semi-submersible floating wind turbine utilizing multi-level semi-active dampers. *Arabian Journal for Science and Engineering*. https://doi.org/10.1007/s13369-025-10507-0

Ding, H., Altay, O., & Wang, J.-T. (2023). Lateral vibration control of monopile supported offshore wind turbines with toroidal tuned liquid column dampers. *Engineering Structures, 286*, Article 116107. https://doi.org/10.1016/j.engstruct.2023.116107

Ding, H., Chen, Y.-N., Wang, J.-T., & Altay, O. (2022). Numerical analysis of passive toroidal tuned liquid column dampers for the vibration control of monopile wind turbines using FVM and FEM. *Ocean Engineering, 247*, Article 110637. https://doi.org/10.1016/j.oceaneng.2022.110637

Ding, H., Ma, R., Wang, J., Bi, K., Fang, X., & Song, J. (2025). Using tuned liquid column dampers (TLCDs) for heave motion mitigation of very large floating structures (VLFSs). *Ocean Engineering, 321*, Article 120391. https://doi.org/10.1016/j.oceaneng.2025.120391

Dong, X., Jia, Y., Li, Y., & Ren, S. (2026). The theoretical analysis and improvement on the vibration reduction effect of particle damping-tuned mass damper for offshore wind turbine structure. *Structures, 84*, Article 111014. https://doi.org/10.1016/j.istruc.2025.111014

Download GWEC's Global Wind Report 2025 | windfair. (n.d.). Retrieved 14 January 2026, from https://w3.windfair.net/wind-energy/news/47450-download-gwec-s-global-wind-report-2025

Elias, S., & Beer, M. (2024). Vibration control and energy harvesting of offshore wind turbines installed with TMDI under dynamical loading. *Engineering Structures, 315*, Article 118459. https://doi.org/10.1016/j.engstruct.2024.118459

Elkodama, A., Ismaiel, A., Abdellatif, A., Shaaban, S., Yoshida, S., & Rushdi, M. A. (2023). Control methods for horizontal axis wind turbines (HAWT): State-of-the-art review. *Energies, 16*(17), 6394. https://doi.org/10.3390/en16176394

Emami, M., Dezvareh, R., & Mousavi, S. A. (2022). Contribution of fluid viscous dampers on fatigue life of lattice-type offshore wind turbines. *Ocean Engineering, 245*, Article 110506. https://doi.org/10.1016/j.oceaneng.2021.110506

Emami, M., Nasab, V. H., Akar, S., & Batako, A. (2023). Experimental investigation into the effect of magnetorheological fluid damper on vibration and chatter in straight turning process. *Journal of Manufacturing Processes, 99*, 825–847. https://doi.org/10.1016/j.jmapro.2023.05.056

Enevoldsen, I., & Mørk, K. J. (1996). Effects of a vibration mass damper in a wind turbine tower. *Mechanics of Structures and Machines, 24*(2), 155–187. https://doi.org/10.1080/08905459608905260

Farahpour, H., & Hejazi, F. (2023). Development of integrated semi-active adaptive vibration control system for bridges subjected to traffic loads. *Structures, 51*, 1773–1794. https://doi.org/10.1016/j.istruc.2023.03.107

Fitzgerald, B., Sarkar, S., & Staino, A. (2018). Improved reliability of wind turbine towers with active tuned mass dampers (ATMDs). *Journal of Sound and Vibration, 419*, 103–122. https://doi.org/10.1016/j.jsv.2017.12.026

Fujino, Y., & Sun, L. M. (1993). Vibration control by multiple tuned liquid dampers (MTLDs). *Journal of Structural Engineering, 119*(12), 3482–3502. https://doi.org/10.1061/(ASCE)0733-9445(1993)119:12(3482)

Gao, Y., Zhai, E., Li, S., Zhang, Z., Xu, Z., Zhang, G., Racic, V., Chen, J., Wang, L., & Zhang, Z. (2024). Integrated design and real-world application of a tuned mass damper (TMD) with displacement constraints for large offshore monopile wind turbines. *Ocean Engineering, 292*, Article 116568. https://doi.org/10.1016/j.oceaneng.2023.116568

Ghaffarzadeh, H., & Younespour, A. (2014). Active tendons control of structures using block pulse functions. *Structural Control and Health Monitoring, 21*(12), 1453–1464. https://doi.org/10.1002/stc.1656

Ghane, M., Rasekhi Nejad, A., Blanke, M., Gao, Z., & Moan, T. (2018). Condition monitoring of spar-type floating wind turbine drivetrain using statistical fault diagnosis. *Wind Energy, 21*(7), 575–589. https://doi.org/10.1002/we.2179

Ha, M., & Cheong, C. (2016). Pitch motion mitigation of spar-type floating substructure for offshore wind turbine using multilayer tuned liquid damper. *Ocean Engineering, 116*, 157–164. https://doi.org/10.1016/j.oceaneng.2016.02.036

Han, D., Li, X., Wang, W., & Su, X. (2025). Vibration suppression performance comparison of floating offshore wind turbines using a tuned liquid column damper and a tuned mass damper. *Ocean Engineering, 330*, Article 121285. https://doi.org/10.1016/j.oceaneng.2025.121285

He, K., Qi, L., Zheng, L., & Chen, Y. (2018). Combined pitch and trailing edge flap control for load mitigation of wind turbines. *Energies, 11*(10), 2519. https://doi.org/10.3390/en11102519

Hemmati, A., Oterkus, E., & Barltrop, N. (2019a). Fragility reduction of offshore wind turbines using tuned liquid column dampers. *Soil Dynamics and Earthquake Engineering, 125*, Article 105705. https://doi.org/10.1016/j.soildyn.2019.105705

Hemmati, A., Oterkus, E., & Khorasanchi, M. (2019b). Vibration suppression of offshore wind turbine foundations using tuned liquid column dampers and tuned mass dampers. *Ocean Engineering, 172*, 286–295. https://doi.org/10.1016/j.oceaneng.2018.11.055

Hochrainer, M. J. (2015). Active tuned liquid column gas damper in structural control. In J. Caicedo & S. Pakzad (Eds.), *Dynamics of civil structures* (Vol. 2, pp. 467–473). Springer International Publishing.

Hochrainer, M. J., & Ziegler, F. (2006). Control of tall building vibrations by sealed tuned liquid column dampers. *Structural Control and Health Monitoring, 13*(6), 980–1002. https://doi.org/10.1002/stc.90

Holierhoek, J. G. (2023). Aeroelastic design of wind turbine blades. In *Advances in Wind Turbine Blade Design and Materials* (pp. 169–189). Elsevier. https://doi.org/10.1016/B978-0-08-103007-3.00010-0

Hosseini Lavassani, S. H., Ebadijalal, M., Shahrouzi, M., Gharehbaghi, V., Noroozinejad Farsangi, E., & Yang, T. Y. (2022). Interpretation of simultaneously optimized fuzzy controller and active tuned mass damper parameters under Pulse-type ground motions. *Engineering Structures, 261*, Article 114286. https://doi.org/10.1016/j.engstruct.2022.114286

Housner, G. W., Bergman, L. A., Caughey, T. K., Chassiakos, A. G., Claus, R. O., Masri, S. F., Skelton, R. E., Soong, T. T., Spencer, B. F., & Yao, J. T. P. (1997). Structural control: Past, present, and future. *Journal of Engineering Mechanics, 123*(9), 897–971. https://doi.org/10.1061/(ASCE)0733-9399(1997)123:9(897)

Hu, Y., & He, E. (2017). Active structural control of a floating wind turbine with a stroke-limited hybrid mass damper. *Journal of Sound and Vibration, 410*, 447–472. https://doi.org/10.1016/j.jsv.2017.08.050

Huang, X., & Yang, B. (2023). Towards novel energy shunt inspired vibration suppression techniques: Principles, designs and applications. *Mechanical Systems and Signal Processing, 182*, Article 109496. https://doi.org/10.1016/j.ymssp.2022.109496

Huang, Y., & Zhao, X. (2025). Wind farm control via offline reinforcement learning with adversarial training. *IEEE Transactions on Automation Science and Engineering, 22*, 12845–12856. https://doi.org/10.1109/TASE.2025.3548565

Hur, S., & Leithead, W. E. (2017). Model predictive and linear quadratic Gaussian control of a wind turbine. *Optimal Control Applications and Methods, 38*(1), 88–111. https://doi.org/10.1002/oca.2244

ISO 20816–21:2025. (n.d.). ISO. Retrieved 26 January 2026, from https://www.iso.org/standard/84280.html

Jahangiri, V., & Sun, C. (2019). Integrated bi-directional vibration control and energy harvesting of monopile offshore wind turbines. *Ocean Engineering, 178*, 260–269. https://doi.org/10.1016/j.oceaneng.2019.02.015

Jahangiri, V., & Sun, C. (2020). Three-dimensional vibration control of offshore floating wind turbines using multiple tuned mass dampers. *Ocean Engineering, 206*, Article 107196. https://doi.org/10.1016/j.oceaneng.2020.107196

Jahangiri, V., & Sun, C. (2022). A novel three dimensional nonlinear tuned mass damper and its application in floating offshore wind turbines. *Ocean Engineering, 250*, Article 110703. https://doi.org/10.1016/j.oceaneng.2022.110703

Jahani, K., Langlois, R. G., & Afagh, F. F. (2022). Structural dynamics of offshore wind turbines: A review. *Ocean Engineering, 251*, Article 111136. https://doi.org/10.1016/j.oceaneng.2022.111136

Jiang, J., Dong, X., Lian, J., & Jia, Y. (2024a). Research on semi-active vibration reduction of offshore wind turbine structure combined eddy current theory with tuned mass dampers. *Renewable Energy, 234*, Article 121185. https://doi.org/10.1016/j.renene.2024.121185

Jiang, Z., Tang, J., Dai, K., Fang, C., & Luo, Y. (2024b). A tuned cable-inerter system for wind turbine blades vibration suppression. *International Journal of Mechanical Sciences, 269*, Article 109030. https://doi.org/10.1016/j.ijmecsci.2024.109030

Jonkman, J., Butterfield, S., Musial, W., & Scott, G. (2009). *Definition of a 5-MW Reference Wind Turbine for Offshore System Development* (NREL/TP-500–38060, 947422; p. NREL/TP-500–38060, 947422). https://doi.org/10.2172/947422

Jonkman, J., & Musial, W. (2010). *Offshore Code Comparison Collaboration (OC3) for IEA Wind Task 23 Offshore Wind Technology and Deployment* (NREL/TP-5000–48191, 1004009; p. NREL/TP-5000–48191, 1004009). https://doi.org/10.2172/1004009

Karakalas, A. A., Manolas, D. I., Machairas, T. T., Riziotis, V. A., & Saravanos, D. A. (2019). Active load alleviation potential of adaptive wind turbine blades using shape memory alloy actuators. *Wind Energy, 22*(5), 620–637. https://doi.org/10.1002/we.2311

Lackner, M. A., & Rotea, M. A. (2011). Structural control of floating wind turbines. *Mechatronics, 21*(4), 704–719. https://doi.org/10.1016/j.mechatronics.2010.11.007

Latif, M., Ambreen, H., Diyin, T., Hassan, F., Imran, M., Hussain Abbasi, S., & Imran, M. (2025). Development of a novel teal-time weighted surrogate model for wind turbine-based on DFIG: Enhancing computational efficiency in power systems with error bound. *IEEE Access, 13*, 50000–50016. https://doi.org/10.1109/ACCESS.2025.3550276

Leng, D., Wang, R., Yang, Y., Li, Y., & Liu, G. (2023). Study on a three-dimensional variable-stiffness TMD for mitigating bi-directional vibration of monopile offshore wind turbines. *Ocean Engineering, 281*, Article 114791. https://doi.org/10.1016/j.oceaneng.2023.114791

Leng, D., Yang, Y., Xu, K., Li, Y., Liu, G., Tian, X., & Xie, Y. (2021). Vibration control of offshore wind turbine under multiple hazards using single variable-stiffness tuned mass damper. *Ocean Engineering, 236*, Article 109473. https://doi.org/10.1016/j.oceaneng.2021.109473

Leon-Medina, J. X., Tibaduiza, D. A., Parés, N., & Pozo, F. (2025). Digital twin technology in wind turbine components: A review. *Intelligent Systems with Applications, 26*, Article 200535. https://doi.org/10.1016/j.iswa.2025.200535

Lesieutre, G. A., Rusovici, R., Koopmann, G. H., & Dosch, J. J. (2003). Modelling and characterization of a piezoceramic inertial actuator. *Journal of Sound and Vibration, 261*(1), 93–107. https://doi.org/10.1016/S0022-460X(02)00943-4

Li, S., Yang, Y., Zhao, Y., & Wang R. (2025). Research on semi-active control of floating offshore wind turbine using MRE-PTMD with collision. *Journal of Vibration and Shock, 44*(16), 1–11.(in Chinese) https://jvs.sjtu.edu.cn/EN/Y2025/V44/I16/1

Li, J.-Y., & Zhu, S. (2018). Versatile behaviors of electromagnetic shunt damper with a negative impedance converter. *IEEE/ASME Transactions on Mechatronics, 23*(3), 1415–1424. https://doi.org/10.1109/TMECH.2018.2813307

Li, J.-Y., Zhu, S., Zhang, J., Ma, R., & Zuo, H. (2024). Vibration control of offshore wind turbines with a novel energy-adaptive self-powered active mass damper. *Engineering Structures, 302*, Article 117450. https://doi.org/10.1016/j.engstruct.2024.117450

Li, L., Liu, Y., Yuan, Z., & Gao, Y. (2019). Dynamic and structural performances of offshore floating wind turbines in turbulent wind flow. *Ocean Engineering, 179*, 92–103. https://doi.org/10.1016/j.oceaneng.2019.03.028

Li, T., Liu, Z., Liu, S., Fan, Y., Yang, Q., & Xiao, H. (2022). Numerical study on passive structural control of semi-submersible floating wind turbine considering non-collinear wind and waves. *Ocean Engineering, 266*, Article 112745. https://doi.org/10.1016/j.oceaneng.2022.112745

Lin, G.-L., Lu, L.-Y., Lei, K.-T., Liu, K.-Y., Ko, Y.-Y., & Ju, S.-H. (2021). Experimental study on seismic vibration control of an offshore wind turbine with TMD considering soil liquefaction effect. *Marine Structures, 77*, Article 102961. https://doi.org/10.1016/j.marstruc.2021.102961

Liu, X., Lu, C., Li, G., Godbole, A., & Chen, Y. (2017). Effects of aerodynamic damping on the tower load of offshore horizontal axis wind turbines. *Applied Energy, 204*, 1101–1114. https://doi.org/10.1016/j.apenergy.2017.05.024

Liu, Y., Waters, T. P., & Brennan, M. J. (2005). A comparison of semi-active damping control strategies for vibration isolation of harmonic disturbances. *Journal of Sound and Vibration, 280*(1–2), 21–39. https://doi.org/10.1016/j.jsv.2003.11.048

Love, J. S., Tait, M. J., & Toopchi-Nezhad, H. (2011). A hybrid structural control system using a tuned liquid damper to reduce the wind induced motion of a base isolated structure. *Engineering Structures, 33*(3), 738–746. https://doi.org/10.1016/j.engstruct.2010.11.027

Lu, L.-Y. (2004). Semi-active modal control for seismic structures with variable friction dampers. *Engineering Structures, 26*(4), 437–454. https://doi.org/10.1016/j.engstruct.2003.10.012

Lu, Y., Chen, H., He, Y., Li, J., Guo, T., Chen, X., Yuan, J., & Zhang, Z. (2025). Shaking table test and simulation study on CS-NES vibration control of offshore wind turbine. *Soil Dynamics and Earthquake Engineering, 197*, Article 109548. https://doi.org/10.1016/j.soildyn.2025.109548

Lu, Z., Zhao, S., Ma, C., & Dai, K. (2023). Experimental and analytical study on the performance of wind turbine tower attached with particle tuned mass damper. *Engineering Structures, 294*, Article 116784. https://doi.org/10.1016/j.engstruct.2023.116784

Luo, M., Jin, X., Yue, Y., & Chen, Y. (2024). Bidirectional vibration control for fully-coupled floating offshore wind turbines. *Ocean Engineering, 292*, Article 116523. https://doi.org/10.1016/j.oceaneng.2023.116523

Ma, R., Jiang, S., Han, C., Bi, K., & Du, X. (2025). Development, testing and characterization of a novel rotational inertia sand damper (RISD) for structural vibration control. *Mechanical Systems and Signal Processing, 225*, Article 112241. https://doi.org/10.1016/j.ymssp.2024.112241

Mani, Y., Veeraragu, J., Sangameshwar, S., & Rangaswamy, R. (2020). Dynamic behavior of smart material embedded wind turbine blade under actuated condition. *Wind and Structures, 30*(2), 211–217. https://doi.org/10.12989/WAS.2020.30.2.211

Martynowicz, P., & Santos, M. (2020). Structural vibration control of NREL 5.0 MW FOWT using optimal-based MR tuned vibration absorber. *IFAC-PapersOnLine, 53*(2), 12378–12383. https://doi.org/10.1016/j.ifacol.2020.12.1253

McMorland, J., Collu, M., McMillan, D., & Carroll, J. (2022). Operation and maintenance for floating wind turbines: A review. *Renewable and Sustainable Energy Reviews, 163*, Article 112499. https://doi.org/10.1016/j.rser.2022.112499

McNamara, D., Pandit, A., & Malekjafarian, A. (2024). Optimized design of multiple tuned mass dampers for vibration control of offshore wind turbines. *Ocean Engineering, 305*, Article 117912. https://doi.org/10.1016/j.oceaneng.2024.117912

Mei, M., Kou, P., Xu, Y., Zhang, Z., Tian, R., & Liang, D. (2025). Improving floating offshore wind farm flow control with scalable model-based deep reinforcement learning. *IEEE Transactions on Automation Science and Engineering, 22*, 18255–18268. https://doi.org/10.1109/TASE.2025.3585016

Moghadam, F. K., & Nejad, A. R. (2022). Online condition monitoring of floating wind turbines drivetrain by means of digital twin. *Mechanical Systems and Signal Processing, 162*, Article 108087. https://doi.org/10.1016/j.ymssp.2021.108087

Moradi, H., & Vossoughi, G. (2015). Robust control of the variable speed wind turbines in the presence of uncertainties: A comparison between H∞ and PID controllers. *Energy, 90*, 1508–1521. https://doi.org/10.1016/j.energy.2015.06.100

Nabhani, A., Tousi, N. M., Coma, M., Bugeda, G., & Bergadà, J. M. (2025). Large-scale horizontal axis wind turbine aerodynamic efficiency optimization using active flow control and synthetic jets. *Energy, 319*, Article 134940. https://doi.org/10.1016/j.energy.2025.134940

Nazokkar, A., & Dezvareh, R. (2022). Vibration control of floating offshore wind turbine using semi-active liquid column gas damper. *Ocean Engineering, 265*, Article 112574. https://doi.org/10.1016/j.oceaneng.2022.112574

Nicoletti, R., & Liebich, R. (2018). Analysis of long wind turbine blades with shape memory alloy wires in super-elastic phase. *Journal of Intelligent Material Systems and Structures, 29*(15), 3108–3123. https://doi.org/10.1177/1045389X18783078

Nigdeli, S. M., & Boduroğlu, M. H. (2013). Active tendon control of torsionally irregular structures under near-fault ground motion excitation. *Computer-Aided Civil and Infrastructure Engineering, 28*(9), 718–736. https://doi.org/10.1111/mice.12046

Ormondroyd, J., & Den Hartog, J. P. (1928). The theory of the dynamic vibration absorber. *Transactions of the American Society of Mechanical Engineers, 49–50*(2), Article 021007. https://doi.org/10.1115/1.4058553

Oueslati, F., & Sankar, S. (1994). A class of semi-active suspension schemes for vehicle vibration control. *Journal of Sound and Vibration, 172*(3), 391–411. https://doi.org/10.1006/jsvi.1994.1182

Padullaparthi, V. R., Nagarathinam, S., Vasan, A., Menon, V., & Sudarsanam, D. (2022). FALCON-FArm Level CONtrol for wind turbines using multi-agent deep reinforcement learning. *Renewable Energy, 181*, 445–456. https://doi.org/10.1016/j.renene.2021.09.023

Park, S., Lackner, M. A., Pourazarm, P., Rodríguez Tsouroukdissian, A., & Cross-Whiter, J. (2019). An investigation on the impacts of passive and semiactive structural control on a fixed bottom and a floating offshore wind turbine. *Wind Energy, 22*(11), 1451–1471. https://doi.org/10.1002/we.2381

Pettas, V., Barlas, T., Gertz, D., & Madsen, H. A. (2016). Power performance optimization and loads alleviation with active flaps using individual flap control. *Journal of Physics: Conference Series, 749*, Article 012010. https://doi.org/10.1088/1742-6596/749/1/012010

Qiu, Z., Han, J., Zhang, X., Wang, Y., & Wu, Z. (2009). Active vibration control of a flexible beam using a non-collocated acceleration sensor and piezoelectric patch actuator. *Journal of Sound and Vibration, 326*(3–5), 438–455. https://doi.org/10.1016/j.jsv.2009.05.034

Qu, W., Sun, J., & Qiu, Y. (2004). Active control of vibration using a fuzzy control method. *Journal of Sound and Vibration, 275*(3–5), 917–930. https://doi.org/10.1016/S0022-460X(03)00795-8

Ren, Y., Li, L., Brindley, J., & Jiang, L. (2016). Nonlinear PI control for variable pitch wind turbine. *Control Engineering Practice, 50*, 84–94. https://doi.org/10.1016/j.conengprac.2016.02.004

Rezaee, M., & Aly, A. M. (2018). Vibration control in wind turbines to achieve desired system-level performance under single and multiple hazard loadings. *Structural Control and Health Monitoring, e2261*(12), 1–31. https://doi.org/10.1002/stc.2261

Rosół, M., & Martynowicz, P. (2016). Implementation of the LQG controller for a wind turbine tower-nacelle model with an mr tuned vibration absorber. *Journal of Theoretical and Applied Mechanics*, 1109. https://doi.org/10.15632/jtam-pl.54.4.1109

Sanchez, H. E., Escobet, T., Puig, V., & Odgaard, P. F. (2018). Health-aware model predictive control of wind turbines using fatigue prognosis. *International Journal of Adaptive Control and Signal Processing, 32*(4), 614–627. https://doi.org/10.1002/acs.2784

Sarkar, S., & Chakraborty, A. (2018). Optimal design of semiactive MR-TLCD for along-wind vibration control of horizontal axis wind turbine tower. *Structural Control and Health Monitoring, 25*(2), Article e2083. https://doi.org/10.1002/stc.2083

Sarkar, S., & Chakraborty, A. (2019). Development of semi-active vibration control strategy for horizontal axis wind turbine tower using multiple magneto-rheological tuned liquid column dampers. *Journal of Sound and Vibration, 457*, 15–36. https://doi.org/10.1016/j.jsv.2019.05.052

Sarkar, S., Fitzgerald, B., & Basu, B. (2021). Individual blade pitch control of floating offshore wind turbines for load mitigation and power regulation. *IEEE Transactions on Control Systems Technology, 29*(1), 305–315. https://doi.org/10.1109/TCST.2020.2975148

Shelley, S. J., Lee, K. L., Aksel, T., & Aktan, A. E. (1995). Active-control and forced-vibration studies on highway bridge. *Journal of Structural Engineering, 121*(9), 1306–1312. https://doi.org/10.1061/(ASCE)0733-9445(1995)121:9(1306)

Shen, W., Hu, Y., Zhu, H., Li, Y., & Luo, H. (2021). Performance enhancement in cable vibration energy harvesting employing inerters: Full-scale experiment. *Structural Control and Health Monitoring, 28*(7). https://doi.org/10.1002/stc.2740

Shi, B., & Zhu, Q. (2025). Uncertainty quantification of dynamic response in wind turbines with a geometric nonlinear tuned mass-damper-inerter. *Ocean Engineering, 337*, Article 121849. https://doi.org/10.1016/j.oceaneng.2025.121849

Sierra-Garcia, J. E., Santos, M., & Pandit, R. (2022). Wind turbine pitch reinforcement learning control improved by PID regulator and learning observer. *Engineering Applications of Artificial Intelligence, 111*, Article 104769. https://doi.org/10.1016/j.engappai.2022.104769

Smith, M. C. & Fu-Cheng Wang. (2003). Performance benefits in passive vehicle suspensions employing inerters. *42nd IEEE International Conference on Decision and Control (IEEE Cat. No.03CH37475), 3*, 2258–2263. https://doi.org/10.1109/CDC.2003.1272954

Soong, T. T., & Spencer, B. F. (2002). Supplemental energy dissipation: State-of-the-art and state-of-the-practice. *Engineering Structures, 24*(3), 243–259. https://doi.org/10.1016/S0141-0296(01)00092-X

Sorge, E., Riascos, C., Caterino, N., Demartino, C., & Georgakis, C. T. (2024). Optimal design of a hinge-spring-friction device for enhancing wind induced structural response of onshore wind turbines. *Engineering Structures, 314*, Article 118305. https://doi.org/10.1016/j.engstruct.2024.118305

Spencer, B. F., Dyke, S. J., Sain, M. K., & Carlson, J. D. (1997). Phenomenological model for magnetorheological dampers. *Journal of Engineering Mechanics, 123*(3), 230–238. https://doi.org/10.1061/(ASCE)0733-9399(1997)123:3(230)

Staino, A., Basu, B., & Nielsen, S. R. K. (2012). Actuator control of edgewise vibrations in wind turbine blades. *Journal of Sound and Vibration, 331*(6), 1233–1256. https://doi.org/10.1016/j.jsv.2011.11.003

Sun, C. (2018a). Semi-active control of monopile offshore wind turbines under multi-hazards. *Mechanical Systems and Signal Processing, 99*, 285–305. https://doi.org/10.1016/j.ymssp.2017.06.016

Sun, C. (2018b). Mitigation of offshore wind turbine responses under wind and wave loading: Considering soil effects and damage. *Structural Control and Health Monitoring, 25*(3), Article e2117. https://doi.org/10.1002/stc.2117

Sun, C., & Jahangiri, V. J. M. S. (2018). Bi-directional vibration control of offshore wind turbines using a 3D pendulum tuned mass damper. *Mechanical Systems and Signal Processing, 105*, 338–360. https://doi.org/10.1016/j.ymssp.2017.12.011

Sun, C., & Jahangiri, V. (2019). Fatigue damage mitigation of offshore wind turbines under real wind and wave conditions. *Engineering Structures, 178*, 472–483. https://doi.org/10.1016/j.engstruct.2018.10.053

Takács, G., & Rohal'-Ilkiv, B. (2014). Model predictive control algorithms for active vibration control: A study on timing, performance and implementation properties. *Journal of Vibration and Control, 20*(13), 2061–2080. https://doi.org/10.1177/1077546313479993

Tang, J., Cai, Q., Dai, K., Liu, Y., Heng, J., & Luo, Y. (2025). A cable-based tuned inerter damper for edgewise vibration control of the wind turbine blade. *Engineering Structures, 323*, Article 119248. https://doi.org/10.1016/j.engstruct.2024.119248

Tang, J., Dai, K., Luo, Y., Bezabeh, M. A., & Ding, Z. (2024). Integrated control strategy for the vibration mitigation of wind turbines based on pitch angle control and TMDI systems. *Engineering Structures, 303*, Article 117529. https://doi.org/10.1016/j.engstruct.2024.117529

Tang, S., Sweetman, B., & Gao, J. (2021). Nonlinear effects and dynamic coupling of floating offshore wind turbines using geometrically-exact blades and momentum-based methods. *Ocean Engineering, 229*, Article 108866. https://doi.org/10.1016/j.oceaneng.2021.108866

Tao, W., Basu, B., & Li, J. (2018). Reliability analysis of active tendon-controlled wind turbines by a computationally efficient wavelet-based probability density evolution method. *Structural Control and Health Monitoring, 25*(3), Article e2078. https://doi.org/10.1002/stc.2078

Tian, H., Soltani, M. N., Yeter, B., & Galván-Pozos, D. E. (2025). Design of a novel tower damping system for semi-submersible floating offshore wind turbines considering fatigue and ultimate limit states. *Ocean Engineering, 320*, Article 120343. https://doi.org/10.1016/j.oceaneng.2025.120343

Wang, C., Liu, Z., Zhang, D., & Yu, Z. (2025a). Nonlinear dynamic characteristics of eddy current tuned rolling cylinder damper and vibration reduction performance on wind turbines. *Structures, 80*, Article 109961. https://doi.org/10.1016/j.istruc.2025.109961

Wang, F.-C., Hong, M.-F., & Lin, T.-C. (2011). Designing and testing a hydraulic inerter. *Proceedings of the Institution of Mechanical Engineers, Part c: Journal of Mechanical Engineering Science, 225*(1), 66–72. https://doi.org/10.1243/09544062JMES2199

Wang, F.-C., & Su, W.-J. (2008). Impact of inerter nonlinearities on vehicle suspension control. *Vehicle System Dynamics, 46*(7), 575–595. https://doi.org/10.1080/00423110701519031
Wang, L., Xiong, C., Wang, X., Li, Y., & Xu, M. (2018). Hybrid time-variant reliability estimation for active control structures under aleatory and epistemic uncertainties. *Journal of Sound and Vibration, 419*, 469–492. https://doi.org/10.1016/j.jsv.2018.01.027
Wang, W., Li, X., Zhao, H., Wang, B., & Li, Y. (2020). Vibration control of a pentapod offshore wind turbine under combined seismic wind and wave loads using multiple tuned mass damper. *Applied Ocean Research, 103*, Article 102254. https://doi.org/10.1016/j.apor.2020.102254
Wang, W., Wang, J., Dai, K., Ba, Z., Wu, M., Zhou, B., Sharbati, R., & El Damatty, A. (2025b). Seismic fragility analysis of near-fault mountainous wind turbines considering source-path-site effects. *Renewable Energy, 254*, Article 123706. https://doi.org/10.1016/j.renene.2025.123706
Wang, W., Zhou, J., Yu, T., Chen, S., Luo, Y., & Hua, X. (2024). A novel tuned hydrodynamic mass damper employing added mass and damping from interaction between heave plate and fluids. *Mechanical Systems and Signal Processing, 220*, Article 111639. https://doi.org/10.1016/j.ymssp.2024.111639
Wang, X., Zhu, H., Su, K., & Zhou, X. (2026). Semi-active vibration control analysis of embedded rock-socketed jacket offshore wind turbines considering time-varying aerodynamic damping. *Ocean Engineering, 344*, Article 123736. https://doi.org/10.1016/j.oceaneng.2025.123736
Xia, K., Sacco, C., Kirkpatrick, M., Saidy, C., Nguyen, L., Kircaliali, A., & Harik, R. (2021). A digital twin to train deep reinforcement learning agent for smart manufacturing plants: Environment, interfaces and intelligence. *Journal of Manufacturing Systems, 58*, 210–230. https://doi.org/10.1016/j.jmsy.2020.06.012
Xie, F., & Aly, A.-M. (2020). Structural control and vibration issues in wind turbines: A review. *Engineering Structures, 210*, Article 110087. https://doi.org/10.1016/j.engstruct.2019.110087
Xu, Q., Sun, R., Li, J., & Zhao, X. (2025). Self-powered electromagnetic shunt damper for semi-active vibration control. *International Journal of Mechanical Sciences, 305*, Article 110761. https://doi.org/10.1016/j.ijmecsci.2025.110761
Xue, M.-A., Dou, P., Zheng, J., Lin, P., & Yuan, X. (2022). Pitch motion reduction of semisubmersible floating offshore wind turbine substructure using a tuned liquid multicolumn damper. *Marine Structures, 84*, Article 103237. https://doi.org/10.1016/j.marstruc.2022.103237
Yang, D., Nagarajaiah, S., Chen, L., Sun, L., & Basu, B. (2025). Dynamic response reduction of floating offshore renewable energy applications with a high-damping mooring system. *Ocean Engineering, 324*, Article 120609. https://doi.org/10.1016/j.oceaneng.2025.120609
Yang, G., Spencer, B. F., Carlson, J. D., & Sain, M. K. (2002). Large-scale MR fluid dampers: Modeling and dynamic performance considerations. *Engineering Structures, 24*(3), 309–323. https://doi.org/10.1016/S0141-0296(01)00097-9
Yang, J. J., & He, E. M. (2020). Coupled modeling and structural vibration control for floating offshore wind turbine. *Renewable Energy, 157*, 678–694. https://doi.org/10.1016/j.renene.2020.05.075
Yanik, A. (2020). Seismic control performance indices for magneto-rheological dampers considering simple soil-structure interaction. *Soil Dynamics and Earthquake Engineering, 129*, Article 105964. https://doi.org/10.1016/j.soildyn.2019.105964
Yao, J., Jin, X., Yue, Y., Huang, X., & Chen, P. (2024). Investigation on vibration control of TetraSpar floating offshore wind turbine. *Ocean Engineering, 311*, Article 118934. https://doi.org/10.1016/j.oceaneng.2024.118934
Yin, W., Sun, F., Cai, X., Zhao, J., Dong, F., Le, Z., Song, J., & Chen, L. (2025). Temperature-frequency-dependent additional damping contributed by constrained layer damping (CLD) treatment for offshore wind towers: Test and calculation methods. *Ocean Engineering, 318*, Article 120213. https://doi.org/10.1016/j.oceaneng.2024.120213
Yucesan, Y. A., & Viana, F. A. C. (2023). Physics-informed digital twin for wind turbine main bearing fatigue: Quantifying uncertainty in grease degradation. *Applied Soft Computing, 149*, Article 110921. https://doi.org/10.1016/j.asoc.2023.110921

Zeng, W., Qiao, H., Guo, C., Ren, W., & Du, Y. (2025). Inerter-based nonlinear energy sink for enhanced vibration reduction in floating wind turbines: Parameter optimization and numerical validation. *Ocean Engineering, 319*, Article 120273. https://doi.org/10.1016/j.oceaneng.2024.120273

Zhang, Z., & Høeg, C. (2020). Dynamics and control of spar-type floating offshore wind turbines with tuned liquid column dampers. *Structural Control and Health Monitoring*, *27*(6). https://doi.org/10.1002/stc.2532

Zhang, D., Liu, Z., Wang, C., & Wang, Y. (2025a). Improved toggle-brace viscous damper for vibration mitigation of wind turbine blade. *Alexandria Engineering Journal, 132*, 40–52. https://doi.org/10.1016/j.aej.2025.10.005

Zhang, J., Heng, J., Dong, Y., Baniotopoulos, C., & Ali, Y. (2025b). A digital twinning framework for intelligent operation and maintenance of floating offshore wind turbines under corrosion-fatigue. *IOP Conference Series: Earth and Environmental Science, 1552*(1), Article 012023. https://doi.org/10.1088/1755-1315/1552/1/012023

Zhang, J., Liang, X., Wang, B., You, P., & Yun, L. (2024a). Vibration reduction and energy harvesting of monopile offshore wind turbines under extreme wind-wave loadings using a novel bidirectional absorber-harvester. *Structures, 66*, Article 106790. https://doi.org/10.1016/j.istruc.2024.106790

Zhang, J., Zhao, X., & Wei, X. (2020). Data-driven structural control of monopile wind turbine towers based on machine learning. *IFAC-PapersOnLine, 53*(2), 7466–7471. https://doi.org/10.1016/j.ifacol.2020.12.1299

Zhang, J., Zhao, X., & Wei, X. (2022). Reinforcement learning-based structural control of floating wind turbines. *IEEE Transactions on Systems, Man, and Cybernetics: Systems, 52*(3), 1603–1613. https://doi.org/10.1109/TSMC.2020.3032622

Zhang, R., Zhao, Z., & Dai, K. (2019). Seismic response mitigation of a wind turbine tower using a tuned parallel inerter mass system. *Engineering Structures, 180*, 29–39. https://doi.org/10.1016/j.engstruct.2018.11.020

Zhang, S. Y., Tan, G., Deng, Y., Zhang, Y., Wang, Y., & Wang, Z. (2024b). Vibration mitigation of spar-buoy floating wind turbines using a nonlinear energy sink. *Ocean Engineering, 304*, Article 117883. https://doi.org/10.1016/j.oceaneng.2024.117883

Zhang, Z. (2022). Understanding and exploiting the nonlinear behavior of tuned liquid dampers (TLDs) for structural vibration control by means of a nonlinear reduced-order model (ROM). *Engineering Structures, 251*, Article 113524. https://doi.org/10.1016/j.engstruct.2021.113524

Zhang, Z., Nielsen, S. R. K., Basu, B., & Li, J. (2015). Nonlinear modeling of tuned liquid dampers (TLDs) in rotating wind turbine blades for damping edgewise vibrations. *Journal of Fluids and Structures, 59*, 252–269. https://doi.org/10.1016/j.jfluidstructs.2015.09.006

Zhao, J., Sun, F., Yin, W., Le, Z., Song, J., Cai, X., & He, Y. (2023). Damping performance and its influencing factors of wind turbine towers with constrained layer damping treatment. *Structures, 57*, Article 105322. https://doi.org/10.1016/j.istruc.2023.105322

Zhao, Z., Dai, K., Lalonde, E. R., Meng, J., Li, B., Ding, Z., & Bitsuamlak, G. (2019). Studies on application of scissor-jack braced viscous damper system in wind turbines under seismic and wind loads. *Engineering Structures, 196*, Article 109294. https://doi.org/10.1016/j.engstruct.2019.109294

Zhu, H., Hao, W., Li, C., Ding, Q., & Wu, B. (2019). Application of flow control strategy of blowing, synthetic and plasma jet actuators in vertical axis wind turbines. *Aerospace Science and Technology, 88*, 468–480. https://doi.org/10.1016/j.ast.2019.03.022

Zhu, S., Shen, W., & Xu, Y. (2012). Linear electromagnetic devices for vibration damping and energy harvesting: Modeling and testing. *Engineering Structures, 34*, 198–212. https://doi.org/10.1016/j.engstruct.2011.09.024

Zuo, H., Bi, K., & Hao, H. (2017). Using multiple tuned mass dampers to control offshore wind turbine vibrations under multiple hazards. *Engineering Structures, 141*, 303–315. https://doi.org/10.1016/j.engstruct.2017.03.006

Zuo, H., Bi, K., & Hao, H. (2020). A state-of-the-art review on the vibration mitigation of wind turbines. *Renewable and Sustainable Energy Reviews, 121*, Article 109710. https://doi.org/10.1016/j.rser.2020.109710

Zuo, H., Zhang, J., Bi, K., Zhu, S., Hao, H., & Ma, R. (2023). Structural vibration control of spar-buoy floating offshore wind turbines. *Engineering Structures, 294*, Article 116732. https://doi.org/10.1016/j.engstruct.2023.116732

Zuo, H., & Zhu, S. (2022). Development of novel track nonlinear energy sinks for seismic performance improvement of offshore wind turbine towers. *Mechanical Systems and Signal Processing, 172*, Article 108975. https://doi.org/10.1016/j.ymssp.2022.108975

Chapter 6
Utilization of Measurement Data for Wind Energy Structures

Rüdiger Höffer, Ika Kurniawati, Simona Bogoevska, and Yasaman Salahshour

Abstract Continuous monitoring of full-scale structures enables improved understanding of aerodynamic and aeroelastic behaviour, supports system identification and facilitates timely detection of structural changes. In wind engineering, the modelling of aerodynamic loads remains particularly challenging due to scaling effects and Reynolds number discrepancies between wind tunnel experiments and real-scale conditions, emphasizing the need for full-scale measurements to provide reliable validation data. Data-driven methods have gained prominence as effective tools for representing complex structural and aerodynamic phenomena. In this context, artificial neural networks offer efficient prediction of wind-induced responses and pressure distributions based on experimental data, while physics-informed neural networks further enhance modelling robustness by incorporating governing physical laws. The chapter presents an integrated perspective on full-scale measurements, data-driven system identification and neural network-based approaches, highlighting their complementary roles in advancing aerodynamic assessment and structural monitoring of wind turbine systems.

R. Höffer (✉) · I. Kurniawati · Y. Salahshour
Faculty of Civil and Environmental Engineering, Ruhr University Bochum, Bochum, Germany
e-mail: ruediger.hoeffer@rub.de

I. Kurniawati
e-mail: ika.kurniawati@rub.de

Y. Salahshour
e-mail: Yasaman.Salahshour@edu.rub.de

S. Bogoevska
Faculty of Civil Engineering, Ss. Cyril & Methodius University, Skopje, North Macedonia
e-mail: simona.bogoevska@gf.ukim.edu.mk

Y. Salahshour
Department of Civil and Environmental Engineering, University of Florence, Florence, Italy

C. Baniotopoulos and E. Marino (eds.), *Advances in Wind Energy in the Era of Artificial Intelligence*, CISM International Centre for Mechanical Sciences 625,
https://doi.org/10.1007/978-3-032-28815-8_6

6.1 Introduction

The ability to observe, assess and reliably interpret the behaviour of engineering structures under operational conditions remains a central challenge in the field of wind energy structures (Gohar et al., 2025; Zhang et al., 2023). This challenge extends beyond structural response monitoring to the accurate identification of environmental and aerodynamic loading, which governs both serviceability and ultimate performance.

Recent advances in sensing technologies and data acquisition systems have enabled the deployment of permanent monitoring systems on full-scale structures, offering continuous streams of measurement data that support improved load identification, model validation and data-aided assessment under real operating conditions (Fang et al., 2025; Mwangi et al., 2024). At the core of such approaches lies the development of mathematical models derived from observed system quantities, which has been enabled by parallel progress in computational capabilities, analytical methods and artificial intelligence for large-scale data processing (Malekloo et al., 2022).

Depending on the underlying assumptions and available information, these models may be classified as physics-based, data-driven, or hybrid. Physics-based models rely on established theoretical formulations, but their applicability may be limited when addressing complex systems governed by interacting physical phenomena (Fig. 6.1).

In such cases, data-driven models constructed directly from observational data may provide a more effective representation of system behaviour. Data-driven models have demonstrated strong potential as reduced-order representations of monitored systems, as they aim to capture the functional relationships between relevant input and output variables without explicitly relying on physical assumptions (Noori et al.,

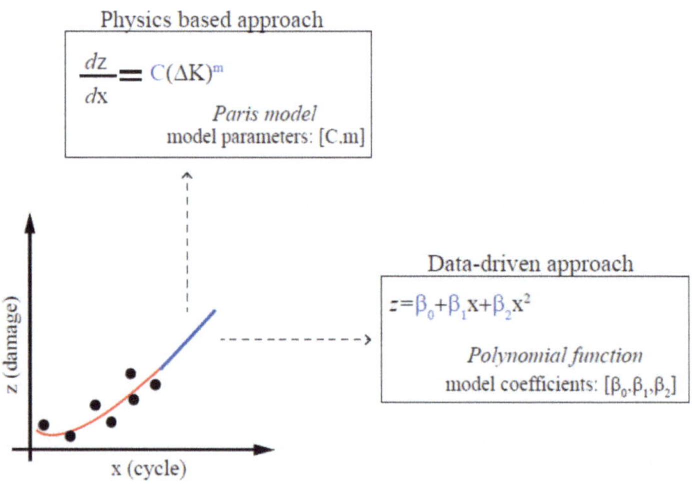

Fig. 6.1 Contrasting the physics-based Paris law with a purely data-driven polynomial formulation (adopted from Bogoevska, 2020)

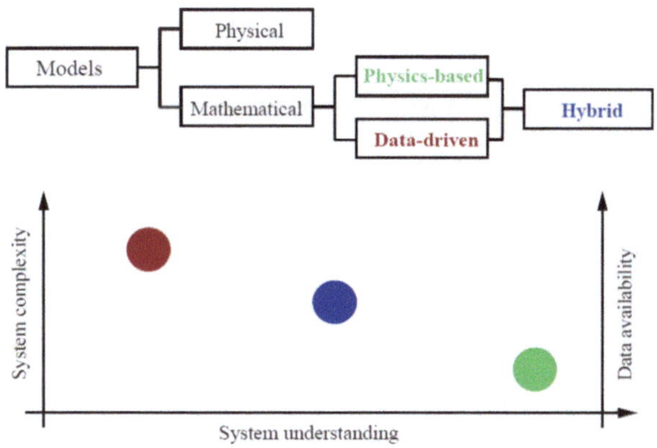

Fig. 6.2 Classification of mathematical models (adopted from Bogoevska, 2020)

2023). Their performance, however, depends critically on the availability and quality of measurement data. Unlike physics-based models, which may be transferable across similar systems, data-driven models are generally valid only within the range of conditions represented in their training data. Moreover, as the dimensionality of the input space increases, the required amount of training data may grow rapidly. Figure 6.2 summarizes the applicability of different modelling approaches in relation to system complexity, level of physical understanding and data availability.

In wind energy applications, data-driven modelling and the use of measurement data form a fundamental basis for both the design and operational assessment of wind turbines throughout their service life. During the design phase, numerous load cases are evaluated and the resulting structural effects must be validated against representative measurements. Data-assisted and data-validated design procedures improve confidence in design assumptions and support the development of efficient turbine configurations. For example, the use of data-driven modelling for design of adaptive aerostructures for blades, as reviewed in Roetzer et al. (2024). In addition, measurement data are indispensable for the calibration and validation of analytical and numerical models used in wind turbine engineering for accurate results (Rodriguez & Jonkman, 2018).

Despite their widespread use, laboratory and wind tunnel measurements of aerodynamic loads are subject to inherent limitations and questions related to geometric scaling (Stathopoulos & Surry, 1983). In particular, discrepancies arise from differences in the Reynolds number (Re) between model-scale experiments and full-scale operating conditions. Wind tunnel tests are typically performed at Reynolds numbers on the order of 104, whereas full-scale wind turbine towers operate at Reynolds numbers approaching 10^6. Since aerodynamic force coefficients depend strongly on the flow regime, measurements obtained at reduced Reynolds numbers cannot be directly extrapolated to full-scale conditions without introducing uncertainty. The application of surface roughness in wind tunnel testing, commonly used to promote critical or transcritical flow regimes at lower Reynolds numbers, further modifies flow separation and wake development, often resulting in aerodynamic coefficients

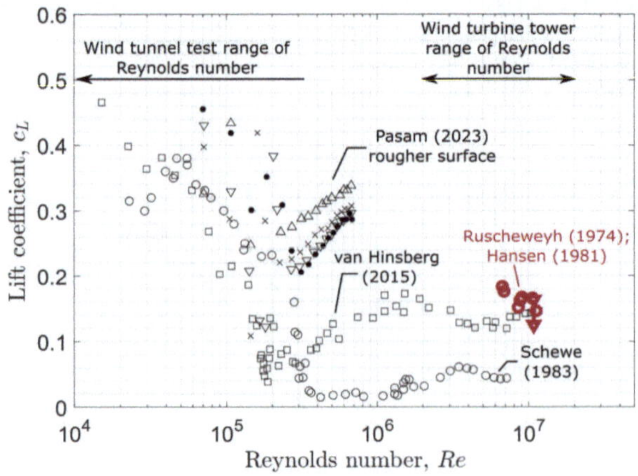

Fig. 6.3 Lift coefficient of circular cylinder in respect to Reynolds number from various literature. Wind tunnel tests: Pasam et al. (2023) with rougher surface of relative roughness $k/D = 1.1 \times 10^{-3}$ to 3×10^{-3}; van Hinsberg (2015) with $k/D = 1.2 \times 10^{-3}$; Schewe (1983) with smooth surface; full-scale measurements (Ruscheweyh, 1974) with $k/D = 3.7 \times 10^{-3}$; Hansen (1981) with concrete surface

that differ from those observed at full scale. This effect is illustrated in Fig. 6.3, which presents the lift coefficient of a circular cylinder representative of a wind turbine tower cross-section as a function of Re. The figure shows that lift coefficients measured in wind tunnel tests at lower Reynolds numbers are significantly higher than those observed at Reynolds numbers representative of full-scale conditions.

Even when surface roughness is applied to shift the flow regime, the resulting lift coefficients remain elevated. In contrast, measurements obtained at higher Reynolds numbers or directly from full-scale structures exhibit consistently lower lift coefficient values. These observations indicate that the aerodynamic design and assessment of wind turbine structures can only be fully validated through real-scale measurements, which avoid Reynolds number mismatch and provide essential scale-bridging information.

Among the data-driven concepts that have gained increasing attention in wind engineering, neural network-based approaches are emerging as powerful tools for modelling complex aerodynamic phenomena (Mostafa et al., 2022). Artificial neural networks offer advanced capabilities for predicting wind-induced responses and pressure distributions on structures using experimental data, thereby complementing traditional wind tunnel testing. Physics-informed neural networks extend this framework by embedding governing physical laws into the learning process, enabling models that are both data-driven and physically consistent. Despite these advantages, such as improved generalization and reduced reliance on exhaustive testing, challenges related to computational cost and the treatment of complex geometries continue to limit their widespread application in engineering practice (Zhao et al., 2024).

While wind tunnel testing provides valuable insights, it is often time-consuming and costly, and the resulting data are typically limited to specific wind directions and structural configurations. Neural network-based approaches address these limitations by learning the complex, non-linear relationships between structural geometry, wind characteristics and pressure response. Once trained, these models enable efficient prediction of pressure coefficients for conditions beyond those explicitly tested, offering a promising pathway toward more flexible and comprehensive aerodynamic assessment frameworks.

In light of the considerations outlined above, the chapter is organized into three main parts. Section 6.2 focuses on the acquisition and evaluation of high-quality full-scale measurement data for assessing aerodynamic and aeroelastic loads, highlighting the role of real-scale measurements in calibrating and validating wind turbine models. Section 6.3 addresses data-driven system identification and monitoring strategies for operational wind turbines, discussing how vibration-based measurements and data fusion approaches support structural assessment and long-term behaviour tracking. Section 6.4 presents neural network-based approaches in wind engineering, covering the fundamentals of artificial neural networks, their integration with wind tunnel data and the application of physics-informed neural networks for combining data-driven learning with physical laws to enhance robustness and predictive capability.

6.2 Measurement Data for Aerodynamic and Aeroelastic Effects at Wind Turbine Towers

This section reports on the use of full-scale measurement data to examine the aerodynamic and aeroelastic parameters of a wind turbine tower, especially related to vortex resonance. It demonstrates how full-scale measurement data can be used to strategically investigate the aerodynamic load and aeroelastic response of wind turbine towers, using the input–output full-scale measurements. The section also provides general guidelines for evaluation practices to ensure high data quality is collected.

6.2.1 Role of Full-Scale Measurement Data

One of the most critical loads on a wind turbine structure, especially its support structure, is due to aeroelastic effects. For example, in the case of a slender, large wind turbine tower, aeroelastic phenomena such as vortex-induced vibrations (VIV) experiencing vortex resonance can occur, creating large cross-wind displacement and cyclic loads that impact the structure's fatigue life. Usually for wind turbine tower, VIV is crucial during installation and free-standing tower. In large and modern wind turbine, 2nd tower-bending mode VIV may occur.

Full-scale measurements can be the strategic tool for investigating VIV as it addresses the correct Reynolds number condition, where the Reynolds number truly reflects the full-scale size of the tower. Furthermore, full-scale and field measurements provide conditions that are representative of natural atmospheric turbulence. This situation is hardly replicated in the normal boundary layer wind tunnel tests or small-scale experiments, as the smaller scale refers to a lower Reynolds number and a laboratory-simulated boundary layer profile.

Full-scale measurement of the wind turbine tower under vortex resonance provides realistic time series data on the tower's displacement. This data can be used to validate the prediction model and provide insights into the response variations that are related to the load cycle for fatigue. To address the aeroelastic loads and their effects specifically, it is necessary to obtain not only the output values (response), but also the input values (force). For example, the surface pressure measurements can be used to determine the force coefficients.

A case of full-scale measurement campaign was performed with two wind turbine towers at the Test Centre in Østerild, Denmark. The turbines are the SG 8.0–167 DD (D8 Tower) and the SG 14–222 DD (SG 14), manufactured by Siemens Gamesa Renewable Energy. Figure 6.4 describes the measurement layout, which consists of response measurements (output quantity) and wind pressure measurements (input quantity). The input–output measurement is specifically for the SG 14 tower.

Through surface pressure measurement, the evaluation can determine the pressure and force coefficients that characterize the vortex-shedding and aerodynamic load input values. The tower's response is also measured, to give information on tower's oscillation characteristics under VIV and wind-induced vibrations. For full-scale measurements intended to address aeroelastic effects and wind-induced loads, information on the surrounding wind conditions is critical and wind parameter measurements should be performed simultaneously. Therefore, wind condition measurements are available on two met masts (MET 1 and MET 2) and through a wind profiling lidar.

6.2.2 Evaluation Practices for Aerodynamic Testing on Wind Turbine Tower

The documentation in this section provides the general guideline to obtain the high-quality data from full-scale measurements related to a specific task of investigating the aerodynamic load and aeroelastic effects.

Full-scale testing of wind turbines, particularly their support structures, primarily involves measuring the structural response (output-only). This type of test is not particularly useful for investigating the aeroelastic effects on a wind turbine tower, since modelling the aeroelastic interaction still requires the deterministic values for the load coefficients and parameters. The coefficients can be calculated by integrating surface pressure along a finite dimension of the body (e.g. the tower's circumference).

To measure the input quantity of the wind force acting on the tower, full-scale pressure measurements were conducted at two different heights on SG 14, as previously outlined in Fig. 6.4. Figure 6.5 shows the placement of the pressure transmitters, which are placed equally along the tower's circumference. JUMO Midas S06

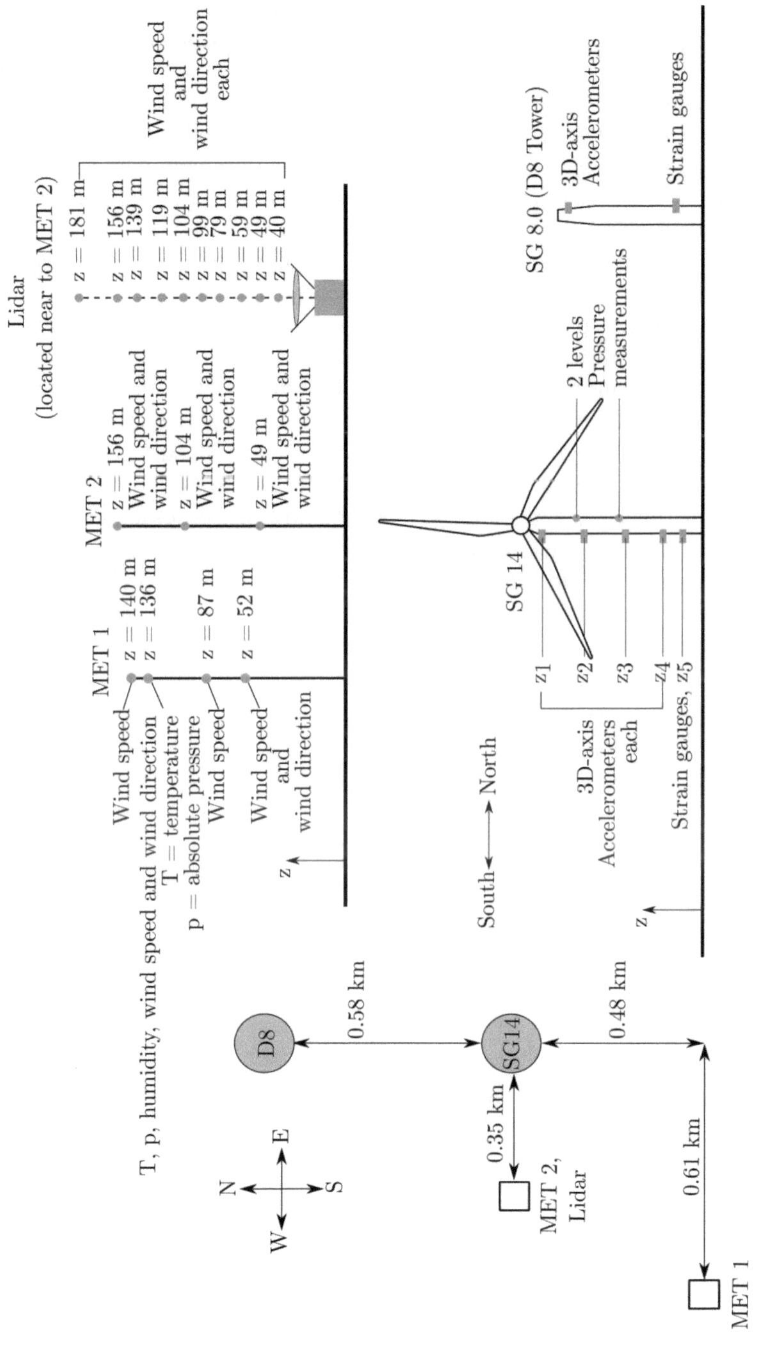

Fig. 6.4 Illustration of the field measurement setup for input–output measurement of wind and aeroelastic effects on wind turbine tower

differential pressure transmitters were used, with 24 sensors placed at each measurement level. The difference between the pressure outside the tower and the reference pressure was measured. The transmitter was mounted in a drilled hole in the tower wall.

The pressure distribution determines the flow development along the body and the flow separation, which affects the aerodynamic coefficients. To determine the vortex shedding load relevant for VIV, the pressure measurement data is evaluated in order to characterise the vortex shedding parameters, such as the lift coefficient and the Strouhal number.

The pressure distribution is associated with the flow regime and depends on the Reynolds number. Full-scale pressure measurements eliminate issues related to differences in the Reynolds number because the data are obtained under 'correct-Re' conditions. Full-scale measurements provide direct load coefficients that remain consistent within the transcritical range of Reynolds numbers and for smooth towers conditions, which cannot be replicated in smaller atmospheric boundary layer wind tunnels, particularly for large wind turbine towers with smooth surface.

Figure 6.6 shows an exemplary pressure coefficients distribution obtained from full-scale pressure measurement on the SG 14. The flow properties are described by the pressure distribution, including the minimum pressure value, the wake pressure value and the location of flow separation. The aerodynamic coefficients are also directly affected by the distribution of positive and negative pressure along the cross-section. To calculate the pressure coefficient (time-signal) for each sensor i at height z, the $c_{P,i}(z,t)$, the measured pressure is normalized by the appropriate reference dynamic pressure $\overline{q}(z)$ according to Eq. (6.1).

$$c_{P,i}(z,t) = \frac{p_i(z,t)}{\overline{q}(z)} \tag{6.1}$$

where

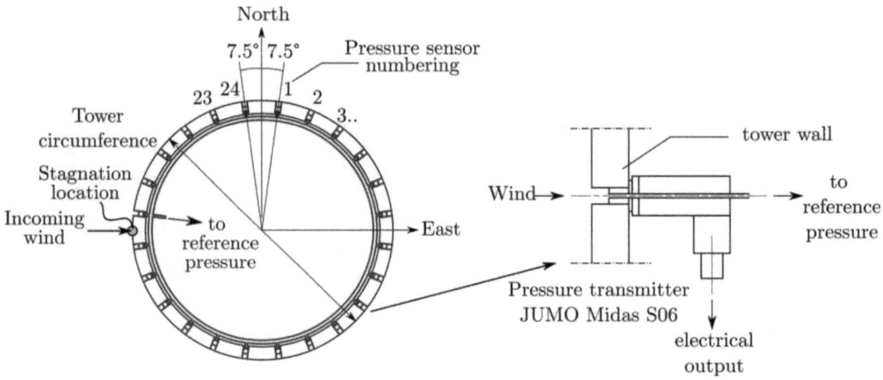

Fig. 6.5 Pressure transmitters installed along the circumference of the tower's cross-section. Adapted from Kurniawati (2025)

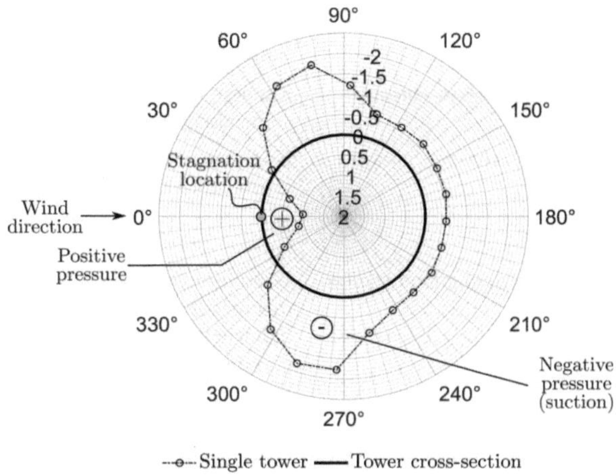

Fig. 6.6 Pressure distribution measured by full-scale surface pressure measurement on SG 14 (tower without nacelle), with illustration of positive and negative pressure. Adapted from Kurniawati (2025)

$$\overline{q}(z)=\frac{1}{2}\rho\overline{u}_q^2(z) \tag{6.2}$$

The appropriate dynamic pressure is supposed to be the free-stream wind velocity pressure at the same measurement height z. The Eq. (6.2) does not use the velocity value from met mast, but the velocity is obtained through the calculation of dynamic pressure q. Thus, a notation subscript 'q' is added in the velocity $\overline{u}_q$.

In the full-scale measurement, it is not easy to obtain the appropriate reference pressure or most of the time, the value of q is unknown. The unknown reference pressure required to normalize the measured pressure is one of the challenges in the full-scale pressure measurement. It is also crucial to ensure that results obtained from a full-scale pressure measurement are plausible, if not accurate. This requires knowledge of certain parameters, which are mostly also unknown, such as the angle of attack that is found to be not the same as the wind direction measured at the met masts.

Table 6.1 summarizes the challenges and mitigations that have been dealt with from the measurement campaign in the Test Centre Østerild. Further information on mitigation approaches is detailed in Kurniawati et al. (2024) and Kurniawati (2025).

One of the challenges that clouded the evaluation process is that the offset value of each pressure sensor is unique and unknown. Therefore, a mitigation or algorithm to find the offset value has been proposed and developed, as reported in Kurniawati et al. (2024).

Full-scale pressure measurement provides much more information if high-quality data can be produced from the evaluation. This is because the measurement is performed continuously over time. Pressure fluctuations are measured over time, providing time histories of load coefficients, such as the drag and lift coefficient. Figure 6.7 shows the exemplary time-histories of pressure coefficient at stagnation, drag coefficient c_D and lift coefficient c_L from full-scale measurement of SG 14, when the tower is not yet without nacelle. An evaluation period of 10-min pressure

Table 6.1 Overview of challenges in full-scale pressure measurement data evaluation

Challenges	Indicators	Mitigation
Offset in pressure sensor	Random fluctuation of pressure value along circumference	Algorithm to find offset value
Unknown reference pressure	Non-plausible pressure coefficient values	Algorithm to recover the reference pressure through difference of base pressure coefficient
Unknown stagnation location	Wind direction measured at the met mast is not accurate to indicate the location of the maximum pressure at stagnation	Calculation of stagnation location through zero-crossing of the pressure distribution

and aerodynamic coefficients is chosen. This makes it possible to obtain the spectral content of the in situ force coefficient. This is particularly important because vortex shedding is centred in a narrow frequency band around the Strouhal frequency, which can be modelled as Gaussian function (Vickery & Clark, 1972). The vortex shedding parameters of the function can be empirically derived by the fitting of the Gaussian function to the spectrum of measured lift coefficient. For further reading of the evaluation of lift coefficient spectrum model and vortex shedding parameters, reader is referred to Kurniawati et al. (2024).

Furthermore, the lift coefficient time history contains information on slower dynamics due to gust fluctuations and large-scale natural turbulence. If pressure

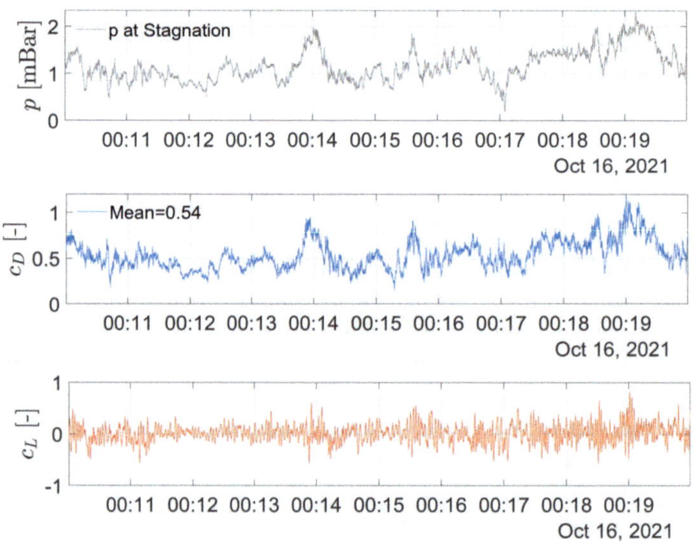

Fig. 6.7 Exemplary time-histories of pressure coefficient at stagnation, drag coefficient c_D and lift coefficient c_L from full-scale measurement of SG 14 (tower without nacelle)

measurements are taken at several spanwise locations, for example at different levels along the tower, the resulting lift coefficient data will also provide the information on the correlation of fluctuating lift, particularly that associated with correlation of vortex shedding. This is an important parameter for vortex resonance load coherence.

To obtain high-quality aerodynamic coefficients data from pressure measurement, curation process might be needed, to reduce the scatter and outliers which helps to seek the information that one investigates. Table 6.2 lists the examples of curation process that may be introduced when evaluating the pressure and aerodynamic coefficients, as well as vortex shedding parameters.

The pressure coefficients are curated to focus on data with a symmetrical pressure distribution between the half-circumference clockwise and counterclockwise of the stagnation point. This avoids non-stationary or bistable data points. Regarding the vortex shedding parameters in relation to the Gaussian lift coefficient model, as well as in characterising the Strouhal frequency, it was necessary to focus on a clear, single peak of vortex shedding in the measurements (Kurniawati et al., 2024). This clear peak indicates a representative Strouhal number for the vortex shedding oscillation of the 10-min period of the lift coefficient time history.

The curation process mentioned in Table 6.2 is not the ultimate guidance, but it demonstrates the practices that may need to be considered when evaluating full-scale pressure measurement data, depending on the evaluation objective. Curation and evaluation procedures differ depending on the research task. For example, when working on a phenomenological model of the lift coefficient, a different procedure or data evaluation task may be required, which should consider all information in the lift coefficient time history.

On that note, it is shown how full-scale pressure measurement or input quantity is highly required for a task of aeroelastic load characterization. The information contained by the full-scale measured parameter, for example lift coefficient, contains

Table 6.2 Curation process in the full-scale pressure measurement evaluation for aerodynamic coefficients and vortex shedding parameters

Curation objective	Variable	Removal
Pressure coefficients		
To obtain good samples of pressure coefficient distribution	Pressure coefficient along the tower circumference	Remove pressure data with asymmetric pressure distribution
To avoid data with alarming large variation of wind direction	Wind direction and stagnation location	Remove pressure data with large difference of wind direction measured at met mast and the calculated stagnation location
Vortex shedding parameters		
To focus on clear single peak of lift coefficient spectrum	Lift coefficient spectrum	Remove data with spectrum of multiple peaks around vortex shedding frequency

richer information by means of realistic fluctuations and effect of slow dynamics by the flow turbulence or turbulent separation on the cylinder. Full-scale pressure measurement data can also give the in situ distribution of the force coefficients for a range of wind velocities and turbulence intensities.

Tower response data is essential for analysing the structural response to aerodynamic phenomena such as VIV. This data can also be used to validate response prediction models. Furthermore, response data obtained from full-scale measurements provides a realistic time history of the signal. This time history is not entirely harmonic oscillation but rather exhibits stochastic variation in the signal.

Response data is obtained by placing accelerometers at the top of the tower or at the location where maximum vibration is expected, depending on the expected vibration mode (e.g. the middle section for the second tower bending mode). Ideally, multiple accelerometers should be placed along the structure's spanwise (e.g. at different heights for a vertical tower), so that the mode shape motion of the structure can be observed.

Alternatively, strain gauges can be installed to measure the bending moment resulted from the tower oscillation. In the full-scale measurements performed in Østerild, the D8 tower was equipped with three-axis accelerometers around the top and strain gauges around the bottom (Fig. 6.8). For SG 14, various response measurement sensors are installed along the height (Fig. 6.9). This is particularly useful as the SG 14 tower has a chance to experience the second mode of tower bending oscillation.

The tower response is to be evaluated using modal displacement ($\hat{y}_n$), which will then be translated into modal bending moment and stress. To obtain the modal displacement for the nth mode, the acceleration signal ($\ddot{y}_n$) is double-integrated according to Eq. (6.3), assuming harmonic oscillation (Meirovitch, 2001). The double integration can be performed in the frequency or time domain. High-pass filtering is applied to the acceleration signal before the integration is performed. The factorized acceleration signal can be calculated for verification purposes and also used

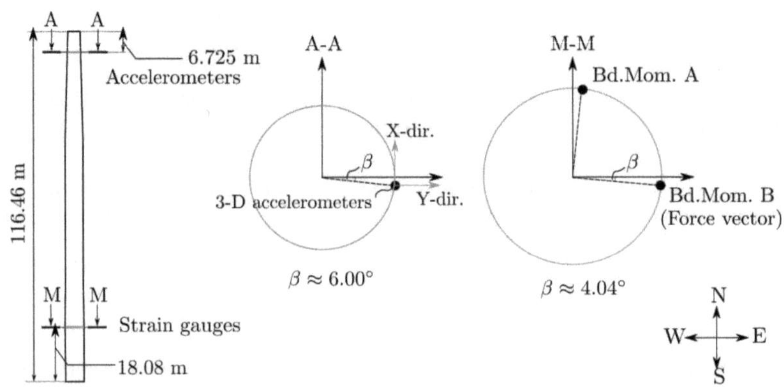

Fig. 6.8 Tower response measurement layout on D8 Tower. Adapted from Kurniawati (2025)

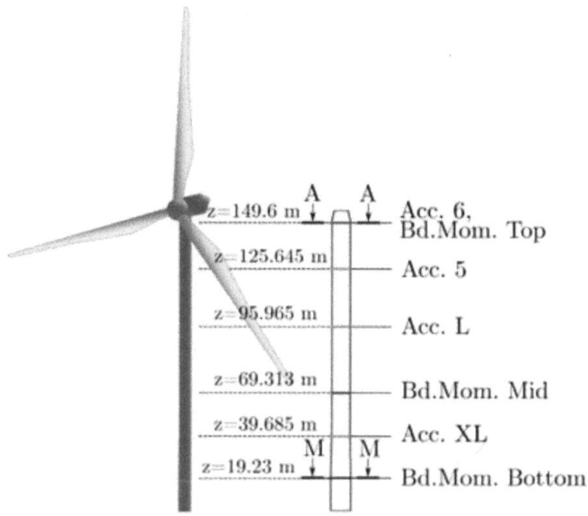

Fig. 6.9 Sensor placements for response measurement on SG 14 Tower. Adapted from Kurniawati (2025)

to estimate displacement according to Eq. (6.4). The factorized acceleration signal is suitable for estimating statistical displacement values such as the maximum and standard deviation.

$$\ddot{y}_n(t) = -\omega^2 y_n(t) \tag{6.3}$$

$$y_n(t) \approx \frac{\ddot{y}_n(t)}{(2\pi f_n)^2} \tag{6.4}$$

where f_n is the natural frequency of the structure i nth mode.

Figure 6.10 shows a comparison of the two methods: double integration and factorized accelerations. As expected, the displacement calculated by double integration is slightly lower, as the high-pass filter removes the peaks of the higher-frequency content. This arguably provides a more realistic displacement.

As shown in Fig. 6.11a, the displacement signal demonstrates that the double integration was successful. The oscillation is primarily in the second mode and the phase shift between the displacement and acceleration signals is confirmed to be approximately 180°.

Without filtering, the resulting displacement signal significantly drifts, as shown by the unrealistic displacement signal in Fig. 6.11b. This demonstrates the importance of applying a high-pass filter before performing integration in the time domain. Without a high-pass filter, the contribution of the low frequency will be linearly extended in velocity (first integration) and squared in acceleration (second integration).

Another option for obtaining the standard deviation of the displacement response is to divide the measured base bending moment (e.g. in standard deviation) by the integrated inertial force along the tower height, described by Eq. (6.5).

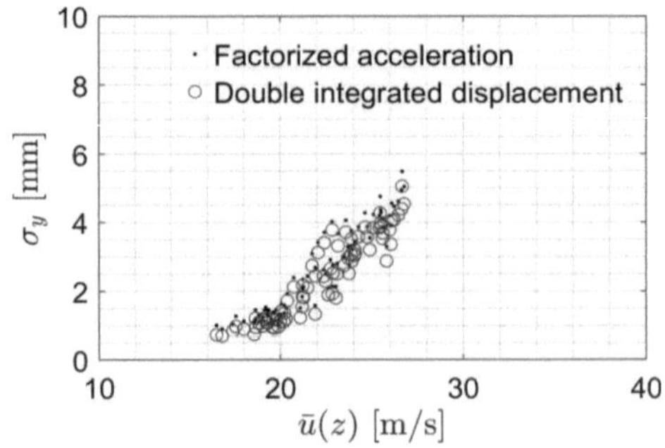

Fig. 6.10 Modal displacement calculated by factorizing acceleration and double integration. Adapted from Kurniawati (2025)

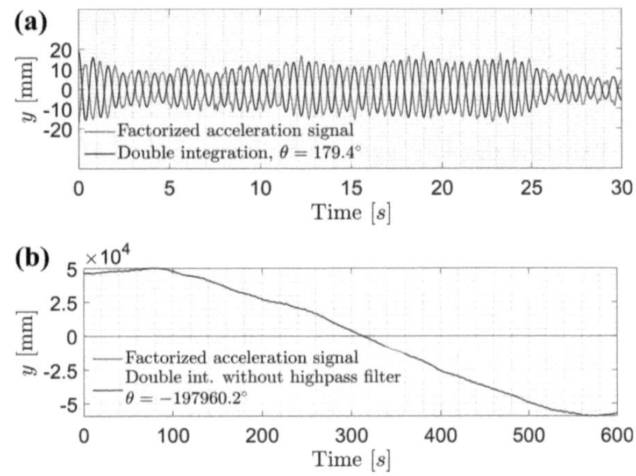

Fig. 6.11 Displacement signal from the double integration compared to the factorized acceleration **a** With high-pass filtering **b** Without high-pass filtering. Adapted from (Kurniawati, 2025)

$$y_n(t) \approx \frac{\sigma_{M_y}}{(2\pi f_n)^2 \int_0^H m(z)\Phi_n(z)zdz} \tag{6.5}$$

where σ_{M_y} is the standard deviation of the base bending moment in the crosswind direction (in Nm). $m(z)$ is the distributed mass along the height, and $\Phi_n(z)$ is the mode shape.

The modal displacement is calculated for an exemplary 10-min time history at different wind velocity ratios relative to the critical velocity ($\overline{u}/u_{cr}$) of VIV. Figure 6.12a, b shows the displacement sample and its distribution for ratios of 1.17 and 1.83, respectively. When the ratio of $\overline{u}/u_{cr}$ is around 1, the incoming wind velocity is around the critical velocity, which causes vortex resonance. When the ratio of $\overline{u}/u_{cr}$ is much larger than 1, the tower should leave the vortex resonance situation. Figure 6.12a and b show two different crosswind displacement characteristics of the tower. Figure 6.12a shows the predominant VIV response, characterised by a maximum distribution around the standard deviation and a bimodal distribution.

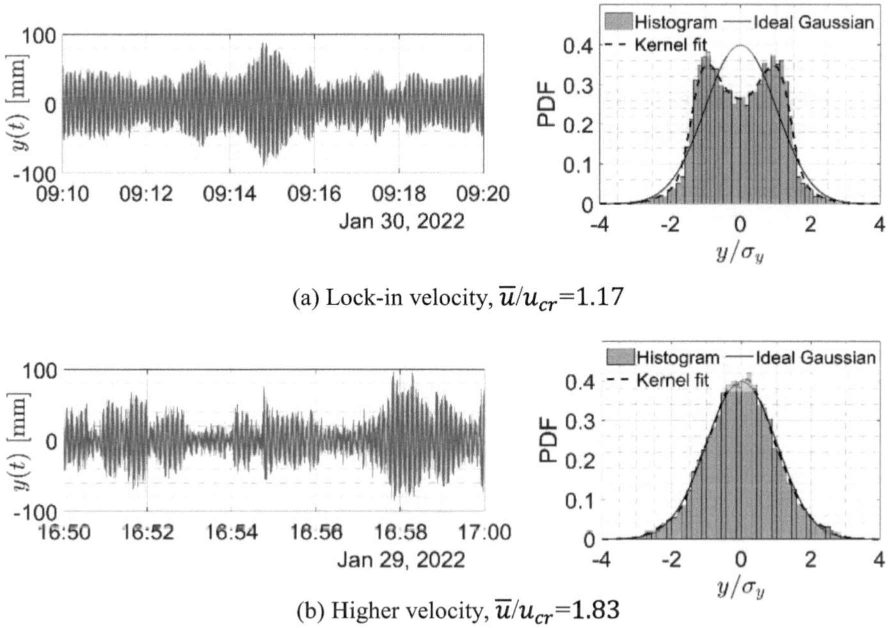

(a) Lock-in velocity, $\overline{u}/u_{cr}$=1.17

(b) Higher velocity, $\overline{u}/u_{cr}$=1.83

Fig. 6.12 PDF of normalized cross-wind oscillation $y(t)/\sigma_y$ from three 10-min cross-wind displacement signals with different velocity ratio $\overline{u}/u_{cr}$. Adapted from Kurniawati (2025)

Figure 6.12b shows a Gaussian distribution, indicating a greater contribution from stochastic components. Such high-quality data on the response can be used to provide field-based evidence of the cross-wind response due to VIV and its potential superposition with the effect of high gust fluctuations. Furthermore, the signal distribution should also be considered, not only for investigating the cross-wind responses, but also for classifying them (Kurniawati, 2025; Kurniawati et al., 2025).

A summary of the general evaluation procedure for the tower's response in relation to the aeroelastic response is presented in Table 6.3. Depending on the evaluation task, the cross-wind and along-wind response signals are calculated, considering the transformation in-line with and parallel to the wind direction. Another evaluation procedure includes double integration and a classification algorithm to characterize the crosswind response signal based on a probability function (e.g. a PDF), which is useful for characterising a VIV or other predominantly signal-dominated process.

Another important aspect of full-scale measurements for analysing aeroelastic loads and responses is that the parameters of the wind conditions (such as speed, direction and temperature) must be measured simultaneously. During the measurement campaign in Østerild, two meteorological masts (MET 1 and MET 2) and one lidar were used. Table 6.4 summarises the sensors used to measure horizontal wind speed and direction, as well as the wind profiling lidar. The lidar was a Vaisala WindCube V2.1 pulsed Doppler lidar covering measurement levels from 40 to 300 m.

Table 6.3 Tower aeroelastic response full-scale measurement evaluation procedure

Evaluation procedure	Variable	Remarks
Calculation of along-wind and cross-wind response signal	Acceleration or displacement	Using transformation in-line to wind direction and parallel to wind direction
Double integration for calculating the displacement	Acceleration	High-pass filter is used. For double integration in time domain, details are presented in Kurniawati (2025)
Classification of cross-wind response using the probability function of the response signal	Displacement or acceleration	See Kurniawati et al. (2025) or Kurniawati (2025)

Table 6.4 Wind conditions sensors and equipment, adapted from Gomez and Yankova (2022) and presented also in Kurniawati (2025)

Sensors	Variable
Thies cup anemometer	0.3–75 m/s
Thies clima Windvane 4.3151.10.020	0–360°
Vaisala WindCube V2.1 Lidar	40 m to 300 m, 0 m/s to 49 m/s, 0–360o, –23 m/s to + 23 m/s (radial wind speed range)

MET 2 is also equipped with air temperature, air pressure, humidity and rain sensors. Air temperature, pressure and humidity are measured at $z = 156$ m using a Risø Pt100 sensor. The Vaisala PTB110 sensor measures air pressure. The Vaisala HMP155 RH/T probe is used for the humidity sensor. The Pronamic Rain-O-Matic Professional Rain Gauge is used as a rain sensor.

Though measurements on met masts and lidar have been provided to ensure that wind condition parameters are available, the distance between the tower, the met masts and the lidar is distant enough to introduce different values being measured at the same time at different locations. This means that the wind conditions just outside the tower are unlikely to be the same as the values measured by the met masts, where non-homogeneous wind conditions are to be expected in the full-scale measurements. This is likely why the wind direction measured at the met mast is not appropriate for the stagnation location on the tower, as discussed previously, affecting the evaluation of full-scale pressure measurements.

During the same 10-min period, the mean wind speed measured by MET 1, MET 2 and lidar can differ, resulting in a different wind profile (Fig. 6.13a). Similar mean wind speeds between two met masts and one lidar refer to more homogeneous wind conditions (Fig. 6.13b).

One of the most important insights from full-scale measurement data is the availability of the distribution of the field-measured wind conditions. This means that, with full-scale testing data, we can achieve a realistic and random variation of the related input conditions. For example, in the assessment of VIV, the variation of

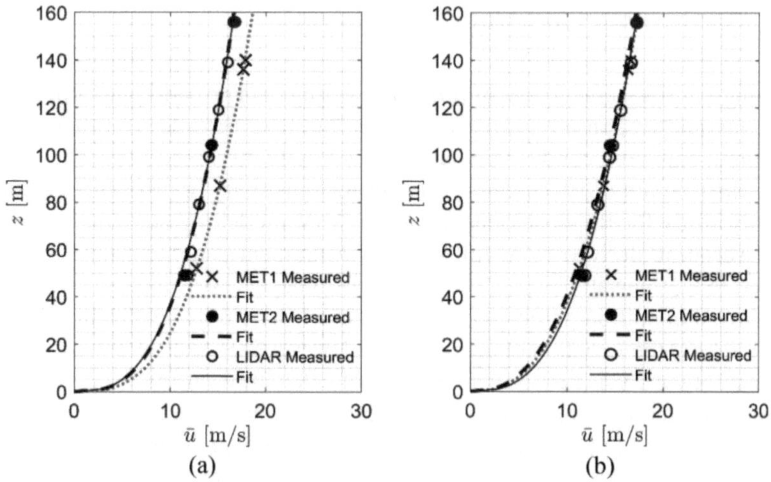

Fig. 6.13 Mean wind speed along the height measured by MET 1, MET 2 and lidar for the same 10-min period and the estimated wind profile lines using the power-law profile **a** On 14-Feb-2022 at 19:40 and **b** on 14-Feb-2022 at 19:30. Adapted from Kurniawati et al. (2025)

the input parameters such as the wind shear exponent α of the wind profile and the turbulence intensity (I_u) is known to affect the tower's response.

Another important point to consider is that, for an aeroelastic phenomenon such as VIV, the critical velocity of vortex resonance usually does not occur at high extreme wind speed for the large structure. For larger wind turbine towers in particular, the critical velocity of VIV is at moderate to smaller wind speeds. At this critical velocity, variations in wind conditions occur as non-neutral atmospheric stability is more often.

For example, at lower wind speeds, the variation in the wind shear exponent, α, shows that the α value does not always correspond to the terrain category. Figure 6.14 displays the distribution of α in relation to wind velocity. The α value is obtained by fitting the measured 10-min mean wind speed data to the met masts or lidar. Only high-quality fits are considered.

Fig. 6.14 Distribution of wind shear exponent in respect to wind velocity from evaluation of wind data of MET 2. Adapted from Kurniawati (2025)

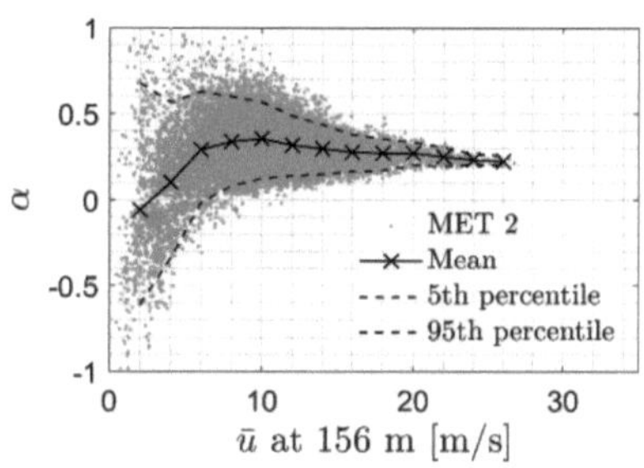

6.2.3 Applications for Model Validation

An example of the strategic use of full-scale testing data providing not only the load coefficients and response data for validation, but also field-derived random variation of input and wind condition parameters is briefly discussed here. The extensive realistic distribution of parameters is useful for sensitivity or probabilistic study. A sensitivity study for the developed advanced VIV modelling adapting the realistic random variation derived from full-scale measurement campaign on D8 Tower is presented in Kurniawati (2025). The sensitivity analysis takes into account the variation of wind condition parameters, such as wind shear exponent α and turbulence intensity I_u, and the variation of measured load parameters, such as Strouhal number. The sensitivity analysis runs a large number of calculations using the developed and proposed VIV model. The VIV model used focuses on the use of 3-D characteristics of negative aerodynamic damping developed based on the measurement data of Lupi et al. (2018). The distributed calculation of parameters is also incorporated in the model, in such a way that the parameters are defined along the height. A realistic replication of calculated output variation compared to the realistic scatters of tower response standard deviation is successfully obtained, as compared in Fig. 6.15. The good match of the simulated lines indicates the advantage of using full-scale measurement data to gain the most realistic insights possible, such as the information contained in the variation of field-measured parameters and its effect to the tower response variation.

6.3 Data-Driven Diagnostics for Large Scale Wind Turbines

Continuous monitoring strategies facilitate the use of data-driven tools, which often offer a more objective and flexible approach in handling the inherent uncertainties and variabilities of real-world systems and environments. This section provides a brief

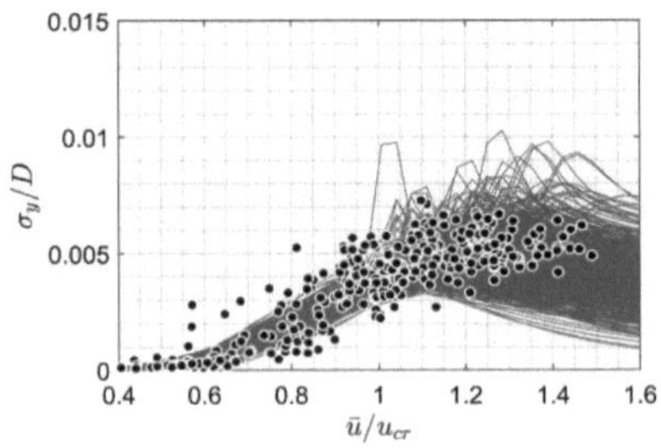

Fig. 6.15 Sensitivity study for the developed VIV prediction model compared to measured standard deviation of tower response, varying over wind speed. The overlapping lines show simulation results, and the points show the realistic measurement scatter of tower response standard deviation. Adapted from Kurniawati (2025)

overview of selected methodologies and applications, with emphasis on key concepts. Given space limitations, the discussion is intentionally concise, while references are provided for the interested reader seeking more detailed theoretical background or extended case-study analyses.

Section 6.3.1 presents representative examples of system identification based on measured vibration data, drawing on the authors' research and selected full-scale applications. Section 6.3.2 discusses the combined use of complementary data-driven approaches and highlights the benefits of data fusion for complex operational systems, with particular emphasis on wind turbines. Section 6.3.3 outlines the potential of these methodologies for tracking structural changes and system behaviour over time, demonstrated through a full-scale operational wind turbine case study.

6.3.1 Structural Dynamics Identification: Evidence from Multiple Case Studies

The attractive potentials in exploiting measured structural responses in operational conditions under natural excitation, introduced in the early 1990s,, marked the beginning of the Operational Modal Analysis (OMA) era. Indeed, the advantages in identifying the actual structural dynamics by conducting fast and relatively cheap experiments have instantly attracted the research and industry communities (Brincker et al., 2002). Since the earliest documented systematic bridge testing in the 1940s, monitoring programmes for bridge infrastructure have historically been utilized for the purpose of understanding the load-structure-response chain, which eventually brings improvements and calibration of numerical models (Brownjohn, 2007). Nowadays, remote control and data retrieving hardware advancements, as well as software solutions for data-driven analysis drive the evolution of traditional bridge monitoring programmes towards more holistic strategies based on output-only SHM systems (Garcia-Ramonda et al., 2025).

The focus of Bogoevska et al. (2020) was to explore data records attained from a short-term output-only testing strategy utilized on two short-span railway bridge structures located in Kosovo. The purpose was to obtain an initial structural identification of the systems' dynamics. Figure 6.16 presents the first vertical bending frequency of one of the monitored bridges. It was successfully identified by employment of the time-domain parametric methods: ARMA method, an approach less sensitive to coupled modes, noise and short data set lengths, and the time-domain SSI method.

The RD-EFDD (Random Decrement-Enhanced Frequency Domain Decomposition) output-only analysis method combines the advantages of the Random Decrement technique and Enhanced Frequency Domain Decomposition (EFDD). This approach enables the estimation of Averaged Normalised Power Spectral Densities (ANPSD) with reduced noise and spectral leakage effects compared with conventional EFDD. As a result, modal parameters can be identified more accurately and

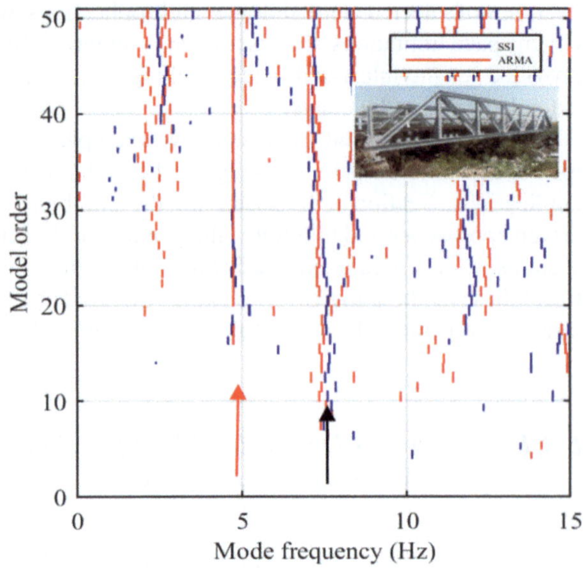

Fig. 6.16 Stabilization plot for Bridge B15 (red arrow is identified harmonic, black arrow is identified structural frequency) adapted from Bogoevska et al. (2020)

reliably. The concept was verified for the case of an out of use steel chimney located on the premises of Ohis Factory in Skopje, Republic of N. Macedonia. In Fig. 6.17, the spectrogram of measured wind induced vibration response, corresponding to the highest installed sensor, clearly exposes the first two natural frequencies of the structure, verifying the result obtained with the amplitude plot of the ANSPD in the same figure. More details on the RD-EFDD methodology can be found in (Bogoevska et al., 2016a). In contrast to response that remains close to stationary, where the standard OMA techniques are commonly employed, the complex dynamic system of operating wind turbine (WT) structures underlines the necessity for utilizing more efficient time-sensitive tools. In order to capture the typical time variable dynamics of WT infrastructure, in Bogoevska et al. (2017), for the case study of a 0.5 MW WT erected in 1997, located in the vicinity of Dortmund, Germany, the SP-TARMA model was successfully implemented. The method is applied on measured acceleration data recorded by triaxial accelerometer installed at the WT structure. The plotted spectrogram (Short Time Fourier Transform) which is provided in the background of the same plot clearly demonstrates the non-stationarity of natural frequencies within the explored time frame (Fig. 6.18).

6.3.2 Fusion of Data-Driven Tools for Complex Systems

Modal parameters are sensitive indicators of structural damage as a result of their correlation with the mass, stiffness and damping properties of systems. However, apart from vibrations, the collected response signals of operating structures carry

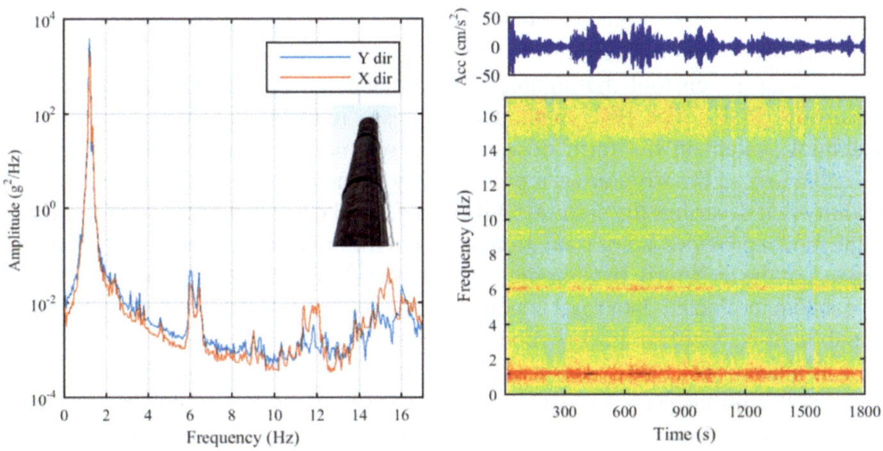

Fig. 6.17 Steel chimney structure stationary dynamics (adapted from Bogoevska et al., 2016a)

Fig. 6.18 WT structure time-varying dynamics (adapted from Bogoevska, 2020)

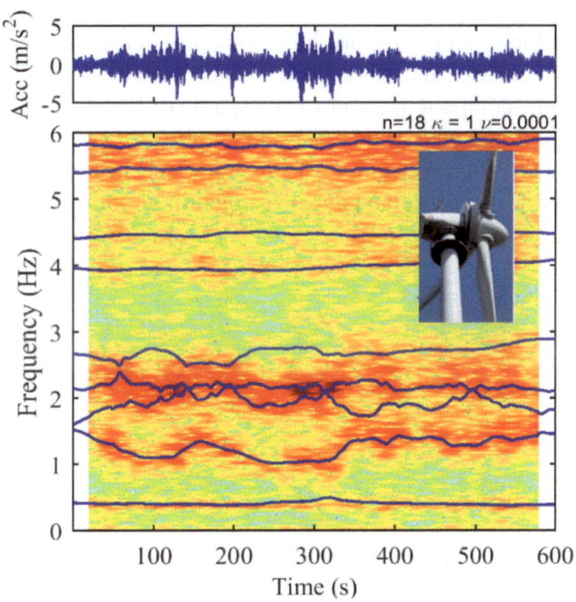

traces of slowly changing quasi-static effects such as diurnal thermal cycles, or effects from other environmental agents such as humidity, wind loading, long-term seasonal temperature varying cycles, operational loading (such as traffic loading for bridges), etc. Unfortunately, for civil engineering structures, these factors can result in modal changes equivalent or even greater than the ones produced by damage. It becomes obvious that a successful practical utilization of SHM frameworks and diagnostic strategies is not possible without an appropriate consideration of these effects (Bogoevska et al., 2016b).

One data-driven tool that enables such consideration is the polynomial chaos expansion (PCE). The PCE method generates a mathematical description of the structural system in the form of a complex polynomial function. The change in the behaviour of the structure is represented by a certain measured or derived output quantity that describes the system, while the selected input database used for training of the model provides a representation of the environment and the operational characteristics of the structure. In Georgijev and Bogoevska (2025) the method is tested for 1-year-long collected data for a concrete arch dam located along the river Treska in Republic of N. Macedonia. The model is developed to reproduce the measured dam crest displacements, using measured temperatures and water levels as input variables (Fig. 6.19). Similar extended studies are reported by Bogoevska and Chatzi (2025).

The applicability of data-driven models spans a wide range of tasks, including clustering and data classification, simulation and data generation, as well as diagnostic and forecasting frameworks. However, with increasing system complexity and associated epistemic and aleatory uncertainties, a multi-target fusion of data-driven tools is required to effectively assess the full spectrum of challenges posed by the problem, which cannot be sufficiently addressed by a single monolithic model. In this context, the work presented in Bogoevska (2020) demonstrates a data-driven, multi-component framework capable of linking measured structural responses with monitored environmental and operational parameters of an in-service wind turbine (WT). The uniqueness of this approach lies in its 'binocular' perspective, which enables the simultaneous investigation of both short-term non-stationary effects and

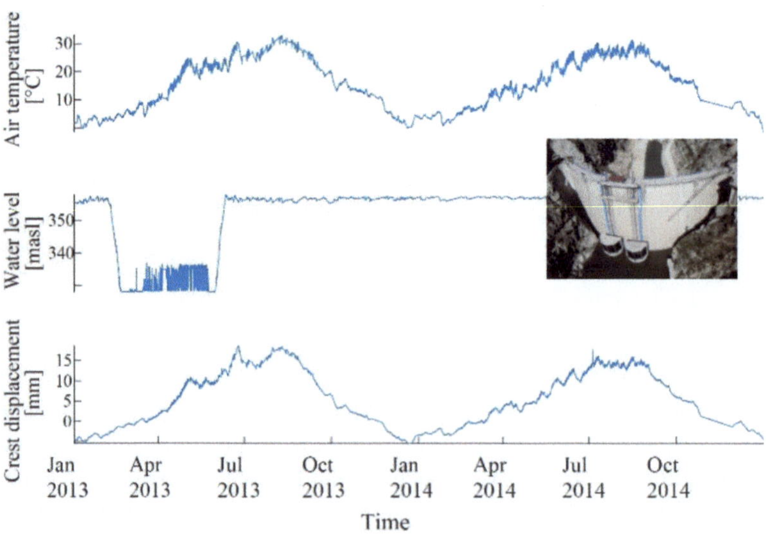

Fig. 6.19 Concrete dam structure: data utilized as PCE input and output (adapted from Georgijev & Bogoevska, 2025)

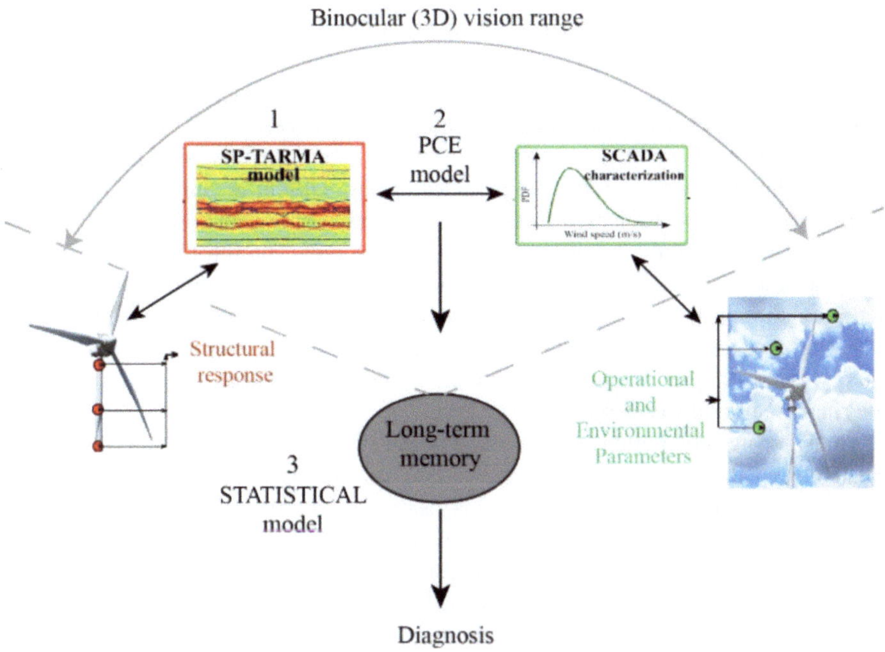

Fig. 6.20 Conceptual model of fusion of data-driven diagnostic tools (adopted from Boĝoevska, 2020)

long-term temporal variability in the WT response data (Fig. 6.20). Such a perspective is particularly well suited to real-life structures whose operational behaviour is inherently complex and non-stationary.

Many real-life structures indeed exhibit complex and/or non-stationary behaviour within their operational spectrum, with wind turbines representing a prototypical and particularly challenging example (Tcherniak et al., 2011). WTs are distinct structures characterized by complex dynamic behaviour and unique operating conditions. They are designed to continuously adapt to rapidly varying environmental factors in order to maximize power production while ensuring structural safety, even under extreme wind conditions. At the same time, the presence of a heavy nacelle mounted atop a slender and flexible mast renders them inherently prone to vibration. Moreover, wind turbines are continuously and heavily excited by varying wind loads, waves (when applicable), earthquakes, rotor rotation, switching regimes and other operational procedures. Structural vibrations have a direct impact on internal stresses, deformations, fatigue accumulation and ultimately the remaining service life of the structure. Consequently, an accurate description of their dynamic behaviour is of critical importance (Oliveira, 2016).

A fully operational wind turbine comprises three rotational subsystems whose behaviour depends on site-specific wind conditions: the rotor, the yaw mechanism that aligns the rotor with the wind direction and the active pitch control system that regulates blade angles. As a result, wind turbines are subjected to a wide range of vibration loads even during normal operation. Beyond stochastic aerodynamic and seismic excitation, these loads include short-term transients, such as extreme gusts, switching events and manoeuvre-induced impulses, as well as cyclic components associated with blade rotation, which must be carefully identified to avoid masking the structure's natural frequencies (Kühn, Gasch, & Sundermann, 2012). The rotating blades, subjected to combined aerodynamic and gravitational forces, substantially modify the loads transmitted to the tower. The rotor-swept area effectively acts as a filter, reshaping the frequency content of the wind-induced forces acting on the mast. Cyclic loads arise from aerodynamic effects, mass imbalance or wave-induced excitation in offshore configurations. Gusts and turbulence redistribute energy toward higher frequencies, producing spectral peaks at integer multiples of the rotor rotational frequency.

Considering the challenges discussed above, fused, multi-target approaches are particularly valuable for simultaneously addressing variability arising from changing environmental and operational conditions, as well as complexity intrinsic to the structure itself, such as non-stationarity and operationally driven dynamics (Bogoevska, 2020). By jointly exploiting structural response data alongside environmental and operational information, these frameworks provide a robust foundation for interpreting measured dynamics and enable reliable condition assessment in complex, real-world systems.

6.3.3 Towards Tracking Structural Changes—A Case Study

To demonstrate the practical application of the previously discussed concept of multi-targeted SHM strategies, the following section focuses on a real, operational wind turbine. The selected case study involves an extensively instrumented and continuously monitored WT over a 4-year period. The presented results highlight how structural response, environmental and operational data can be jointly exploited to assess the behaviour of the structure, and concise results from the fused monitoring framework are presented to illustrate its effectiveness in capturing both environmental variability and intrinsic structural complexity.

The structure under study is a 500-kW gearless horizontal-axis wind turbine erected in 1997 near Dortmund, Germany. It consists of a conical steel tower 63 m in height and a 40.3 m diameter rotor. The tower is composed of three separate 21 m-long steel segments, joined by bolted, prestressed flanged connections and anchored on a reinforced concrete foundation. The tower wall thickness increases from 10 mm at the top to 22 mm at the bottom, with corresponding outer diameters of 1.2 m, 2.15 m and 2.99 m at the three segment levels. The rotor comprises three 19.13 m fiberglass-reinforced epoxy (GRP) blades, rotating clockwise at variable speeds of 18–36 rpm on the upwind side of the structure.

In 2010, the turbine was equipped with a comprehensive monitoring system as part of a DFG-funded research project (Höffer et al., 2016). The sensing system includes six triaxial accelerometers (PCB-3713D1FD3G MEMS) mounted at five heights along the tower, six inductive displacement transducers (HBM-W5K) at two levels, three single-axis piezoelectric seismic accelerometers (PCB-393B12) at the foundation and ten temperature sensors (Pt100) distributed along the tower shaft for both measurement and displacement compensation. A 3D ultrasonic anemometer monitors horizontal and vertical wind components, wind direction and temperature, while SCADA data continuously records power output, rotor speed, nacelle orientation and blade pitch angle over the full monitoring period. Figure 6.21 summarizes the instrumentation layout; additional details regarding the monitoring system can be found in (Lachmann et al., 2009).

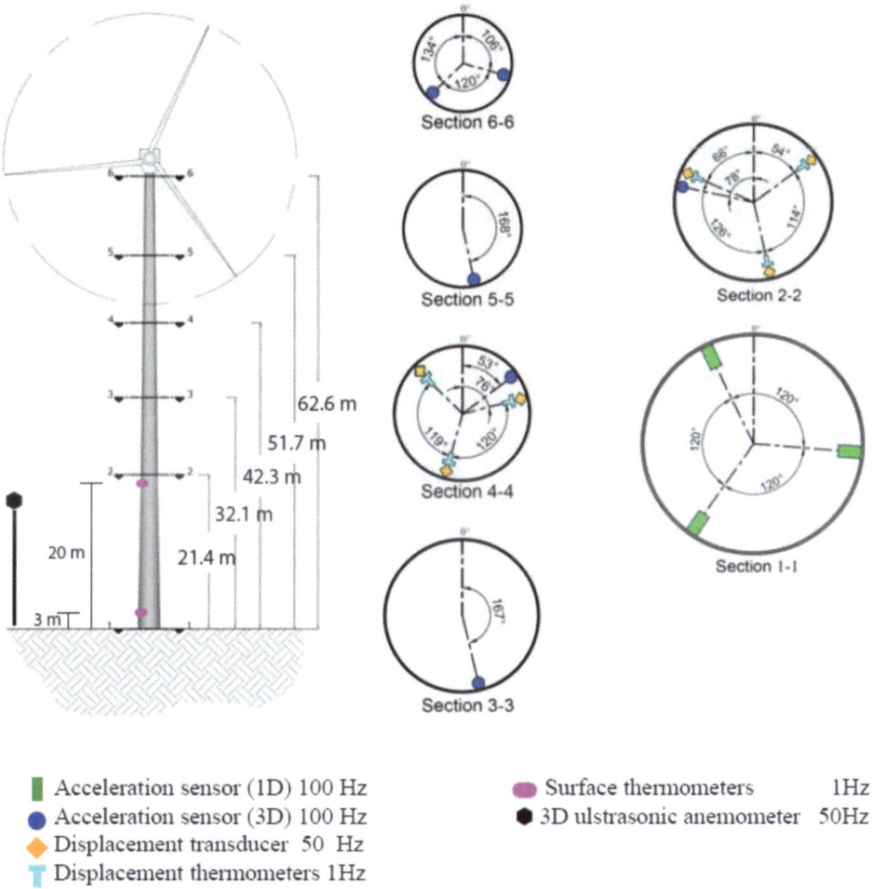

Fig. 6.21 Installed monitoring system and sampling frequencies (adopted from Höffer, Tewolde, & Bogoevska, 2017)

In the Dortmund WT case study, statistical methods can be applied to the Damage Indicator (DI) derived from trained data-driven models to support anomaly detection and long-term tracking. Among these, control charts are widely used due to their simplicity and suitability for automated, continuous monitoring of operational systems. The underlying concept, in historical context often considered as the formal beginning of the utilization of statistical quality control strategies, was developed in year 1924 by Walter A. Shewhart of the Bell Telephone Laboratories (Montgomery, 2007).

The schematic graph illustrated in Fig. 6.22 represents a typical control chart which displays a quality characteristic, measured or computed from a certain sample set, versus the sample number or time, marked at the horizontal axis. The average value of the monitored parameter, which corresponds to the normal or in-control state, is plotted as a central line. The upper control limit and the lower control limit lines are selected to mark the area where the process is assumed to be in control. As soon as the monitored feature performs outside of the control limits further investigation and according action should be undertaken. In order to attain an easier visualization of the feature evolution in time, the points are usually connected as straight-line segments. When designing a control chart, definition of the control limits is the most critical task. The 'distance' of the control limits from the centre line, expressed in standard deviation units, represents the percentage of observations which will fall between the control limits when the monitored system is in good condition. The standard deviation multiplier is usually selected as three, corresponding to approximately 99.7% of the observations falling between the limits under the assumption of a normal distribution. This choice is supported by practical experience, as three-sigma control limits generally deliver good results in practice.

Shewhart charts are an extremely advantageous approach when an out-of-control process results in large shifts of the tracked feature. In addition, as a result of the straightforwardly visualized patterns they can be very useful in identifying the nature or source of the system change. On the other hand, a big disadvantage of the approach

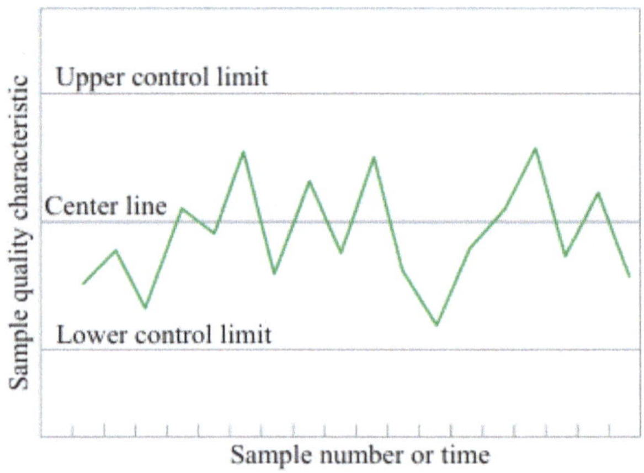

Fig. 6.22 Schematic graph of control chart (adopted from Bogoevska, 2020)

is related to its limited use of information about the process. More precisely, the chart utilizes information about the monitored process only from the last sample observation. As a result of this feature, namely ignoring information related to the complete sequence of points, the control chart is relatively insensitive to small process shifts. In this context, in a scenario when the process tends to operate within limit lines, with acceptable estimates of the process mean and standard deviation, and the sources of fluctuation do not result in large disturbances, the Shewhart chart becomes less useful. A second inferiority is related to its application for multivariable processes with correlated variables.

In addition to control charts, another approach is to utilize indicators based on statistical dissimilarity measures (Sugiyama et al., 2013). Statistical distance and similarity tools are based on concepts which resemble the familiar approach of 'template matching'. They have fundamental importance in the field of statistics and are commonly utilized tools in many areas of data analysis for a long time (Fig. 6.23). For example, the Euclidean distance and the linear correlation coefficient are frequently applied representatives of the two metrics, respectively.

In the context of the present case study, Fig. 6.24 illustrates the performance of a control chart applied to a DI obtained from a relatively short training period of 2 months (extract from Bogoevska, 2020). The input data time histories were analysed using univariate outlier detection based on the commonly used discordancy measure. In Fig. 6.24, the validation set of the estimated DI and statistical outlier analysis of the input data time histories illustrate that index values exceeding the $\pm$ 3std thresholds (99.7% confidence intervals calculated for the fitted Gaussian distribution of the estimation set of the obtained DI) can be linked to novel data ranges of the measured influencing agents, more precisely temperature and RPM values between months March and November in the year 2012, as well as April and September in the year 2013.

Further increase of the training period (Fig. 6.25) results in a decrease in the outlier percentages (SCADA variables with zero percent are not included in the plots) and correspondingly, the DI values become significantly reduced. In the case

Fig. 6.23 Typical distance and divergence metrics (adopted from Bogoevska, 2020)

Metric	Formulation
Euclidean distance	$L^2(P,Q) = \sqrt{\sum_{x\in X} (P(x) - Q(x))^2}$
Total Variation distance	$\delta(P,Q) = \frac{1}{2} \sum_{x\in X} \mid P(x) - Q(x) \mid$
Komogorov-Smirnov distance	$KS(P,Q) = \sup_{x\in X} \mid P(x) - Q(x) \mid$
Hellinger distance	$H(P,Q) = \frac{1}{\sqrt{2}} \sqrt{\sum_{x\in X} (\sqrt{P(x)} - \sqrt{Q(x)})^2}$
Kullback–Leibler divergence	$KLD(P\|Q) = \sum_{x\in X} P(x)\, log(\frac{P(x)}{Q(x)})$

* where P, Q are discrete probability distribution functions, and X is a discrete random variable that can take on values $x1, x2, \ldots$..

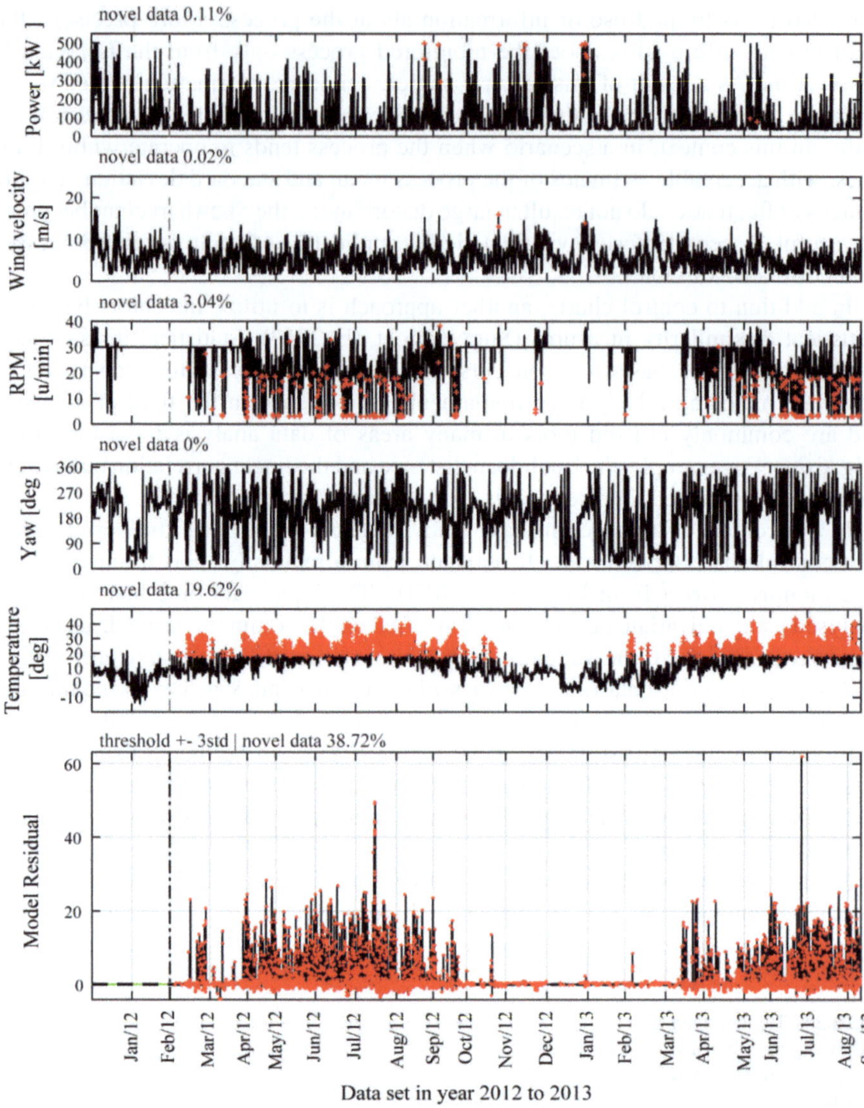

Fig. 6.24 Two-month training set. Identified novel data (red points) within time history of 10-min mean values of measured SCADA and control chart of the DI (model residual) (adopted from Bogoevska, 2020)

of the 12-month period of training, the outlier percentages drop below 0.2% and the DI distribution pattern of the testing set is evidently improved, with substantially less points above the threshold values. More results are presented in Bogoevska (2020).

As most sensitive to outliers, in this case study the Kullback–Leibler (K-L) divergence measure is utilized. When an estimated threshold limit value of the

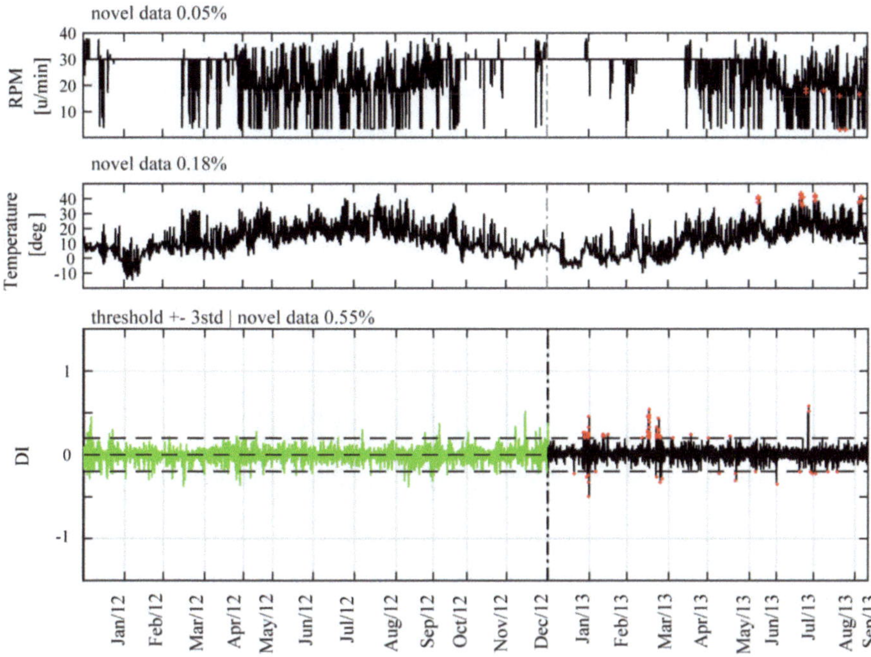

Fig. 6.25 Twelve-month training set. Identified novel data (red points) within time history of 10-min mean values of measured SCADA and Control chart of the DI (adopted from Bogoevska, 2020)

K-L divergence measure is exceeded, an alarm should trigger a proper retraining action. A 1-year setting with 1-month training set is presented in Fig. 6.26. When the resulting values exceed the pre-set threshold, for a certain predefined amount of time, a retraining procedure starts (see time history in Fig. 6.26, marked pink). This evidently improves the model performance for the following months, May to December 2012.

The results from the K-L divergence, applied daily on the obtained DI, are presented in Fig. 6.27. The sensitivity of the DI values to unfamiliar EOPs fluctuations is tested for 2-month, 4-month and 12-month training periods. The red dots, marked at the plot, are the first alarming values (indicating changes in the system) delivered by the tested statistical model. Furthermore, the estimated K-L divergence value changes in correlation with utilized training lengths (improved training results in lower values). A detailed discussion and analysis of these effects can be found in Bogoevska (2020).

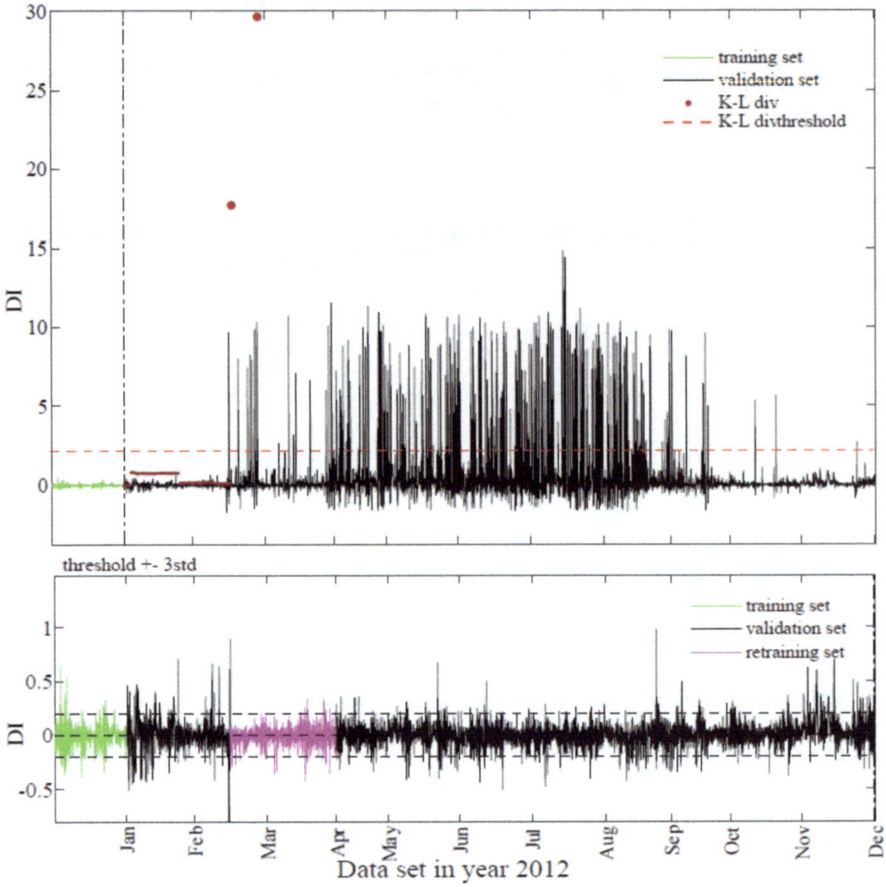

Fig. 6.26 DI-based K-L divergence illustrated on 1-year sample with proper retraining (adopted from Bogoevska, 2020)

6.4 Application of Artificial Neural Networks in Wind Engineering

Artificial neural networks (ANNs) have emerged as powerful tools in various scientific and engineering disciplines, demonstrating remarkable ability to learn complex patterns from data. A foundational understanding of these networks is aimed for in this document, starting with their fundamental building blocks. The core components of an artificial neural network will first be delved into, exploring the concept of artificial neurons and how they are organized into interconnected layers: the input layer, hidden layers and the output layer. Following this, the flow of data through the network will be described, illustrating the process by which input data is transformed through these layers to produce an output.

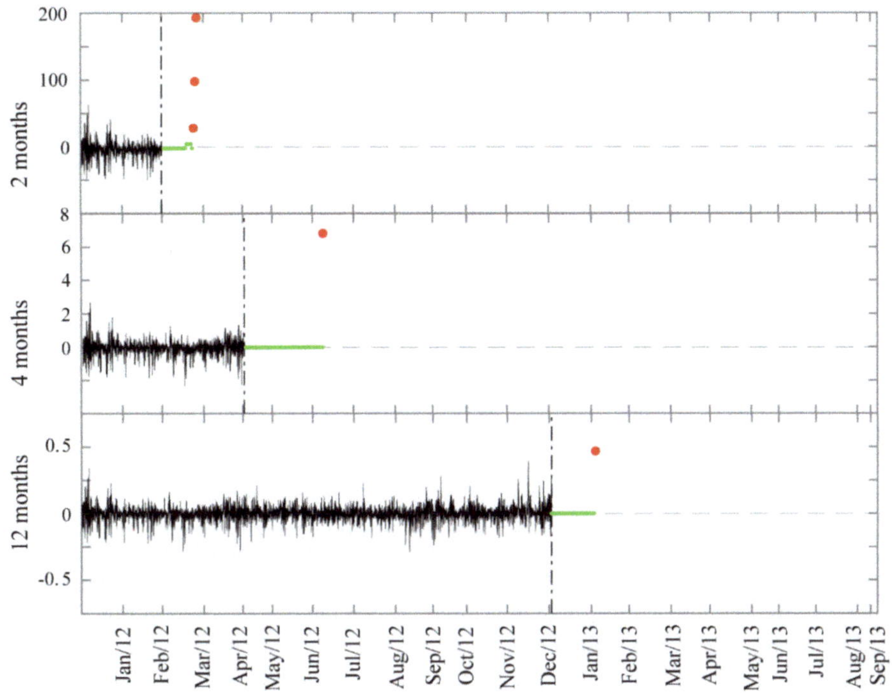

Fig. 6.27 Kullback–Leibler divergence applied on the obtained DI with cumulative assessment of one DI value per day, and tested for 2, 4 and 12 months of training (adopted from Bogoevska, 2020)

Following this foundational overview, a practical example of utilizing ANNs will be presented and explained in detail. Finally, recent advancements in this field by focusing on physics-informed neural networks (PINNs) will be briefly touched upon.

Its objective is to provide experts in the field of wind engineering with the opportunity to apply their knowledge and experience in neural networks. This could be a step forward in overcoming some challenges in this specialized field.

6.4.1 Fundamentals of Neural Networks

Artificial intelligence (AI) is the broad concept of creating machines capable of tasks requiring human intelligence, like learning and problem-solving. Machine Learning (ML) is a subset of AI that focuses on enabling computers to learn from data without explicit programming, identifying patterns to make predictions or decisions. Neural networks (NN), in turn, are a specific type of Machine Learning model inspired by the human brain's structure, excelling at recognizing patterns in complex data. Therefore, AI is the overarching goal, ML is a method to achieve AI by learning from data and NNs are a particular type of ML model used for pattern recognition.

Fig. 6.28 The schematic of an artificial neural network

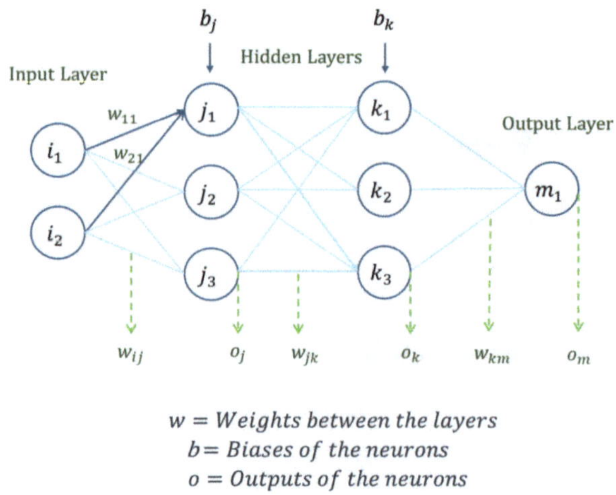

At its core, a neural network processes data through interconnected layers. Data enters via an input layer, where each neuron receives a specific feature. This information then flows through one or more hidden layers as shown in Fig. 6.28.

For a single neuron in a hidden layer, let $x=[x_1, x_2, ..., x_n]$ be the vector of inputs to a neuron, and $w=[w_1, w_2, ..., w_n]$ be the vector of synaptic weights. Let 'b' be the bias term. The weighted sum X is calculated as follows:

$$X=\sum_{i=1}^{n}(w_i x_i)+b=w^T x+b \tag{6.6}$$

In each neuron of these hidden layers, the weighted sum (X) is then passed through an activation function to produce the neuron's output ($o=\sigma(X)$). Figure 6.29 shows the process taking place in each neuron. This function introduces non-linearity, which is crucial for the network to learn and represent complex, real-world patterns that linear models cannot. Common activation functions include Sigmoid, Tanh and ReLU.

The network's result is produced by the output layer. The learning capability of neural networks stems from the adjustment of these weights and also biases during a process called training. This begins with a forward pass, where data propagates through the network to generate a prediction. The accuracy of this prediction is then measured by a loss function, which quantifies the discrepancy between the network's output and the true target value. Mean Squared Error (MSE) is a frequently used loss function due to its differentiability, which is essential for gradient-based optimization. For a given set of N training examples, the difference between the predicted output 'o' and true value 't' is calculated, squared, and then the average of these squared differences is calculated as follows:

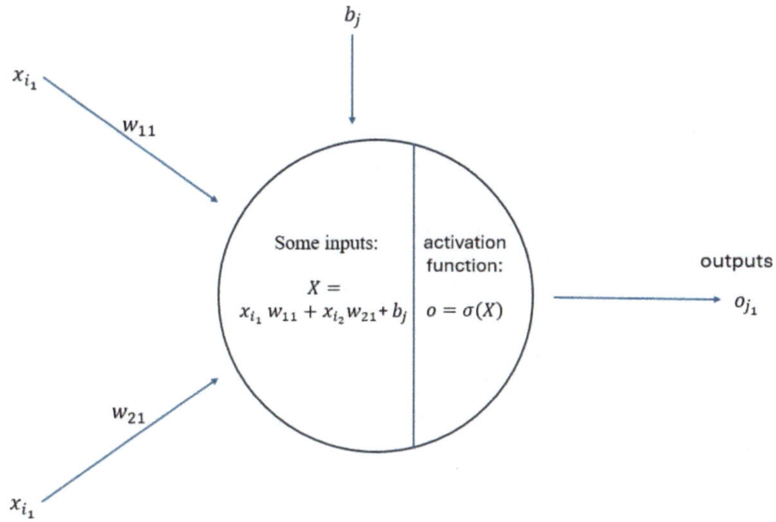

Fig. 6.29 The schematic of one neuron

$$MSE = \frac{\sum_{m=1}^{N} (t_m - o_m)^2}{N} \tag{6.7}$$

To minimize this loss, the network employs optimization algorithms, primarily Gradient Descent. This method iteratively adjusts weights and biases in the direction opposite to the gradient of the loss function, effectively moving towards a minimum error. Backpropagation is the systematic algorithm used to calculate these gradients and propagate the error signal backward through the layers, allowing for efficient weight updates. More advanced optimizers like Adam (Kingma & Ba, 2015) and L-BFGS (Zhu & Byrd, 1997) further enhance this process.

Effective training also relies on proper weight initialization, such as Glorot initialization (Glorot & Bengio, 2010), which helps prevent issues like vanishing or exploding gradients. Additionally, normalizing input data is vital; it ensures faster and more stable convergence during training by bringing features to a similar scale. Various network hyperparameters, such as the learning rate, the number of hidden layers and the number of neurons per layer, significantly influence performance. These settings are not learned but are tuned through methods like Bayesian optimization to achieve optimal results.

6.4.2 *Integrating Wind Tunnel Data with Artificial Neural Networks*

Artificial neural networks have been increasingly utilized within the field of wind engineering, particularly in tasks involving the prediction of wind pressures on structures based on experimental data.

Wind tunnel tests are a standard methodology for measuring wind pressures on scaled structural models. While these tests yield valuable data, they can be both time-consuming and financially intensive. Furthermore, data is typically collected for specific wind orientations and structural geometries, making it challenging to predict pressures for conditions not directly tested.

Neural networks possess the ability to learn the complex, non-linear relationships between various input parameters, such as building geometry, wind characteristics (like wind direction and velocity) and the specific location on the structure and the resultant pressure coefficients. This learned relationship allows ANNs to serve as powerful predictive tools.

To illustrate the potential of ANNs in this domain, consider a scenario where wind tunnel test data of the rectangular building for different wind angles has been collected. This dataset includes the location of pressure taps on the structure (x, y coordinates) and the measured mean pressure coefficients (C_p) and fluctuating pressure coefficients ($\grave{C}_p$) at these locations for each tested wind angle.

A typical workflow to leverage ANNs for pressure coefficient prediction in this context could involve the following steps:

1. **Data Acquisition**: Obtain wind tunnel test data encompassing different building dimensions and wind angles. The dataset should include pressure tap coordinates and corresponding measured pressure coefficients (C_p and $\grave{C}_p$).
2. **Data Preparation**

   - **Feature Identification**: Define the input features for the neural network. These would typically include parameters influencing the pressure coefficient, such as follows:

     - Normalized x-coordinate (x/B).
     - Normalized y-coordinate (y/L).
     - Wind angle.

   - **Target Identification**: Define the target variables to be predicted. In this case, these are the mean pressure coefficient (C_p) and fluctuating pressure coefficient ($\grave{C}_p$

     - **Data Integration**: Combine the data from different wind angles and aspect ratios into a structured dataset. Each data point would represent a specific measurement condition (x, y, Angle) and the corresponding C_p and $\grave{C}_p$

   - **Normalization**: Normalize the x and y coordinates using the building's width (B) and length (L), respectively. This is a critical step for ensuring feature consistency across varying structural dimensions.

3. **Model Construction**: Design a neural network architecture. This typically involves defining an input layer (matching the number of features), one or more hidden layers (to learn complex relationships) and an output layer (configured to predict C_p and $\grave{C}_p$) as shown in Fig. 6.30.
4. **Training Process**: Train the neural network using the prepared dataset. During training, the network iteratively adjusts its internal weights and biases to minimize the difference between its predicted pressure coefficients and the actual measured values. Optimization algorithms, such as Adam or L-BFGS, are commonly used to guide this process.
5. **Prediction Generation**: Once trained, the model can be used to predict pressure coefficients for new, unseen conditions. This could involve generating a new set of features for another building and the subsequent use of the trained model to predict the corresponding coefficient pressure values.

The predictions generated by the ANN can then be evaluated against any available validation data or compared to results from computational fluid dynamics (CFD) simulations. A well-trained neural network can effectively serve as a predictive tool, offering estimated pressure coefficients for various structural geometries and wind angles, potentially reducing the need for extensive additional physical wind tunnel tests or computationally expensive simulations for every possible scenario. This highlights the significant potential of ANNs to complement traditional methods in wind engineering practice.

The work presented here focuses on a single building size. However, this approach can be made more general. To predict pressure coefficients for buildings of various shapes and sizes, more measurement data from different building geometries would need to be collected. If this is done, the ANN model can be taught to handle these variations. In such a case, the building's dimensions (like its length and width) should be added as new input features for the ANN, alongside other information like wind angles.

6.4.3 Application of Physics-Informed Neural Networks in Wind Engineering

Physics-informed neural networks (PINNs) (Raissi et al., 2019) represent an innovative approach to solving problems governed by physical laws, often expressed as partial differential equations (PDEs). Unlike traditional neural networks that primarily learn from data, PINNs integrate the underlying physics directly into the training process. This allows them to not only leverage available data but also to discover solutions that are consistent with fundamental physical principles, such as

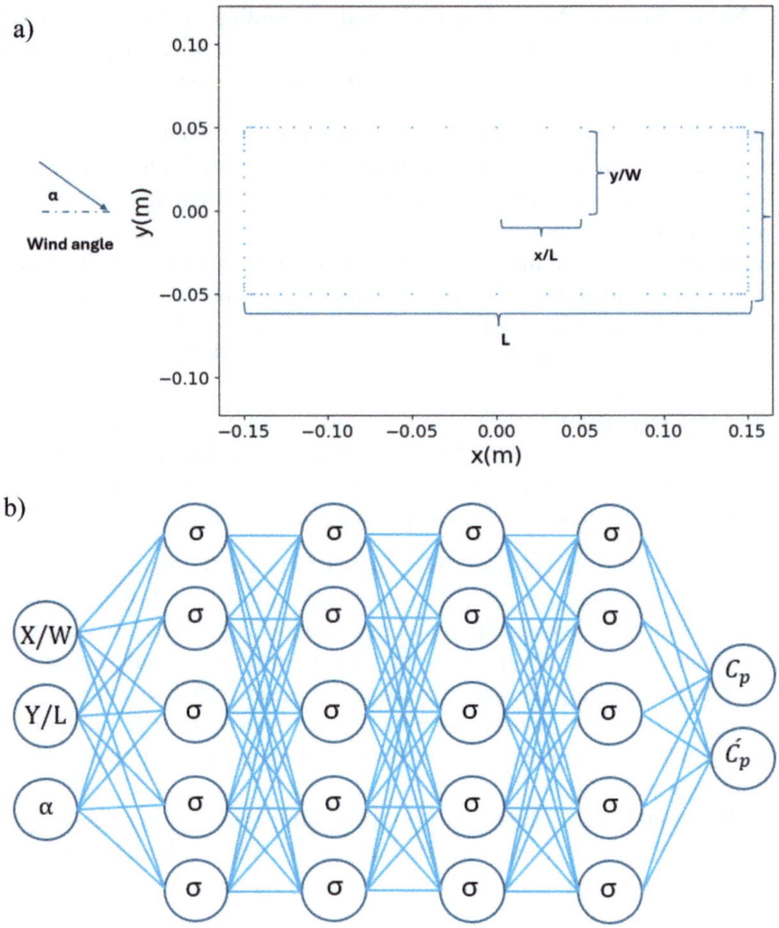

X, Y : Spatial coordinates of the building roof

B, L: Building's dimension

C_p: Mean coefficient pressure

α : Wind angle

$\acute{C}_p$: Fluctuating wind pressure coefficient

σ : Activation function

Fig. 6.30 **a** Neural network input features on the building plan **b** Schematic of ANN for predicting coefficient pressure on building roof

those describing heat transfer or fluid dynamics. To fully appreciate the mechanics of PINNs, automatic differentiation (AD) (Güne et al., 2018) is a key concept. AD efficiently calculates exact derivatives of complex functions, essential for neural network training by computing gradients during backpropagation.

As mentioned above, in PINNs, the loss function is designed to ensure that the neural network's output not only matches any available training data but also satisfies the underlying physical laws described by partial differential equations (PDEs). This is achieved by incorporating the PDE residuals into the loss function.

The loss function in a PINN typically consists of multiple components, such as data loss, physics-informed loss and boundary and initial condition loss. Data loss (L_{data}) is a standard data-driven loss, like those used in traditional neural networks. It measures the discrepancy between the network's predictions at locations where training data is available and the actual observed data values. For example, this could be the Mean Squared Error (MSE) between the predicted solution and the measured data points. Physics-informed loss ($L_{physics}$) is the crucial component that incorporates the physical laws. It is calculated based on the residuals of the governing PDEs. In PINNs, the neural network's output is treated as an approximate solution to the PDE. PINNs also typically include loss components that enforce the boundary and initial conditions of the problem ($L_{boundary/initial}$). These measure the discrepancy between the network's output at the boundaries and initial time points and the specified boundary and initial condition values. The total loss function (L_{total}) is usually a weighted sum of these individual loss components:

$$L_{total}=w_{data}L_{data}+w_{physics}L_{physics}+w_{boundary/initial}L_{boundary/initial} \tag{6.8}$$

where w_{data}, $w_{physics}$ and $w_{boundary/initial}$ are weighting factors that control the relative importance of each loss component during training (Wang et al., 2020). By minimizing this total loss function, the PINN is trained to find a solution that not only fits the available data but also satisfies the physical laws (PDEs) and boundary/initial conditions of the system. Automatic differentiation is essential for calculating the $L_{physics}$ component by efficiently computing the derivatives needed to evaluate the PDE residuals.

In the context of wind engineering, PINNs hold a potential for predicting complex flow fields around buildings and structures. By incorporating the fundamental equations of fluid dynamics (such as the Navier–Stokes equations) into the neural network's loss function (Fig. 6.31), PINNs can learn to predict flow variables (like velocity and pressure) while simultaneously satisfying the physical laws governing the fluid behaviour. This physics-informed approach can lead to more accurate and physically consistent predictions, especially in scenarios with limited experimental data.

Challenges of Physics-Informed Neural Networks for Wind Engineering

The application of physics-informed neural networks (PINNs) to wind engineering problems, such as predicting wind flow around structures, holds significant promise. However, several inherent challenges must be addressed for their widespread and reliable implementation in this domain. Concurrently, these challenges highlight fertile ground for future research and development.

The inherent complexity of the governing equations poses a significant challenge in accurately modelling wind flow. Specifically, the Navier–Stokes equations, which describe fluid motion, are a set of highly non-linear partial differential equations. Integrating these complex equations both accurately and stably into the PINN loss

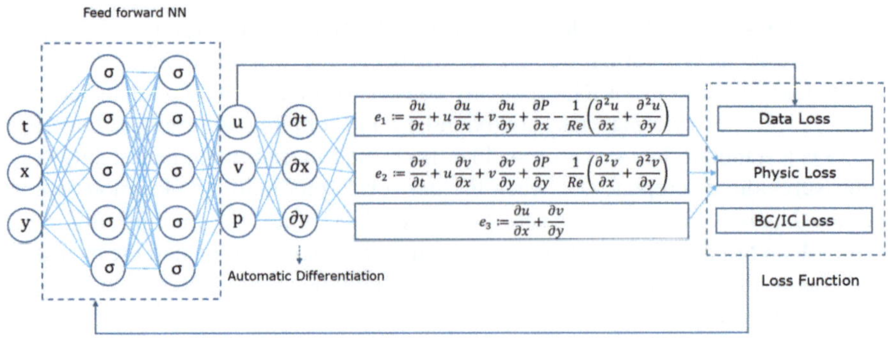

Fig. 6.31 Schematic of PINN for solving Navier–Stokes equation

function becomes particularly challenging, especially when dealing with the turbulent flows commonly encountered in wind engineering applications. Furthermore, the computational cost associated with training PINNs is considerable, particularly for intricate geometries and large computational domains typical of wind engineering. This demanding nature is largely due to the necessity of computing derivatives of the neural network output through automatic differentiation at numerous collocation points. Another technical hurdle lies in effectively handling complex geometries. Wind flow simulations frequently involve sophisticated building shapes and urban environments. Representing these intricate geometries accurately within the PINN framework and ensuring precise enforcement of physical boundary conditions are crucial.

It is important to carefully consider how to include limited and imperfect experimental data. PINNs can use data from sources like wind tunnel tests or field studies. However, it is very important to find a balance between reducing errors with this data and making sure the model follows physical laws. This balance is essential for getting a good model that performs well and can be applied to new situations. Lastly, it is still hard for PINNs to correctly show turbulence, which is a key part of how wind moves. Future research must work on using existing ways to model turbulence or create new PINN-based methods to solve this important problem.

Open Research Avenues and Potential Impact of PINNs in Wind Engineering

Despite the aforementioned challenges, the application of physics-informed neural networks (PINNs) in wind engineering remains a nascent field offering substantial opportunities for impactful research and development. Advancements in this domain are poised to yield significant benefits. Firstly, PINNs have the potential to reduce preprocessing efforts compared to traditional Computational Fluid Dynamics (CFD) methods. These conventional techniques often necessitate extensive mesh generation, whereas PINNs are mesh-free, which could streamline problem setup, particularly for geometries of high complexity. Secondly, once trained, PINN models are capable of providing predictions of wind flow variables considerably faster than iterative CFD simulations. This acceleration is particularly valuable for applications demanding rapid analysis, such as real-time wind environment assessments or preliminary design evaluations. Thirdly, by integrating fundamental physical laws, PINNs demonstrate enhanced generalization capabilities when applied to novel wind scenarios or slightly modified geometries, especially under conditions where experimental data is limited. Furthermore, in instances of complex wind phenomena where precise governing equations are uncertain, PINNs could serve as a valuable tool for discovering or refining these physical laws from available data. Finally, the PINN framework offers a flexible approach for integrating and solving coupled multiphysics problems, which is crucial for addressing wind engineering challenges involving interactions between different physical domains, such as fluid–structure interaction and aeroacoustics.

The use of physics-informed neural networks (PINNs) in wind engineering is a new area with great potential for important research and development. Advances in this field can bring many benefits. For instance, PINNs can reduce the effort needed for preparing data because they do not require complex mesh generation, which is a common feature of traditional computational fluid dynamics (CFD) methods. This can save time and effort, especially for complex shapes. Furthermore, once trained, PINN models can predict wind flow variables much quicker than standard CFD simulations. This speed is very useful for tasks that need quick analysis, such as real-time evaluation of wind conditions or early design checks. PINNs also show promise in adapting to new wind situations or slightly different designs, even when there is not much experimental data. This is because they incorporate basic physical laws. Moreover, in complex wind situations where the exact governing equations are not fully known, PINNs could help to find or improve these physical rules using existing data.

6.5 Conclusions

This chapter demonstrates how measurement data can play a strategic complementary role, but highly essential, in the development of wind energy structures. Continuous full-scale measurements provide reliable insight into the aerodynamic and aeroelastic behaviour of wind turbine systems under real operational conditions. Through the use of input–output field measurements and data evaluation procedures, high-quality datasets can be utilized to support load identification, model calibration and the validation of prediction models for tracking behaviour or phenomena (such as vortex-induced vibrations) in wind turbine towers.

Recent advances in sensing technologies, data acquisition systems and computational capabilities have further enabled the deployment of permanent monitoring systems, offering continuous data streams that support structural assessment, long-term behaviour tracking and data-informed decision-making throughout the structural lifecycle. Within this framework, data-driven and hybrid modelling approaches complement traditional physics-based models by capturing complex, interacting phenomena and natural parameter variability that are difficult to represent analytically. Neural network-based methods, including physics-informed neural networks, enhance robustness by combining data-driven learning with governing physical laws. Collectively, the approaches presented in this chapter highlight how high-quality measurement data not only enable advanced and realistic prediction models and probabilistic analyses of aeroelastic effects such as VIV, but also provide a valuable foundation for broader research tasks, modelling efforts and the sustainable operation of wind turbine structures.

Acknowledgements Part of the research is funded by the Deutsche Forschungsgemeinschaft (DFG, German Research Foundation)—project number 493357786 which is gratefully acknowledged by the authors. The authors would also like to thank SIEMENS Gamesa Renewable Energy Deutschland GmbH for providing the measurement data and measurement system as well as for all further support.

References

Bogoevska, S., Dumova-Jovanoska, E., & Höffer, R. (2016a). Response-based modal identification of a real steel chimney. *Scientific Journal of Civil Engineering*. ISSN 1857–839X.

Bogoevska, S., Spiridonakos, M., Chatzi, E., Dumova-Jovanoska, E., & Höffer, R. (2016b). A data-driven framework for comprehensive identification of operational wind turbines under uncertainty. In *Proceedings of ISMA 2016–International Conference on Noise and Vibration Engineering and USD2016–International Conference on Uncertainty in Structural Dynamics.*

Bogoevska, S., Nakov, D., Arangjelovski, T., & Markovski, G. (2020). Monitoring-based structural identification of two railway bridges. *Scientific Journal of Civil Engineering, 9*(1), 77–82. ISSN 1857–839X.

Bogoevska, S. (2020). *A holistic framework for data-driven diagnostics of operational wind turbines,* Doctoral dissertation, Ruhr-Universität Bochum, Germany.

Bogoevska, S., & Chatzi, E. (2025). Robust data-driven health monitoring of full-scale concrete dam structures via stochastic ML schemes. In *Proceedings of the 11th International Operational Modal Analysis Conference (IOMAC 2025).*

Bogoevska, S., Spiridonakos, M., Chatzi, E., Dumova-Jovanoska, E., & Höffer, R. (2017). A data-driven diagnostic framework for wind turbine structures: A holistic approach. *Sensors, 17*(4), 720. https://doi.org/10.3390/s17040720

Brincker, R., Ventura, C. E., & Andersen, P. (2002). Why output-only testing is a desirable tool for a wide range of practical applications. In *Proceedings of the 20th International Modal Analysis Conference (IMAC-XX).*

Brownjohn, J. M. W. (2007). Structural health monitoring of civil infrastructure. *Philosophical Transactions of the Royal Society A, 365*, 589–622.

Fang, X., Wang, H., Liu, G., & Xie, Y. (2025). A review of structural health monitoring for fixed offshore wind turbines driven by multisource heterogeneous data. *Structural Health Monitoring, 0*(0). https://doi.org/10.1177/14759217251371673

Garcia-Ramonda, L., Bogoevska, S., Reuland, Y., Martakis, P., Dertimanis, V., & Chatzi, E. (2025). Ponte Moesa Campagnola: A bridge benchmark for structural identification under controlled damage progression. *Journal of Structural Engineering, 151*(12), 04025218.

Georgijev, V., & Bogoevska, S. (2025). Application of the polynomial chaos expansion method for forecasting structural response of two full-scale case studies. *Building Materials and Structures, 68*(1), 13–32.

Glorot, X., & Bengio, Y. (2010). Understanding the difficulty of training deep feedforward neural networks. http://www.iro.umontreal.

Gohar, I., Yew, W. K., Halimi, A., & See, J. (2025). Review of state-of-the-art surface defect detection on wind turbine blades through aerial imagery: Challenges and recommendations. *Engineering*

Applications of Artificial Intelligence, 144, Article 109970. https://doi.org/10.1016/j.engappai.2024.109970

Gomez, P., & Yankova, G. G. (2022). SG 14.X-DD-222 Measurement System & Calibration Report. DTU Wind Energy WTT I-1334(EN)-R0. Denmark.

Güne¸, A., Baydin, G., Pearlmutter, B. A., & Siskind, J. M. (2018). Automatic differentiation in machine learning: A survey. *In Journal of Machine Learning Research, 18*.

Hansen, S. O. (1981). Cross-wind vibrations of a 130-m tapered concrete Chimney. *In Journal of Wind Engineering and Industrial Aerodynamics, 8*, 145–155.

Höffer, R., Tewolde, S., Bogoevska, S., Baitsch, M., Zimmermann, S., & Barbanti, G. (2016). Monitoring-based identification for structural life cycle management of wind energy converters. In *Europe and the Mediterranean towards a sustainable built environment (SBE16)*, vol. 1, pp. 509–520.

Höffer, R., Tewolde, S., & Bogoevska, S. (2017). Fatigue assessment and damage detection of wind turbine structures by continuous health monitoring. In *IABSE Symposium Report (Vol. 109). 58th International Association for Bridge and Structural Engineering*, vol. 809–816.

Kingma, D. P., & Ba, J. (2015). Adam: A method for stochastic optimization. http://arxiv.org/abs/1412.6980

Kühn, M., Gasch, R., & Sundermann, B. (2012). Structural dynamics. In R. Gasch & J. Twele (Eds.), Wind power plants (Chap. 8). Berlin: Springer.

Kurniawati, I. (2025). Cross-Wind Vibrations of Wind Turbine Towers. PhD Thesis, Ruhr-Universität Bochum.

Kurniawati, I., Lupi, F., Seidel, M., Höffer, R., & Niemann, H. J. (2025). Vortex-induced vibrations and post-lock-in cross-wind oscillations of wind turbine tower based on field measurements. *Journal of Wind Engineering and Industrial Aerodynamics, 258*, Article 106010.

Kurniawati, I., Lupi, F., Seidel, M., Niemann, H. J., & Höffer, R. (2024). Vortex shedding load on a full-scale wind turbine tower at transcritical Reynolds numbers. *Journal of Wind Engineering and Industrial Aerodynamics, 252*, Article 105821.

Lachmann, S., Höffer, R., Hartmann, D., & Baitsch, M. (2009). Structural lifetime prediction for wind energy converters basedon health monitoring and system identification. In *5th European & African conference on wind engineering: Florence Italy*, July 19th-23rd 2009, conference proceedings.-(Proceedings e report; 51) (pp. 1000–1004). Firenze University Press.

Lupi, F., Niemann, H. J., & Höffer, R. (2018). Aerodynamic damping model in vortex-induced vibrations for wind engineering applications. *Journal of Wind Engineering and Industrial Aerodynamics, 174*, 281–295.

Malekloo, A., Ozer, E., AlHamaydeh, M., & Girolami, M. (2022). Machine learning and structural health monitoring overview with emerging technology and high-dimensional data source highlights. *Structural Health Monitoring, 21*(4), 1906–1955.

Meirovitch, L. (2001). Fundamentals of Vibrations. McGraw-Hill Higher Education. McGraw-Hill. https://books.google.de/books?id=u358QgAACAAJ.

Montgomery, D. C. (2007). *Introduction to statistical quality control*. John Wiley & Sons.

Mostafa, K., Zisis, I., & Moustafa, M. A. (2022). Machine learning techniques in structural wind engineering: A state-of-the-art review. *Applied Sciences, 12*(10), 5232.

Mwangi, A., Sahay, R., Fumagalli, E., Gryning, M., & Gibescu, M. (2024). Towards a software-defined industrial IoT-edge network for next-generation offshore wind farms: State of the art, resilience, and self-X network and service management. *Energies, 17*(12), 2897. https://doi.org/10.3390/en17122897

Noori, M., Rainieri, C., Domaneschi, M., & Sarhosis, V. (2023). *Data-driven methods for civil structural health monitoring and resilience: Latest developments and applications*. CRC Press.

Oliveira, G. (2016). *Vibration-based structural health monitoring of wind turbines*, Doctoral dissertation, University of Porto, Portugal.

Pasam, A., D. Tudball Smith, J. D. Holmes, D. Burton, & M. C. Thompson. (2023). The influence of surface roughness on postcritical flow over circular cylinders revisited. *In Journal of Fluid Mechanics 975*, A36. Cambridge Core. https://doi.org/10.1017/jfm.2023.846.

Raissi, M., Perdikaris, P., & Karniadakis, G. E. (2019). Physics-informed neural networks: A deep learning framework for solving forward and inverse problems involving nonlinear partial differential equations. *Journal of Computational Physics, 378*, 686–707. https://doi.org/10.1016/j.jcp.2018.10.045

Rodriguez, C., & Jonkman, J. (2018). Introductory Note. In *Meeting Proceeding IEA Wind Task 11-Topical expert meeting. Topical Expert Meeting #88 on Three-Way Verification and Validation between Data, High-Fidelity Models and Engineering Models.*

Roetzer, J., Li, X., & Hall, J. (2024). Review of data-driven models in wind energy: demonstration of blade twist optimization based on aerodynamic loads. *Energies, 17*(16), 3897.

Ruscheweyh, H. (1974). Beitrag Zur Windbelastung Hoher Kreiszylinderähnlicher Schlanker Bauwerke Im Natürlichen Wind Bei Reynoldszahlen Bis Re=1.4 x 107. Phd, Technical University Aachen.

Schewe, G. (1983). On the force fluctuations acting on a circular cylinder in Crossflow from subcritical up to transcritical reynolds numbers. *In Journal of Fluid Mechanics, 133*, 265–285.

Stathopoulos, T., & Surry, D. (1983). Scale effects in wind tunnel testing of low buildings. *In Journal of Wind Engineering and Industrial Aerodynamics, 13*(1), 313–326. https://doi.org/10.1016/0167-6105(83)90152-6

Sugiyama, M., Kanamori, T., Suzuki, T., du Plessis, M. C., Liu, S., & Takeuchi, I. (2013). Density-difference estimation. *Neural Computation, 25*(10), 2734–2775.

Tcherniak, D., Chauhan, S., & Hansen, M. H. (2011). Applicability limits of operational modal analysis to operational wind turbines. In Structural dynamics and renewable energy (Vol. 1, pp. 317–327). Berlin: Springer.

van Hinsberg, N. (2015). The reynolds number dependency of the steady and unsteady loading on a slightly rough circular cylinder: from subcritical up to high transcritical flow state. *In Journal of Fluids and Structures, 55*, 526–539. https://doi.org/10.1016/j.jfluidstructs.2015.04.002

Vickery, B. J., & Clark, A. W. (1972). Lift or across-wind response of tapered stacks. *Journal of the Structural Division, 98*(1), 1–20.

Wang, S., Teng, Y., & Perdikaris, P. (2020). Understanding and mitigating gradient pathologies in physics-informed neural networks. http://arxiv.org/abs/2001.04536

Zhang, P., He, Z., Cui, C., Xu, C., & Ren, L. (2023). An edge-computing framework for operational modal analysis of offshore wind-turbine tower. *Ocean Engineering, 287*, Article 115720. https://doi.org/10.1016/j.oceaneng.2023.115720

Zhao, C., Zhang, F., Lou, W., Wang, X., & Yang, J. (2024). A comprehensive review of advances in physics-informed neural networks and their applications in complex fluid dynamics. In *Physics of Fluids*, vol. 36, no. 10. American Institute of Physics. https://doi.org/10.1063/5.0226562

Zhu, C., & Byrd, R. H. (1997). Algorithm 778: L-BFGS-B: Fortran Subroutines for Large-Scale Bound-Constrained Optimization.

Chapter 7
Knowledge-Enhanced Machine Learning in Wind Engineering

Teng Wu, Xinyang Yu, and Shaopeng Li

Abstract Recent studies have demonstrated the domain knowledge could be leveraged for advanced design and training of machine learning (ML) models along with improved selection of their inputs and outputs in engineering applications. This chapter first briefly introduces the knowledge representations (e.g., physical laws, empirical models, domain-specific guidance, and domain-related solutions) and effective approaches to incorporating various knowledge types into ML pipelines (e.g., modifying loss functions, model architectures, and training process). Then, a systematic review on knowledge-enhanced machine learning (KEML) in wind engineering is presented, with simulation applications in windstorms under climate change, bluff-body aerodynamics and aeroelasticity, structural dynamics, structural response mitigation, and post-hazard recovery (i.e., from traditional Alan G. Davenport Wind Loading Chain to windstorm resilience). Finally, the challenges and prospects are discussed to highlight the promising future of KEML in improving data efficiency, extrapolation capability, and model trustworthiness for wind engineering applications.

7.1 Introduction

The escalating climate crisis, driven primarily by greenhouse gas emissions, underscores the importance of seeking clean and renewable energy (IPCC, 2021). Among the available options, wind energy has gained prominence due to its vast resource,

T. Wu (✉) · X. Yu
Department of Civil, Structural and Environmental Engineering, University at Buffalo, Buffalo, NY, USA
e-mail: tengwu@buffalo.edu

X. Yu
School of Civil Engineering, Central South University, Changsha, Hunan, China

S. Li
Department of Civil and Environmental Engineering, University of Louisiana at Lafayette, Lafayette, LA, USA

C. Baniotopoulos and E. Marino (eds.), *Advances in Wind Energy in the Era of Artificial Intelligence*, CISM International Centre for Mechanical Sciences 625,
https://doi.org/10.1007/978-3-032-28815-8_7

fast-developing technology and efficient land use (Van Kuik et al., 2016; GWEC, 2021). Yet the advancement of wind energy systems depends not only on resource exploitation but also on addressing the fundamental engineering challenges associated with their exposure to the atmospheric environment (Pryor & Barthelmie, 2010; Willis et al., 2018; Veer et al., 2019). Essentially, the performance and integrity of the large-scale flexible wind turbines, support towers, and transmission infrastructures are governed by wind-structure interactions. This topic is the core of the wind engineering discipline (Cermak, 1975; Davenport, 2002), which spans from atmospheric flow and aerodynamic loading to structural dynamics and response control, ultimately the post-event recovery (Kareem, 2020). While wind energy provides the critical motivation, this chapter focuses specifically on the wind engineering perspective that underpins the structural safety and reliability of wind energy systems.

Climate change further complicates the problem of wind-structure interaction, because it alters the large-scale wind climate and increases the frequency and intensity of extreme wind events in a complex fashion (Leckebusch & Ulbrich, 2004; Meucci et al., 2020). Furthermore, human activities-associated urbanization and construction reshape the terrain and introduce additional perturbations through the atmosphere-structure feedbacks (Blocken, 2014). These combined effects on wind environment propagate to civil structures through nonlinear aerodynamics. In addition to aeroelastic instability (Chen & Kareem, 2001), the extreme winds may push systems beyond their original design envelopes to the region dominated by geometric and material nonlinearities, leading to difficult-to-predict responses and potential cascading failures (ASCE, 2019). The abovementioned uncertainties may compound at each stage, resulting a nonlinear amplification of overall risks.

This dire situation necessitates a shift in design philosophy, from traditional prescriptive codes toward performance-based design (PBD) (Scott et al., 2023). The PBD framework aims to explicitly satisfy the performance objectives that directly reflect stakeholder expectations, with the dual aim of enhancing safety and considerably reducing building lifecycle costs (ASCE, 2019). However, the practical implementation of PBD is computationally formidable, as it requires evaluation of structural performance through a large number of nonlinear dynamic analyses for reliable uncertainty quantification (Gidaris et al., 2017). The ability to efficiently and accurately simulate wind-structure interactions thus becomes the bottleneck for realizing the full promise of PBD (Spence & Kareem, 2014).

Concurrently, rapid advances in artificial intelligence (AI), driven by the availability of large datasets, breakthroughs in fundamental theory, and the associated computational resources, are transforming engineering disciplines (Jordan & Mitchell, 2015). In a broad sense, AI aims to create systems enabling perform tasks that traditionally require human intelligence, including recognition, reasoning, and decision-making (Russell & Norvig, 1995). Marked by the rise of machine learning (ML), AI has evolved from rule-based systems to data-driven approaches (Bishop, 2006), which focuses on developing algorithms to learn from data without explicit programming (Murphy, 2012). Particularly, the development of artificial neural networks (ANNs) has led to the emergence of deep learning (DL). DL is a branch of ML characterized by multi-layered neural architectures capable of automatic, hierarchical feature extraction from high-dimensional data, which cannot be solved by conventional ML models (Goodfellow et al., 2016; LeCun et al., 2015).

Depending on the purpose, ML models can be classified into three categories, namely supervised learning for regression/classification, unsupervised learning for pattern recognition/dimension reduction), and reinforcement learning (RL) for decision-making/closed-loop control (Murphy, 2012; Sutton & Barto, 1998). In the context of wind engineering, these methods have proven invaluable in many applications, from surrogate modeling of atmospheric flows, aerodynamic loads, structural response to the optimal control of structural systems (Wu & Snaiki, 2022). However, this data-driven promise is compromised by several challenges such as overfitting, limited extrapolation capabilities, and lack of physical interpretability (Chen et al., 2021; Karpatne et al., 2017; Willard et al., 2020). These deficiencies stem primarily from their agnosticism to the physical mechanism, making them particularly ill-suited for scenarios with scarce or biased data. This issue is critical in wind engineering. In addition to dealing with data scarcity of extreme wind events, reliable machine models are expected to comply with fundamental conservation laws (mass, momentum, and energy) and established engineering principles (e.g., structural dynamics and stability criteria) (Karniadakis et al., 2021; Von Rueden et al., 2021). These may resort to domain knowledge accumulated over the development of the research field. This gap highlights the need for a more robust approach of knowledge-enhanced machine learning (KEML).

KEML provides a systematic framework for integrating diverse forms of knowledge, either explicit or tacit (Psichogios & Ungar, 1992; Dissanayake & Phan-Thien, 1994; Raissi et al., 2019; Li et al., 2020a; Li et al., 2020b), into ML pipelines. By providing guidance through loss functions, ML architectures or the training process, KEML can improve generalizability, reduce reliance on large datasets, and enhance trustworthiness in various applications. KEML has created promising opportunities for wind engineering through a spectrum of strategies, tailored by the availability and nature of the knowledge. For tasks governed by explicit physical laws, such as wind field and aerodynamic load modeling, KEML can enforce these equations through loss functions to enhance physical consistency and data efficiency (Snaiki & Wu, 2019; Wang & Wu, 2020; Yu & Wu, 2023). KEML can also leverage domain intuition/heuristics to guide data preprocessing or model architecture design for enhanced computation efficiency (Li & Wu, 2023, 2025, 2025a; Yu & Wu, 2023). Furthermore, even in the absence of explicit laws or domain intuition/heuristics, KEML can treat the learned representations from low-cost, related tasks as transferable knowledge via transfer learning (Pan & Yang, 2009), thereby enhancing model performance for high-fidelity learning tasks (Li et al., 2021a; Yu & Wu, 2023).

The objective of this chapter is to establish a systematic review and perspective on KEML in wind engineering. Unlike prior reviews that focus broadly on ML applications (Wu & Snaiki, 2022), this work emphasizes the role of embedding knowledge into ML pipelines, with examples from wind field simulation to structural recovery. By mapping how KEML principles can be applied in different scenarios, this chapter aims to outline a cohesive framework that advances both the applicability and reliability of ML in wind engineering.

7.2 Framework of KEML: From Data-Driven Modeling to Knowledge-Assisted Learning

The transformative potential of ML in wind engineering has been widely acknowledged and comprehensively reviewed through its successful applications in a wide range of problems (Kareem, 2020; Wu & Snaiki, 2022). Most of these ML applications leverage supervised learning, targeting input–output mappings from the given data, such as training a surrogate to predict aerodynamic responses under different inflow conditions (Kim et al., 2021; Meddage et al., 2024). Unsupervised learning, where ML models draw hidden patterns from unlabeled data, also appears in dimensionality reduction for high-dimensional wind field data (Mylonas et al., 2019). Reinforcement learning, where an agent (typically a neural network) learns a policy to maximize cumulative rewards, has been gaining attention in wind engineering community as an efficient way to obtain optimal aerodynamic shape (Li et al., 2021a) and control scheme for wind-sensitive structures (Li et al., 2025a; Renn & Gharib, 2022; Xie et al., 2023). These ML-based solutions offer a powerful alternative to traditional methods, providing significant computational efficiency for tasks like surrogate modeling and real-time prediction/control/decision making.

Three interconnected barriers need to be overcome before reliably deploying these powerful models in wind engineering. The primary obstacle is data scarcity. High-fidelity data are inherently difficult/costly to obtain, whether through field measurements, wind tunnel experiments, or computational fluid dynamics (CFD). Complex ML models possess millions of parameters to be determined, and can overfit when trained on sparse data. These models then perform poorly on new, statistically similar scenarios, i.e., exhibiting poor generalizability. This issue is particularly critical for extreme events like tropical cyclones, where data is inherently rare. Secondly, purely data-driven models may suffer from physical inconsistency. The underlying physics in wind-structure interaction is governed by the Navier–Stokes equations coupled with structural dynamics, where the continuity of mass and momentum, turbulence dynamics, and aeroelastic feedbacks impose strong constraints. A ML model that is agnostic to these principles may produce spurious flow features, non-conservative force predictions, or unstable structural responses, undermining their reliability for safety–critical applications. The third and most fundamental challenge stems from the intrinsic need for extrapolation. Unlike conventional ML applications where ML operates within the bounds of its training data, engineering design and safety evaluation constantly demand predictions for untested configurations and unprecedented extreme events. A purely data-driven model, trained to interpolate within known scenarios, perform poorly for extrapolation. Its failure for reliable extrapolation threatens the core motivation for using ML in wind engineering, underscoring the necessity for models embracing physical principles, not just statistical patterns.

To overcome these barriers, pure data-driven approaches have evolved toward a more synergistic mode of KEML. While a closely related topic, the Physics-Informed Neural Networks or Physics-Informed Machine/Deep Learning (Karniadakis et al., 2021; Raissi et al., 2019), has gained significant attention in science and engineering applications, the scope of KEML is broader. The term "knowledge" is defined not only as first-principles physics (e.g., conservation of mass and momentum) but as any prior, domain understanding that can be used to constrain, guide, or improve the learning process. This encompasses empirical models, expert intuition, structural constraints, and tacit or implicit knowledge distilled from closely related but low-cost tasks. This chapter adopts "knowledge-enhanced" to capture this full spectrum of integration, while acknowledging "physics-informed" methods as a critical subset.

To establish a cohesive framework for KEML in wind engineering and present a systematic review of its pioneering applications, the rest of this section first introduces the typology of knowledge and the corresponding methodologies for enhancing machine learning. Section 7.3 then presents a detailed examination of KEML implementations, from windstorm modeling and aerodynamic simulation to structural dynamics, response control, and post-hazard recovery. Finally, Sect. 7.4 builds upon this review to discuss existing challenges and perspectives on research avenues for KEML, highlighting emerging areas where KEML could have a significant impact on wind engineering.

7.2.1 An Anatomy of Knowledge in Wind Engineering

It is essential to systematically clarify the forms and sources of domain knowledge in wind engineering that can be incorporated into machine learning pipeline. As summarized in Table 7.1, five primary types of knowledge are identified, namely, physical laws, empirical models, statistical relations, domain-specific guidance, and domain-related solutions. This anatomy builds conceptually upon the foundations of physics-informed and theory-guided learning frameworks (Karniadakis et al., 2021; Karpatne et al., 2017; Raissi et al., 2019) while emphasizing the context of wind engineering. It offers a structured lens to analyze how different forms of knowledge are integrated into data-driven models, a theme that will be explored through wind engineering applications in Sect. 7.3; however, the knowledge presented here is not intended to form a complete and disjoint knowledge spectrum in wind engineering field. There exist some overlaps in how these types of knowledge are represented. For example, the full-order and reduced-order Navier–Stokes equations, respectively, correspond to high- and low-fidelity knowledge representation of flow momentum conservation.

Table 7.1 Anatomy of knowledge in wind engineering

Knowledge category	Description	Representative examples in wind engineering
Physical laws	Fundamental governing equations and conservation laws derived from first principles	Navier–Stokes equations; structural dynamics equations; conservation laws (e.g., mass or energy)
Empirical models	Simplified physical models and empirical relationships	Logarithmic law of the wall; aerodynamic admittance functions; quasi-steady theory
Statistical relations	Statistics from data distribution	Probabilistic density function (PDF) of wind pressure
Domain-specific guidance	Procedural knowledge, operational heuristics, and design rules derived from engineering practice	Engineering design heuristics; expert judgment on flow separation; code-based safety margins
Domain-related solutions	Patterns and features learned by models from data-rich source tasks	Latent features/weights from low-fidelity simulations (e.g., low-resolution CFD) or pre-trained models on related systems

7.2.2 *Methodologies for Enhancing Machine Learning with Knowledge*

The schemes of integrating knowledge into the ML workflow are summarized in Table 7.2. Enhancement through the loss function is the most common approach. The KEML strategy augments the standard data-mismatch loss L_D with a knowledge-based loss term L_k, resulting in a composite objective $\mathrm{L} = L_D + \lambda\, L_k$. It represents a flexible approach where knowledge acts as a soft constraint, where effectiveness can be sensitive to the weighting parameter λ. The loss function-based constrains can be extracted from equation-based knowledge, including the physical laws, empirical models, and statistical relations. It should be noted that embedding equations in loss function do not guarantee an improved learning performance, considering that gradient descent-based algorithm for minimizing loss function is not effective for complex equations such as full-order Navier–Stokes equations under high Reynolds number.

In contrast, enhancement through model architecture encodes knowledge as a hard structural prior. Following this strategy, knowledge can be used, for example, to tailor recurrent units of long-short term memory (LSTM) networks according to "fluid memory" or select grid structures of convolutional neural networks (CNNs) according to the spatiotemporally varying flow field. These modified model architectures are often more data-efficient and inherently satisfy certain physical properties by design.

Table 7.2 Methodological framework for KEML in wind engineering

Enhancement pathway	Mechanism	Knowledge type or source (from Table 7.1)	Exemplary wind engineering applications (to be detailed in Sect. 7.3)
Loss function-based constrains	Incorporating residuals of physical equations/ empirical relations as regularization terms in the loss function	(1) Physical laws; (2) Empirical models; (3) Statistical relations	Loss function derived from reduced-order Navier–Stokes equations in simulation of tropical cyclone winds (Snaiki & Wu, 2019); Loss function derived from equations of motion for structural dynamics (Wang & Wu, 2020); Loss function derived from non-Gussian PDF for wind pressure simulations (Li & Wu, 2026)
Model architecture-based constrains	Hard-coding knowledge like symmetries or conservation laws as structural priors into the network design	(1) Physical laws; (2) Empirical Models	Tailored LSTM unit for unsteady aerodynamics modeling (Li & Wu, 2023); Non-uniform grid for CNN-based flow field simulations (Yu & Wu, 2023)
Training process-based constrains	Leveraging knowledge to guide the learning algorithm itself, e.g., through data sequencing or model initialization	(3) Domain-specific guidance; (4) Domain-related solutions	Transfer learning from low- to high-fidelity simulation for shape optimization (Li et al., 2021a); Design auxiliary reward for RL in post-hazard recovery decision making (Li & Wu, 2025, 2025a)

Finally, enhancement through the training process leverages knowledge to guide the learning algorithm itself. One strategy is based on transfer learning, where knowledge extracted from low-cost, data-rich tasks is reused to accelerate learning on a data-scarce target (Pan & Yang, 2009). This transfer may occur at the level of model parameters, initializing the model with pretrained weights to have a warm-start optimization. Or, the transfer learning can occur at the level of latent representations, where intermediate features learned in the source domain are preserved and selectively adapted to support new tasks. By reusing informative priors, the model converges more rapidly and exhibits improved generalization without requiring exhaustive data from scratch. This pathway is particularly powerful for leveraging existing, low-fidelity data to bootstrap models for complex, data-scarce target tasks.

7.3 Applications of KEML in Wind Engineering

7.3.1 Windstorm Modeling Under Climate Change

Accurate simulation of near-ground wind flows, particularly for extreme wind events, constitutes the first critical component in the wind-loading chain. Climate change has generally increased the frequency and intensity of extreme wind events, elevating the associated risks of the built environment (Peterson, 2000; Walsh et al., 2016). Effective risk assessment and mitigation rely on the ability to generate numerous, physically realistic wind field scenarios for probabilistic analysis, which poses significant computational challenges.

As one of the most hazardous extreme wind events, tropical cyclone (TC) are responsible for inflicting severe economic losses and casualties over regions spanning hundreds of kilometers through strong wind, heavy rain, and high surge (Pielke et al., 2008; Rappaport, 2014). The strong wind in the atmospheric boundary layer is an important driving force to TC-related damage. Caused by pressure difference, earth rotation and surface frictional forcing, TC is a 3D rotating flow system. Accurate and efficient modeling of the wind field is critical to effective mitigation of TC-related hazards. While numerical weather prediction models (e.g., Weather Research and Forecasting model) based on solving Navier–Stokes equations can obtain high-accuracy wind fields, they are computationally prohibitive for rapid scenario-based simulations in uncertainty quantification. Empirical or analytical models (e.g., Holland, 1980; Kepert, 2001; Snaiki & Wu, 2017a, 2017b) are computationally efficient but rely on simplifications or predefined profile functions, often failing to fully capture the complex nonlinear TC dynamics. ML approaches have emerged as promising alternatives. Huang and Xu (2013) utilized the ANN to consider topography effects in the simulation of directional typhoon wind field at the Stonecutters Bridge site in Hong Kong. Stiles et al. (2014) developed an ANN model that utilized SeaWinds scatterometer measurements to predict sea surface wind speed with a 12.5 km resolution. Wei (2015) evaluated Support Vector Machine (SVM) models with various kernel functions, successfully predicting surface wind speeds during TCs up to 6 h ahead for offshore islands near Taiwan using 13 features, including central pressure and geographic location. Similarly, Zhang et al. (2017) employed Least Squares SVM with and Radial Basis Function Neural Networks to predict the critical radii of 34-knot and 50-knot winds based on parameters such as TC age and center latitude. However, the required large amount of high-fidelity dataset to extend the point-wise model toward a spatially resolved wind field model could be a great challenge due to the high cost of data from simulation, experiment or field measurement (Swischuk et al., 2019).

Snaiki and Wu (2019) confronted this challenge by introducing a KEML framework that effectively integrates equation-based physical principles into an ANN-based model for predicting TC wind field. The reduced-order Navier–Stokes equations were employed as part of the loss function to facilitate the training of the neural networks. The automatic differentiation was employed to compute the partial derivatives used in the equation (e.g., the derivative of velocity field with respect to the three axes x, y and z). The developed model can be used to predict the spatial distribution of TC boundary-layer wind speed from essential parameters of center location, central pressure difference, radius of maximum winds, translational speed, approach angle and surface roughness. These storm parameters are provided by the National Hurricane Center. The developed KEML for TC boundary-layer wind simulation can be utilized in conjunction with the TC risk analysis tools and also can be implemented as part of the early warning system. Empirical insights were strategically employed to derive a simplified form of the governing equations, which were then enforced via the loss function. This approach effectively balances physical consistency with computational feasibility, avoiding the cost of a full-order Navier–Stokes solution (knowing that embedding the full-order equations in the loss function is too difficult for gradient descent-based learning). It stands as a clear example of how domain expertise can guide and optimize the integration of physics/empirical models into ML.

Unlike TC winds that are associated with large-scale (horizontal scales of thousands of kilometers and time scales of days) meteorological systems, non-synoptic wind events (e.g., thunderstorms, downbursts and tornadoes) exhibit limited spatial scales (100–1000 m) and short lifetimes (minutes-hours) (Chowdhury & Wu, 2021). The transient and small-scale nature of non-synoptic windstorms make them difficult to be detected, measured, and modeled (Solari, 2020). Numerical studies of thunderstorms employ both mesoscale and microscale models (Hawbecker, 2021). Mesoscale models capture relevant physics but lack fine spatiotemporal details, whereas microscale models resolve important small-scale structures but are constrained by small computational domains and are sensitive to selected boundary conditions. Thunderstorms induce intense winds with localized damage, including downbursts with intense radial outflows, as well as tornadoes with characteristic rotating columns (Hao & Wu, 2016, 2020; Solari, 2020; Yang et al., 2018). Laboratory experiments and CFD-based studies have been conducted to reproduce both downbursts and tornadoes (e.g., Aboshosha et al., 2015; Ishihara et al., 2011; Junayed et al., 2019; Li et al., 2024; Liu & Ishihara, 2015; Mason et al., 2009; Refan & Hangan, 2016; Tang et al., 2018). They are either labor-intensive or computationally expensive, limiting their use for large-scale scenario generation or probabilistic risk assessment. The rapid development of field-measurement networks (e.g., THUNDERR project at University of Genova (Solari et al., 2018)) and laboratory studies in recent years facilities (e.g., WindEEE at Western University (Hangan, 2014)) have improved understanding of non-synoptic wind systems. These advances offer valuable data, and hence provide potentials in ML applications to non-synoptic winds.

However, across all three event types, current ML development for non-synoptic winds remains preliminary, with applications mainly on predicting event occurrence or likelihood rather than reconstructing the detailed wind field. For thunderstorms, ML models generally use atmospheric thermodynamic parameters (e.g., temperature, humidity) and wind measurements (from radiosonde observations, reanalysis dataset and surface networks) for prediction, and these ANN/CNN-based models demonstrate acceptable to reliable forecasting skill at lead times up to about 15 h (Chakrabarty et al., 2013; Chen & Lombardo, 2020; Litta et al., 2012; McCann, 1992). Efforts on downbursts and tornadoes follow a similar pattern, where ML is typically used for classification or coarse regression of hazard probability or intensity (e.g., Lagerquist et al., 2017; Medina et al., 2019; Smith et al., 2004). Very limited studies attempt to directly model or forecast spatially resolved wind fields. In fact, the ML models generally require a large number of input variables, reflecting physical complexity but also reducing operational usability. Furthermore, selection of model input is often ad-hoc, and performance metrics vary substantially among studies, which hinders systematic comparison. Progress has been constrained by (1) limited understanding of triggering mechanisms and multiscale coupling, (2) high cost of obtaining numerical/experimental data to complement the sparse and heterogeneous observational data, particularly near the ground, and (3) difficulties in identifying a suitable set of input variables that balances physical relevance and availability. While ML has demonstrated potential to supplement traditional modeling, efforts needs to move from coarse event prediction to lifecycle-resolved, wind field-resolved modeling. Advancing ML for non-synoptic winds thus requires shifting its role from a black-box forecasting toward a knowledge/physics-guided prediction.

7.3.2 Aerodynamics and Aeroelasticity Modeling

The bluff-body aerodynamics and aeroelasticity play a critical role in the safe and cost-effective design of wind-sensitive structures. Unsteady wind flows with significant separation around the structures introduce complexities in predicting wind effects. High-fidelity CFD can resolve the complex vortex shedding and transient pressure fields that dominate the structural aerodynamics. However, the computational cost presents a major bottleneck for applications requiring fast iterations, including the preliminary structural design or aeroelastic assessments of extended time periods.

In recent years, ML has been frequently adopted to approximate bluff-body aerodynamics through flow field reconstruction and wind load prediction. Early studies employed ANNs or SVMs to predict aerodynamic coefficients, flutter derivatives, and vortex-induced vibrations (Chen et al., 2002; Fu et al., 2006, 2007; Jung et al., 2004; Turkkan & Srivastava, 1995; Zhang & Zhang, 2004). More advanced architectures such as CNNs have been used to recover pressure and velocity fields from geometry or sparse sensors (Bhatnagar et al., 2019; Guo et al., 2016), with incorporation of recurrent structures for unsteady wake dynamics (Hasegawa et al., 2019, 2020).

These approaches demonstrated notable speedup compared with CFD, enabling rapid aerodynamic assessment under varying Reynolds numbers or bluff-body shapes. Nevertheless, most existing ML surrogates still rely heavily on large datasets from CFD, struggle to generalize across flow regimes or body geometries, and have limited guarantees of physical consistency. Their performance deteriorates markedly when extrapolating in Reynolds number or incident flow angles beyond the training dataset. The lack of embedded physical constraints often results in non-conservative or unrealistic predictions. The goal of employing ML is usually to avoid the prohibitive cost of expensive CFD simulations, yet obtaining sufficient data to train a generalizable ML model may itself require an even larger computational investment. It remains questionable regarding the practical utility of such a purely data-driven surrogate in engineering applications.

To overcome the data-efficiency bottleneck, Yu and Wu (2023) introduced a knowledge-enhanced CNN that strategically integrates domain knowledge at three levels. First is the equation-free expertise from CFD modeling in data preprocessing. Recognizing that flow fields exhibit extreme gradients within the thin boundary layer, the authors empirically designed a non-uniform grid for the CNN's input and output. This embeds fundamental aerodynamic intuition directly into the model's data structure, focusing computational resources on regions of highest physical significance. Second is the experiential latent representations for model construction. The idea that by a low-cost simulation can provide an initial flow field to accelerate the convergence in CFD inspired a two-phase CNN training scheme. In the first phase, a core encoder-decoder aimed at predicting baseline flow fields. In the second phase, the model employed high-level features extracted by the trained modules as a refined initial condition for subsequent time steps. This transfer-learning strategy enhances data efficiency and temporal generalization with minimal additional data. Third is the equation-based physical principle in training. Mass and momentum conservation laws from unsteady Reynolds-Averaged Navier–Stokes (URANS) equations were embedded as soft constraints in the loss function. The mass conservation-based loss function penalized the total mass flux differences between predicted and reference flow variables in each CNN pixel. Similarly, the momentum-based loss function enforced consistency for convection, pressure gradient, and shear stress terms. The Reynolds stress and unsteady terms were omitted since the relevant information was not needed in the training dataset. The culmination of this holistic knowledge enhancement is an end-to-end, pixel-to-pixel CNN capable of near real-time, high-fidelity simulation of unsteady flows and aerodynamic pressures around a bluff body. This approach transcends being a mere surrogate; it represents a new design for integrating multi-faceted domain knowledge to build trustworthy, efficient, and scalable solutions for wind engineering applications.

While these advances demonstrate the power of KEML in reconstructing complex aerodynamic flows, a challenging exists in extending to aeroelastic systems, where fluid–structure coupling and aerodynamic dynamic instability must be captured. A typical problem is the prediction of flutter, which is of critical importance for long-span bridges. Flutter is a dangerous self-excited vibration caused by the interaction of aerodynamics and structural dynamics. Since the collapse of the original Tacoma

Narrows Bridge in 1940, the onset of flutter for long-span bridges under strong winds has been systematically investigated (Scanlan & Tomko, 1971). Current codes typically prevent linear flutter by ensuring that critical flutter speed exceeds design wind speed, which implicitly restrict consideration to small-amplitude responses. With the rise of performance-based design emphasizing explicit performance targets and life-cycle considerations (ASCE, 2019), attention has shifted to wind performance after the start of flutter, motivating the need to characterize nonlinear flutter. Traditional reduced-order models such as the original/improved quasi-steady theory (Diana et al., 1993), hybrid models combining nonlinear quasi-steady and linear unsteady theories (Chen & Kareem, 2003), Volterra-based models (Wu & Kareem, 2014), and describing functions (Zhang et al., 2020), despite their efficiency, fail to fully capture the complex nonlinear dynamics of wind-bridge interactions (Wu & Kareem, 2015). While fully coupled 3D fluid–structure interaction (FSI) simulations, combining CFD and computational structural dynamics (CSD), are promising, their computational cost is prohibitive for the extensive scenario analysis required in practice (De Risi, 2022; Kim & Yhim, 2014).

Recognizing these challenges, ANNs, particularly the LSTM networks with strong ability in modeling long-time dependencies, present a promising path forward (Perez-Ramirez et al., 2019; Xu et al., 2021). The input to a LSTM network is typically the node motion output from a finite element model (FEM) of the structure, and the aerodynamic forces input to each FEM node are fed back as the output to the corresponding network. LSTM-based flutter analysis typically only simulates a 2D rigid segment of the whole structure (Li et al., 2020a, 2020b). However, 2D flutter analysis fails to capture multi-mode coupling effects and the dependence of aeroelastic forces on the spanwise amplitude variation, two key nonlinear aerodynamic features that are crucial for bridge flutter analysis (Ge & Tanaka, 2000; Wu & Kareem, 2013; Wu et al., 2020). To accurately characterize and quantify the oscillation amplitudes and frequencies in the post-flutter stage, 3D modeling of the entire bridge is needed. LSTM is data-hungry network since it contains a large number of learnable parameters. To ensure the generalization ability, a large dataset is needed to cover a wide range of bridge motion frequencies and amplitudes. However, obtaining high-fidelity training data is expensive.

To reduce the required training data and ensure the generalization ability, a knowledge-enhanced LSTM framework have been proposed (Li & Wu, 2023). To customize a lightweight LSTM for aerodynamic analysis, both the general and domain-specific expertise from aerodynamics were integrated. According to the underlying mechanism of LSTM, the redundant input gate and sequenced functions designed for its original application, i.e., text sequence predictions, are eliminated for the bridge aerodynamics application. A more efficient form of activation function was used as an approximation of original sigmoid nonlinear activation function to alleviate the computational costs. According to the domain-specific expertise, the forget gate is eliminated by pre-determining the duration of the fluid memory effect through a unit-impulse function. All biases were also removed to allow a zero-input (motion) zero output (wind force) system, i.e., static wind forces are constant values for a motionless structure. Learnable parameters in knowledge-enhanced LSTM unit are less than 50% of those in a standard LSTM unit, which remarkably improve the training efficiency (Vapnik & Chervonenkis, 2015). In addition, the training scheme was streamlined to further accelerate the learning process. An inverse-free form of mBFGS method (Chen & Jiao, 2007) enabled by the Sherman-Morrison formula (Sherman & Morrison, 1950) has been developed. The proposed training algorithm further improved the training efficiency of knowledge-enhanced LSTM by around 50%. The knowledge-enhanced LSTM model demonstrates a paradigm where the deep integration of structural, aerodynamic, and algorithmic knowledge yields a tool that is simultaneously more data-efficient, computationally faster, and physically more credible than its purely data-driven form.

In addition to the deterministic relation-based knowledge, the statistics knowledge can be also leveraged to facilitate the learning process in structural aerodynamics. Actually, several studies suggested that training a typical time-series ML model (e.g., recurrent neural network) for chaotic system predictions can lead to unbounded gradients (Mikhaeil et al., 2022; Tang et al., 2024). Li and Wu (2026) recently developed a statistics knowledge-informed deep learning for simulations of non-Gaussian wind pressure field. Inspired by the universal approximation theorem for stochastic systems, a deep neural network (DNN) is trained in this study on the statistics in terms of PDF and auto-correlation function (ACF). This one-step learning method (i.e., simultaneously matching both PDF and PSD) has shown great promise for high simulation accuracy by integrating statistics knowledge (i.e., both spectral and probabilistic contents) into loss function consideration.

7.3.3 Structural Dynamics Modeling

Structural dynamics governs how wind-induced forces propagate through complex structural systems, dictating serviceability, fatigue, and ultimate performance. Capturing these nonlinear, path-dependent, and uncertain behaviors is essential for resilience-oriented performance-based design (Chen, 2013; Di Matteo et al., 2014; Zhang & Xu, 2001). While the high-fidelity finite element model (FEM) provides accurate simulations, the computational cost, due to its small time steps and long simulation duration, prohibits the extensive simulations required for uncertainty quantification in the performance-based wind engineering.

Physics-based reduced-order models have well provided efficient alternatives, but the numerical and/or experimental identification of their parameters remains quite challenging. This limitation motivates the adoption of ML models, which have shown promise in modeling nonlinear structural response the field of earthquake engineering without tedious parameters identification (Facchini et al., 2014; Jiang & Adeli, 2005; Linse & Stengel, 1993). Early studies employed relatively simple ML models and standard training schemes to predict nonlinear dynamic response in bridges and tall buildings under wind loads (Oh et al., 2017; Vyavahare et al., 2012; Wu & Kareem, 2011). Recently, the rapid growth of DL further accelerated the progress with providing more advanced network architectures like DNNs with recurrent architecture. Meanwhile, post-disaster reconnaissance programs such as Natural Hazards Reconnaissance Facility (known as the "RAPID") and Structural Extreme Events Reconnaissance (StEER) Network launched in 2018 by National Science Foundation (NSF) provide a great opportunity for ML models with increasingly available field data characterizing structural damages under strong wind events.

The LSTM became a straightforward choice because structural response output at each time step is essentially determined by the system status of the previous time step and the excitation of the current time step. To train the LSTM with limited data, Wang and Wu (2020) introduced the KEDL framework that integrates the domain knowledge at multiple stages. The enhancement was twofold, targeting both computational efficiency and physical consistency. Prior to training, the structural response and external forces were projected to the wavelet domain. As a domain-specific knowledge that guided data preprocessing, the rationale was that structural dynamics are often dominated by a limited number of significant frequency components. By filtering out negligible frequencies, it drastically reduces the dynamics complexity and accelerates network training. Another innovation was the design of a hybrid loss function. In addition to standard data-based loss function, a knowledge-based loss function derived from the equations of motion for structural dynamics system was designed. Accordingly, the conventional data-based loss function was enhanced by the knowledge-based loss function to deal with force/response wavelet coefficients. The knowledge-based loss function acted as a powerful regularizer, which effectively prevents model overfitting to the limited data. To demonstrate the performance of this KEDL model, a multi-degree-of-freedom (MDOF) system was used to investigate the simulation accuracy for nonlinear structural dynamics. Trained on only 20 samples, it achieved accuracy comparable to a high-fidelity 4th-order Runge–Kutta solver while requiring only 3% of the computational time. More importantly, it demonstrated exceptional robustness to noise-contaminated training data, a common reality in engineering practice. This robustness stems directly from the knowledge-based loss function, which remain uncorrupted by noise in the responses. This research is a representative example of KEML in structural dynamics. It moves beyond simply using a neural network, demonstrating that by strategically injecting knowledge about data representation (wavelet transform) and system mechanics (force-response relationship) into the ML pipeline, one can create a model that is more data-efficient and reliable than its purely data-driven counterpart.

7.3.4 Structural Response Mitigation

Structural response mitigation often focusses on reducing wind-induced vibrations through shape optimization or structural control. Shape optimization is the primary approach to reduce aerodynamic loading. Traditional aerodynamic design relies on limited candidate geometries guiding by engineering intuition, which restricts the design space and risks missing unconventional yet efficient configurations. The need for a systematic and automated process capable of exploring high-dimensional shape spaces has motivated the integration of optimization algorithms with aerodynamic performance evaluation (Ding & Kareem, 2018; Skinner & Zare-Behtash, 2018). While those essentially hand-design optimization approaches are frequently used, the determination of their parameters for a specific application is usually based on manual trial-and-error process (Andrychowicz et al., 2016).

Elshaer et al. (2016, 2017) pioneered the use of ANN-based surrogate models with genetic algorithms for aerodynamic shape optimization of tall buildings. More recently, the RL has emerged as a model-free alternative (Hu et al., 2020; Silver et al., 2017) of optimization schemes. In the RL framework, an agent, typically represented by neural networks, learns a policy to seek globally optimal shapes through trial-and-error interactions with a flow environment (Sutton & Barto, 1998). The application of deep RL (DRL) has been demonstrated in both experimental and simulated active flow control for bluff bodies (Fan et al., 2020). However, this capability comes at a steep price of obtaining sufficient data through interacting with the environment. As purely data-driven models without integrated prior knowledge, DRL still requires large datasets and lead to inefficient optimization process.

Unlike KEML studies in modeling TC or extratropical cyclone wind field (Snaiki & Wu, 2019, 2022) and structural dynamics (Wang & Wu, 2020), optimization problem lacks explicit governing equations for domain knowledge integration (Min et al., 2017). In addition, the no-free-lunch theorem (Wolpert and Macready, 2002) prohibits a universal optimizer that performs well for all scenarios. To enhance the efficiency without explicit equations, latent experiential knowledge was incorporated into the DRL optimizer developed by Li et al. (2021a). Specifically, domain-related knowledge extracted from the low-cost source tasks were integrated into the DRL shape optimizer for its applications to high-cost target tasks. The knowledge is incorporated via the transfer-learning technique (Pan & Yang, 2009; Yan et al., 2019). Policy weights were first trained using low-fidelity RANS simulations and transferred to high fidelity large-eddy simulation (LES) environments. Knowledge can also be imported through meta-learning technique based on a group of inexpensive tasks generated from a common probability distribution (e.g., multivariate Gaussian distribution) that reflects important high-level structures of the current optimization problem (e.g., Finn et al., 2017; Zhou et al., 2017). Numerical experiments, focusing on the aerodynamic shape optimization of a tall building cross-section, provided compelling validation (Li et al., 2021a). The knowledge-enhanced DRL optimizers consistently outperformed conventional gradient-based and gradient-free algorithms, both in the final quality of the discovered design and the speed of convergence. This work serves as a demonstration of how knowledge be leveraged when lacking explicit governing equations. Beyond its practical success, this study highlights a conceptual advance, demonstrating that knowledge need not be confined to explicit equations but can be encoded as transferable inductive biases. The demonstrated framework also provides a scalable route toward self-evolving aerodynamic design systems, where learned experience across tasks accelerates innovation and enhances resilience of wind-sensitive structures.

Beyond modifying external aerodynamic shape for wind load reduction, response mitigation in terms of semi-active and active control reflects broader efforts toward intelligent and resilient structures (Fisco & Adeli, 2011). The mitigation of vibrations in stay cables, a critical component of long-span bridges, is a long-existing challenge in wind engineering (Fujino & Siringoringo, 2013; Xu, 2013). These cables are vulnerable to various wind-induced vibrations, such as vortex-induced vibration and rain-wind-induced vibration, causing fatigue and serviceability issues (Jafari et al., 2020; Zuo & Jones, 2010). Magnetorheological dampers (MRD) have shown great promise due to their adjustable damping properties and low energy consumption (Duan et al., 2006; Li et al., 2007). The core challenge is in real-time control, which involves computing the desired input current from sensor measurements. The behavior of MRD are dependent on multiple physical processes, including fluid dynamics and electromagnetics governing the behavior of magnetorheological fluids under magnetic fields (Zhu et al., 2012).

Traditional control methods often struggle to balance accuracy and computational efficiency. For example, some easy-to-implement models (Stanway et al., 1987) lacks accuracy under varying strain rates, while more sophisticated models (Spencer et al., 1997; Parlak et al., 2022; Hu et al., 2022) capture hysteresis but are too complex for real-time control. LSTM with robust potential in addressing challenges in modeling nonlinear time series offer a promising solution to the MRD control. Li et al. (2025b) developed a LSTM-based semi-active control of MRD to mitigate the cable vibration of cable-stayed bridges. Recognizing that cable vibration is a temporal process with historical dependence, the selection of LSTM is straightforward, considering that control decisions must be based on sequential data. This approach bypassed the need for an explicit physical model of the magnetorheological damper. Instead, it leveraged the strength of LSTM in modeling dynamics from data, resulting in a more adaptive and efficient solution than traditional methods. While current applications show promise in learning direct control policies from data, the integration of KEML remains largely unexplored. The primary challenge hindering the application of KEML in this domain stems from the intrinsic complexity of MRD behavior, which involves the tight coupling of multiple physical processes. This multi-physics nature makes it difficult to define a single, governing equation for knowledge embedding. Instead of seeking a single overarching model, future research might pivot toward decomposing and constraining the multi-physics problem. A promising direction is the development of physics-informed hybrid architectures. For instance, one could design a network where different sub-modules are implicitly regularized by specific physical principles, such as embedding fluid-mechanical constraints in layers processing velocity inputs, and electromagnetic relationships in layers handling current commands.

Recent advances in active flow control also demonstrated the potential of ML approaches to facilitate transient wind generation at the laboratory scales (Li et al., 2021b; Yang et al., 2024). Reproducing transient wind fields in multiple-fan wind tunnels can be seen as a typical reference tracking problem, which requires an optimal controller to minimize the difference between the target wind field and the simulated wind field. The core idea parallels that of structural load mitigation, in which the control objective is to minimize the discrepancy between the target and controlled aerodynamic states through automated, model-free learning. Conventional wind tunnel control methods rely on proportional-integral-derivative (PID)-like or fuzzy logic controllers that assume a simplified, often linear relationship between control signals and the resulting flow (Nishi et al., 1999; Khan & Rapal, 2006; Smith et al., 2012). These controllers require extensive manual tuning and cannot adequately reproduce the complex transient features as in downbursts. The developed DRL-based controller in Li et al. (2021b), by contrast, learns an optimal control policy through trial-and-error interaction between the agent (controller) and the environment (wind tunnel). The DNN serves as the functional core of the agent, mapping high-dimensional state information, such as fan speeds, flow velocities, and temporal histories, to optimal control actions (commands for fan rotation). By interacting directly with the complex physical system, the DRL agent implicitly resolved the hidden dynamics of the wind tunnel into an effective control policy, establishing a powerful alternative to hand-tuned controllers. Yang et al. (2024) later extended the framework to the control of digital wind tunnel. The numerical simulation based in a digital wind tunnel enables a more comprehensive analysis of the complex mixing process between the airflow generated by each fan. A DRL agent that can interact with a two-dimensional flow field environment with a Reynolds number of 200 was trained and learned to generate boundary control conditions based on desired wind speeds. The DRL-based controller has been recently extended to the active flutter control of long-span bridges (Li et al., 2025a; Wu et al., 2023). Although these DRL-based controllers do not explicitly incorporate knowledge, these studies also point toward a broader vision where knowledge-enhanced learning could bridge different scales of wind/structural control, from fan-level actuation to flow/structural-level feedback. By embedding fluid–structure interaction knowledge or control priors into the learning loop, future systems could achieve physically consistent co-adaptation between wind turbulence generation and structural response control.

7.3.5 *Post-Hazard Recovery*

For the effective post-event recovery in wind engineering, the damage analysis and repairing need to go beyond individual structures to interdependent infrastructure systems such as power grids, transportation networks, and telecommunication lines. Along with pre-event preparedness and during-event emergency response, post-event recovery is critical to infrastructure resilience. This phase involves relatively long duration (weeks to months) and intense governmental intervening, allowing for collaborative decision-making among diverse stakeholders.

Effective recovery in this context hinges on two complementary fronts, information management and physical restoration. The rapid growth of communication platforms provides unprecedented volumes of disaster-related data. Many studies have used ML classifiers to extract and disseminate actionable information during rescue and relief operations (Ashktorab et al., 2014; Imran et al., 2016; O'Neal et al.,

2018; Robertson et al., 2019; Manna & Nakai, 2019; Devaraj et al., 2020). Although concerns persist due to inconsistent data quality from social-media sources, these applications demonstrate considerable potential for improving situational awareness and timely decisions.

Complementary to situation awareness is selecting optimal strategies from various alternatives to create an effective recovery schedule. This optimization ensures the most efficient allocation of repair resources to damaged components, facilitating the rapid restoration of infrastructure functionality (e.g., Zou & Chen, 2019). Traditional scheduling approach that selects the best recovery policy by evaluating multiple ones prescribed by domain experts may lead to suboptimal schedules. Formulating post-disaster recovery as a stochastic sequential decision problem enables systematic solution exploration. Static (open-loop) optimization methods, like binary particle swarm optimization (Zou & Chen, 2019) and deep-ensemble-assisted active learning (Zou & Chen, 2021), seek a fixed repair sequence but struggle with the dynamic uncertainty. In contrast, dynamic (closed-loop) optimization adapts to time-varying states, offering a more promising approach. The Markov decision process (MDP) provides a fundamental model for dynamical optimization (e.g., Memarzadeh & Pozzi, 2019; Nozhati et al., 2020a, 2020b; Papakonstantinou & Shinozuka, 2014). The MDP could be optimized via dynamic programming (DP), which requires explicit system dynamics. However, the interconnected infrastructure systems in post-hazard recovery are typically modeled through "black-box" simulators.

This gap has motivated data-driven approaches. A review by Gerolimos et al. (2025) confirms the growing role of autonomous decision-making in disaster management. Many studies have employed RL to prioritize the recovery of transportation network (Ghannad et al., 2021), large-scale power distribution systems (Dehghani et al., 2021), and road network under various hazards (Fan et al., 2023). The model-free nature of RL for optimization, offers a method to find optimal solutions through a trial-and-error approach, interacting with these "black-box" simulators with an advantage of no need for explicit system dynamics.

Considering the trial-and-error exploration of the action space in a deep RL (DRL)-based approach can be quite time-consuming, it is imperative to employ advanced learning techniques to improve training efficiency. Li and Wu (2025, 2025a) addressed this challenge by developing a knowledge-enhanced DRL framework that integrates the domain knowledge in terms of procedural constraints of repair tasks (e.g., nonreversible sequential order) to improve the training efficiency of conventional DRL-based schemes. Specifically, two mechanisms namely knowledge-guided exploration (e.g., Memarzadeh & Pozzi, 2019) and knowledge-shaped reward (e.g., Hu et al., 2020) were utilized to integrate the domain knowledge. By modifying the probabilities of actions occurring, knowledge-guided exploration promotes favorable actions that have the potential to yield high rewards while discouraging unfavorable actions that are less likely to result in high rewards. Taking the traffic network as an example, the role of knowledge-guided exploration is to regenerate the exploration action if the action suggests repairing the road segment without first fixing the road links whose functionalities are the prerequisites. By designing an auxiliary reward, knowledge-shaped exploration gives bonus to desirable actions and/or penalties to undesirable ones. For the traffic network, its purpose is to penalize exploration actions that involve restoring a road link before full recovery of its "prerequisites." The knowledge-enhanced DRL framework exemplifies how embedding domain-specific procedural and relational knowledge can significantly enhance the learning efficiency and physical validity of data-driven decision models. By constraining exploration within a knowledge-informed action space, it bridges the gap between theoretical reinforcement learning and real-world recovery management.

7.4 Prospects and Future Directions

7.4.1 From Knowledge Embedding to Knowledge Discovery

While current KEML applications have successfully demonstrated their values in enhancing predictive accuracy and data efficiency of surrogate models, their transformative potential may extend beyond being mere accelerators for simulations. It is expected to evolve from a consumer of established knowledge into a generator of new scientific insight, actively contributing to our fundamental understanding of wind-structure interactions (Butler et al., 2018). This evolution encompasses three key pathways.

Knowledge Extraction and Model Interpretation: The internal representations learned by sophisticated KEML models, especially those trained under multi-faceted physical constraints, may encode latent patterns and simplified relationships that are not immediately obvious from first principles or empirical data alone (Udrescu & Tegmark, 2020). Future research must develop advanced explainable AI (XAI) techniques tailored for physics-constrained networks. The goal is to move toward symbolic regression or rule extraction methods that can distill the network's learned behavior into human-comprehensible formulas, conceptual models, or scaling laws. For instance, analyzing a successfully trained KEML model for nonlinear flutter under turbulent inflow conditions might reveal a previously unrecognized role of aerodynamic damping due to multi-scale wind-structure interactions, or a KEML-based shape optimizer may be interrogated to extract a new, non-empirical rule for aerodynamic efficiency.

Active Hypothesis Generation and Exploration: KEML systems can be architected to combine generative models (e.g., generative adversarial networks and diffusion models) with physics-based discriminators. These systems can propose novel, physically-plausible configurations or scenarios that maximize a desired objective or explore the boundaries of known physics (Sanchez-Lengeling & Aspuru-Guzik, 2018). A KEML-driven "digital wind tunnel" may, for example, autonomously search for and characterize previously undocumented aeroelastic instability mechanisms by intelligently perturbing structural and flow parameters in simulation. In aerodynamic design, this approach may shift the paradigm from optimizing within a known design space to discovering entirely new regions of high-performance shapes that is against human intuition.

Closing the Loop with Autonomous Discovery Frameworks: An ultimate expression of KEML as a scientific tool could be its integration into a closed-loop, autonomous discovery pipeline (Shields et al., 2021). This framework would iteratively (1) propose a hypothesis (e.g., a new control law or a material distribution); (2) test it using a high-fidelity KEML surrogate or by guiding a targeted physical experiment; (3) assimilate the results to update knowledge base; and (4) formulate a new and refined hypothesis. Such a system could dramatically accelerate the discovery of resilient structural forms and control strategies, systematically navigating the vast, non-convex design spaces that are currently intractable for human-led exploration.

By pursuing these directions, KEML may transcend its current role as a sophisticated interpolation tool and become an indispensable partner in unraveling the complex, multi-scale physics that govern wind-sensitive systems, ultimately leading to groundbreaking advances in both fundamental knowledge and engineering practice. The improved understanding of fundamental physics and other types of domain knowledge underlying each subfield of wind engineering also provides an opportunity for the KEML to develop itself adaptively. Hence, it is important to enhance its ability of continual learning to incrementally acquire, update, accumulate, and exploit the newly discovered domain knowledge (Wang et al., 2024).

7.4.2 From Isolated Surrogates to Integrated Systems

The current KEML predominantly focuses on building high-fidelity surrogate models for individual components of the wind loading chain. Simply concatenating these independently developed models in a sequential workflow for full-chain analysis, from wind field acquisition to post-hazard recovery, inevitably leads to the error accumulation and propagation. The predictive bias from an upstream component serves as the input for the downstream component, where it can be nonlinearly amplified. For instance, a minor systematic bias in the simulated wind field could become a significant error in the final recovery strategy assessment, after being processed through nonlinear aerodynamic load calculation and structural response analysis. Therefore, future research needs to go from "isolated component optimization" to "system-level co-design." This advance may require breakthroughs in the following three directions.

Developing End-to-End Differentiable Co-simulation Frameworks: An ambitious future direction is to construct a global KEML framework where the entire chain of physical processes (from fluid to structure) is modeled as differentiable components (de Avila Belbute-Peres et al., 2018). This would allow for joint optimization of all-component models' parameters through backpropagation across the entire chain, explicitly minimizing the cumulative error on the final decision objective (e.g., life-cycle cost or recovery time), rather than optimizing the fitting accuracy of each intermediate step in isolation.

Quantifying and Propagating Uncertainty: Next-generation KEML models should not only provide point estimates but also inherently quantify the uncertainty in their predictions. By adopting probabilistic models (e.g., Bayesian neural networks or deep ensembles), each surrogate can output a predictive distribution (Lakshminarayanan et al., 2017; Perdikaris et al., 2017). This uncertainty information can then be fed as input to the next component, enabling forward propagation of uncertainty analysis throughout the chain. Ultimately, the uncertainty propagation and quantification provide decision-makers with a risk distribution of the overall system performance, rather than a single deterministic value.

Establishing System-Level Performance-Oriented Loss Functions: Accurate risk assessment for structures and infrastructure systems under extreme windstorms requires back-propagating system-level engineering constraints (e.g., allowable displacements and acceleration limits) to inform the training of upstream surrogates, guiding them to learn the features most critical for the final system performance, rather than pursuing mean squared error minimization on all features indiscriminately. For example, instead of training a wind field model solely to minimize wind speed error, the multi-objective, system-level loss functions that also consider the impact of wind prediction error on the accuracy of downstream structural response simulations should be designed.

Addressing the challenge of error accumulation is the prerequisite for elevating KEML from "proof-of-concept demonstration" to "practical application." It demands closer collaboration between wind engineering and ML researchers to re-examine and design learning architectures from a systems perspective.

7.5 Concluding Remarks

This chapter has systematically depicted the paradigm of knowledge-enhanced machine learning (KEML) and its transformative potential across the entire Alan G. Davenport Chain in wind engineering. It has been delineated how different forms of knowledge can be embedded into machine learning models in five research fields: windstorms under changing climate, bluff-body aerodynamics and aeroelasticity, structural dynamics, structural response mitigation, and post-hazard recovery. Collectively, these studies illustrated that KEML enables not only improved predictive capability and data efficiency but also enhanced physical interpretability and extrapolation reliability, which are essential for resilient infrastructure design under climate change conditions.

Despite the rapid progress, significant gaps still exist toward a fully integrated end-to-end KEML framework. Existing studies are largely limited to some isolated components in the wind loading chain. The coupling among these components introduces new sources of uncertainty, and the errors of individual surrogate models can accumulate through the workflow. Developing true end-to-end tools therefore requires new strategies for cross-stage consistency, dynamic knowledge transfer, and uncertainty propagation. Future efforts should also emphasize multi-scale and multi-fidelity integration, enabling the physical and statistical coherence of KEML models across engineering meteorology, aerodynamics, and structural and infrastructural domains, through jointly re-envisioning learning architectures within wind engineering and machine learning communities. Finally, the knowledge discovery and the ability of continual learning make the KEML a promising next-generation AI tool to facilitate the performance-based design, hazard risk assessment, and infrastructure systems resilience enhancement under extreme windstorms.

References

Aboshosha, H., Bitsuamlak, G., & El Damatty, A. (2015). Turbulence characterization of downbursts using LES. *Journal of Wind Engineering and Industrial Aerodynamics, 136*, 44–61.
Andrychowicz, M., Denil, M., Gómez, S., Hoffman, M. W., Pfau, D., Schaul, T., Shillingford, B., & de Freitas, N. (2016). Learning to learn by gradient descent by gradient descent. In *Proceedings of 30th conference on neural information processing systems (NIPS 2016)*, Barcelona, Spain.
ASCE (American Society of Civil Engineers). (2019). *Prestandard for performance-based wind design*. American Society of Civil Engineers.
Ashktorab, Z., Brown, C., Nandi, M., & Culotta, A. (2014, May). Tweedr: Mining twitter to inform disaster response. In *Proceedings of 11th international ISCRAM conference* (pp. 269–272). University Park, PA, USA.
Bhatnagar, S., Afshar, Y., Pan, S., Duraisamy, K., & Kaushik, S. (2019). Prediction of aerodynamic flow fields using convolutional neural networks. *Computational Mechanics, 64*(2), 525–545.
Bishop, C. M. (2006). *Pattern recognition and machine learning*. Springer.
Blocken, B. (2014). 50 years of computational wind engineering: Past, present and future. *Journal of Wind Engineering and Industrial Aerodynamics, 129*, 69–102.
Butler, K. T., Davies, D. W., Cartwright, H., Isayev, O., & Walsh, A. (2018). Machine learning for molecular and materials science. *Nature, 559*(7715), 547–555.
Cermak, J. E. (1975). Applications of fluid mechanics to wind engineering. *Journal of Fluids Engineering, 97*(1), 9–38.
Chakrabarty, H., Murthy, C. A., & Gupta, A. D. (2013). Application of pattern recognition techniques to predict severe thunderstorms. *International Journal of Computer Theory and Engineering, 5*(6), 850–855.
Chen, G., & Lombardo, F. T. (2020). An automated classification method of thunderstorm and non-thunderstorm wind data based on a convolutional neural network. *Journal of Wind Engineering and Industrial Aerodynamics, 207*, Article 104407.
Chen, L., & Jiao, B. (2007). The generalized quasi-newton method for nonconvex unconstrained optimization problems. *Advances in Mathematics, 1*, 81–85.
Chen, X. (2013). Estimation of stochastic crosswind response of wind-excited tall buildings with nonlinear aerodynamic damping. *Engineering Structures, 56*, 766–778.
Chen, X., & Kareem, A. (2001). Nonlinear response analysis of long-span bridges under turbulent winds. *Journal of Wind Engineering and Industrial Aerodynamics, 89*(14–15), 1335–1350.
Chen, X., & Kareem, A. (2003). Aeroelastic analysis of bridges: Effects of turbulence and aerodynamic nonlinearities. *Journal of Engineering Mechanics, 129*(8), 885–895.
Chen, Y., Kopp, G. A., & Surry, D. (2002). Interpolation of wind-induced pressure time series with an artificial neural network. *Journal of Wind Engineering and Industrial Aerodynamics, 90*(6), 589–615.
Chen, Z., Liu, Y., & Sun, H. (2021). Physics-informed learning of governing equations from scarce data. *Nature Communications, 12*(1), 6136.

Chowdhury, J., & Wu, T. (2021). Aerodynamic loading due to non-synoptic wind systems. In *The Oxford handbook of non-synoptic wind storms* (p. 337). Oxford University Press.

Davenport, A. G. (2002). Past, present and future of wind engineering. *Journal of Wind Engineering and Industrial Aerodynamics, 90*(12–15), 1371–1380.

de Avila Belbute-Peres, F., Smith, K., Allen, K., Tenenbaum, J., & Kolter, J. Z. (2018). End-to-end differentiable physics for learning and control. *Advances in Neural Information Processing Systems, 31*.

De Risi, R. (2022). A computational framework for finite element modeling of traveling loads on bridges in dynamic regime. *Computer-Aided Civil and Infrastructure Engineering, 37*(4), 470–484.

Dehghani, N. L., Jeddi, A. B., & Shafieezadeh, A. (2021). Intelligent hurricane resilience enhancement of power distribution systems via deep reinforcement learning. *Applied Energy, 285*, Article 116355.

Devaraj, A., Murthy, D., & Dontula, A. (2020). Machine-learning methods for identifying social media-based requests for urgent help during hurricanes. *International Journal of Disaster Risk Reduction, 51*, Article 101757.

Di Matteo, A., Iacono, F. L., Navarra, G., & Pirrotta, A. (2014). Experimental validation of a direct pre-design formula for TLCD. *Engineering Structures, 75*, 528–538.

Diana, G., Bruni, S., Cigada, A., & Collina, A. (1993). Turbulence effect on flutter velocity in long span suspended bridges. *Journal of Wind Engineering and Industrial Aerodynamics, 48*(2–3), 329–342.

Ding, F., & Kareem, A. (2018). A multi-fidelity shape optimization via surrogate modeling for civil structures. *Journal of Wind Engineering and Industrial Aerodynamics, 178*, 49–56.

Dissanayake, M. G., & Phan-Thien, N. (1994). For solving partial differential equations. *Communications in Numerical Methods in Engineering, 10*(3), 195–201.

Duan, Y. F., Ni, Y. Q., & Ko, J. M. (2006). Cable vibration control using magnetorheological dampers. *Journal of Intelligent Material Systems and Structures, 17*(4), 321–325.

Elshaer, A., Bitsuamlak, G., & El Damatty, A. (2016, June). Aerodynamic shape optimization of tall buildings using twisting and corner modifications. In *8th International colloquium on bluff body aerodynamics and applications*. Northeastern University, Boston, MA, USA.

Elshaer, A., Bitsuamlak, G., & El Damatty, A. (2017). Enhancing wind performance of tall buildings using corner aerodynamic optimization. *Engineering Structures, 136*, 133–148.

Facchini, L., Betti, M., & Biagini, P. (2014). Neural network based modal identification of structural systems through output-only measurement. *Computers & Structures, 138*, 183–194.

Fan, D., Yang, L., Wang, Z., Triantafyllou, M. S., & Karniadakis, G. E. (2020). Reinforcement learning for bluff body active flow control in experiments and simulations. *Proceedings of the National Academy of Sciences, 117*(42), 26091–26098.

Fan, X., Zhang, X., Wang, X., & Yu, X. (2023). A deep reinforcement learning model for resilient road network recovery under earthquake or flooding hazards. *Journal of Infrastructure Preservation and Resilience, 4*(1), 8.

Finn, C., Abbeel, P., & Levine, S. (2017). Model-agnostic meta-learning for fast adaptation of deep networks. In *Proceedings of the 34th International conference on machine learning, PMLR* (pp. 1126–1135).

Fisco, N. R., & Adeli, H. (2011). Smart structures: Part I—active and semi-active control. *Scientia Iranica, 18*(3), 275–284.

Fu, J. Y., Li, Q. S., & Xie, Z. N. (2006). Prediction of wind loads on a large flat roof using fuzzy neural networks. *Engineering Structures, 28*(1), 153–161.

Fu, J. Y., Liang, S. G., & Li, Q. S. (2007). Prediction of wind-induced pressures on a large gymnasium roof using artificial neural networks. *Computers and Structures, 85*(3–4), 179–192.

Fujino, Y., & Siringoringo, D. (2013). Vibration mechanisms and controls of long-span bridges: A review. *Structural Engineering International, 23*(3), 248–268.

Ge, Y. J., & Tanaka, H. (2000). Aerodynamic flutter analysis of cable-supported bridges by multi-mode and full-mode approaches. *Journal of Wind Engineering and Industrial Aerodynamics, 86*(2–3), 123–153.

Gerolimos, N., Alevizos, V., Edralin, S., Xu, C., Priniotakis, G., Papakostas, G. A., & Yue, Z. (2025). Autonomous decision-making enhancing natural disaster management through open world machine learning: A systematic review. *Human-Centric Intelligent Systems*, 1–16.

Ghannad, P., Lee, Y. C., & Choi, J. O. (2021). Prioritizing postdisaster recovery of transportation infrastructure systems using multiagent reinforcement learning. *Journal of Management in Engineering, 37*(1), 04020100.

Gidaris, I., Padgett, J. E., Barbosa, A. R., Chen, S., Cox, D., Webb, B., & Cerato, A. (2017). Multiple-hazard fragility and restoration models of highway bridges for regional risk and resilience assessment in the United States: State-of-the-art review. *Journal of Structural Engineering, 143*(3), 04016188.

Goodfellow, I., Bengio, Y., Courville, A., & Bengio, Y. (2016). *Deep learning* (vol. 1(2)). MIT Press.

Guo, X., Li, W., & Iorio, F. (2016, August). Convolutional neural networks for steady flow approximation. In *Proceedings of the 22nd ACM SIGKDD international conference on knowledge discovery and data mining* (pp. 481–490).

GWEC (Global Wind Energy Council). (2021). *Global wind report 2021* (p. 80). Global Wind Energy Council.

Hangan, H. (2014). The wind engineering energy and environment (WindEEE) dome at Western University, Canada. *Wind Engineers, 39*(4), 350–351.

Hao, J., & Wu, T. (2016). Tornado-induced effects on aerostatic and aeroelastic behaviors of long-span bridge. In *Proceeding of the 2016 world congress on advances in civil environmental & materials research*, Jeju, Korea.

Hao, J., & Wu, T. (2020). Numerical analysis of a long-span bridge response to tornado-like winds. *Wind and Structures, 31*(5), 459–472.

Hasegawa, K., Fukami, K., Murata, T., & Fukagata, K. (2019). Data-driven reduced order modeling of flows around two-dimensional bluff bodies of various shapes. In *Fluids engineering division summer meeting*. American Society of Mechanical Engineers, San Francisco, CA.

Hasegawa, K., Fukami, K., Murata, T., & Fukagata, K. (2020). Machine-learning-based reduced-order modeling for unsteady flows around bluff bodies of various shapes. *Theoretical and Computational Fluid Dynamics, 34*(4), 367–384.

Hawbecker, P. (2021). Mesoscale, microscale, and numerical models. In *The Oxford Handbook of Non-Synoptic Wind Storms* (pp. 239). Oxford University Press.

Holland, G. J. (1980). An analytic model of the wind and pressure profiles in hurricanes. *Monthly Weather Review, 108*(8), 1212–1218.

Hu, G., Wu, L., Deng, Y., Yu, L., & Li, G. (2022). Optimal design and performance analysis of magnetorheological damper based on multiphysics coupling model. *Journal of Magnetism and Magnetic Materials, 558*, Article 169527.

Hu, Y., Wang, W., Jia, H., Wang, Y., Chen, Y., Hao, J., Wu, F., & Fan, C. (2020). Learning to utilize shaping rewards: A new approach of reward shaping. *Advances in Neural Information Processing Systems, 33*, 15931–15941.

Huang, W. F., & Xu, Y. L. (2013). Prediction of typhoon design wind speed and profile over complex terrain. *Structural Engineering and Mechanics, 45*(1), 1–18.

Imran, M., Mitra, P., & Castillo, C. (2016). Twitter as a lifeline: Human-annotated Twitter corpora for NLP of crisis-related messages. In *10th international conference on language resources and evaluation* (pp. 1638–1643). European Language Resources Association (ELRA).

IPCC (Intergovernmental Panel on Climate Change). (2021). *Climate change 2021: The physical science basis.* Contribution of Working Group I to the Sixth Assessment Report of the Intergovernmental Panel on Climate Change, Technical Summary.

Ishihara, T., Oh, S., & Tokuyama, Y. (2011). Numerical study on flow fields of tornado-like vortices using the LES turbulence model. *Journal of Wind Engineering and Industrial Aerodynamics, 99*(4), 239–248.

Jafari, M., Hou, F., & Abdelkefi, A. (2020). Wind-induced vibration of structural cables. *Nonlinear Dynamics, 100*(1), 351–421.

Jiang, X., & Adeli, H. (2005). Dynamic wavelet neural network for nonlinear identification of high rise buildings. *Computer-Aided Civil and Infrastructure Engineering, 20*(5), 316–330.

Jordan, M. I., & Mitchell, T. M. (2015). Machine learning: Trends, perspectives, and prospects. *Science, 349*(6245), 255–260.

Junayed, C., Jubayer, C., Parvu, D., Romanic, D., & Hangan, H. (2019). Flow field dynamics of large-scale experimentally produced downburst flows. *Journal of Wind Engineering and Industrial Aerodynamics, 188*, 61–79.

Jung, S., Ghaboussi, J., & Kwon, S. D. (2004). Estimation of aeroelastic parameters of bridge decks using neural networks. *Journal of Engineering Mechanics, 130*(11), 1356–1364.

Kareem, A. (2020). Emerging frontiers in wind engineering: Computing, stochastics, machine learning and beyond. *Journal of Wind Engineering and Industrial Aerodynamics, 206*, Article 104320.

Karniadakis, G. E., Kevrekidis, I. G., Lu, L., Perdikaris, P., Wang, S., & Yang, L. (2021). Physics-informed machine learning. *Nature Reviews Physics, 3*(6), 422–440.

Karpatne, A., Atluri, G., Faghmous, J. H., Steinbach, M., Banerjee, A., Ganguly, A., Shekhar, S., Samatova, N., & Kumar, V. (2017). Theory-guided data science: A new paradigm for scientific discovery from data. *IEEE Transactions on Knowledge and Data Engineering, 29*(10), 2318–2331.

Kepert, J. (2001). The dynamics of boundary layer jets within the tropical cyclone core: Part I: Linear theory. *Journal of the Atmospheric Sciences, 58*(17), 2469–2484.

Khan, A. A., & Rapal, N. (2006, April). Fuzzy PID controller: design, tuning and comparison with conventional PID controller. In *IEEE international conference on engineering of intelligent systems* (pp. 1–6). IEEE.

Kim, B. C., & Yhim, S. S. (2014). Buffeting analysis of a cable-stayed bridge using three-dimensional computational fluid dynamics. *Journal of Bridge Engineering, 19*(11), 04014044.

Kim, B., Lee, D. E., Preethaa, K. S., Hu, G., Natarajan, Y., & Kwok, K. C. (2021). Predicting wind flow around buildings using deep learning. *Journal of Wind Engineering and Industrial Aerodynamics, 219*, Article 104820.

Lagerquist, R., McGovern, A., & Smith, T. (2017). Machine learning for real-time prediction of damaging straight-line convective wind. *Weather and Forecasting, 32*(6), 2175–2193.

Lakshminarayanan, B., Pritzel, A., & Blundell, C. (2017). Simple and scalable predictive uncertainty estimation using deep ensembles. *Advances in Neural Information Processing Systems, 30*.

Leckebusch, G. C., & Ulbrich, U. (2004). On the relationship between cyclones and extreme windstorm events over Europe under climate change. *Global and Planetary Change, 44*(1–4), 181–193.

LeCun, Y., Bengio, Y., & Hinton, G. (2015). Deep learning. *Nature, 521*(7553), 436–444.

Li, H., Liu, M., Li, J., Guan, X., & Ou, J. (2007). Vibration control of stay cables of the Shandong Binzhou yellow river highway bridge using magnetorheological fluid dampers. *Journal of Bridge Engineering, 12*(4), 401–409.

Li, K., Liu, M., & Wu, T. (2025a). Active control of aerodynamic flaps for long-span bridge flutter using proximal policy optimization-based deep reinforcement learning. *IABSE Structural Engineering International*, 1–11.

Li, S., & Wu, T. (2025). Optimizing post-hurricane recovery of interdependent infrastructure systems via knowledge-enhanced deep reinforcement learning. *Advances in Wind Engineering, 2*(1), Article 100039.

Li, S., Catarelli, R. A., Phillips, B. M., Bridge, J. A., & Gurley, K. R. (2024). Physical simulation of downburst winds for civil structures: A review. *Journal of Wind Engineering and Industrial Aerodynamics, 254*, Article 105900.

Li, S., Snaiki, R., & Wu, T. (2021a). A knowledge-enhanced deep reinforcement learning-based shape optimizer for aerodynamic mitigation of wind-sensitive structures. *Computer-Aided Civil and Infrastructure Engineering, 36*(6), 733–746.

Li, S., Snaiki, R., & Wu, T. (2021b). Active simulation of transient wind field in a multiple-fan wind tunnel via deep reinforcement learning. *Journal of Engineering Mechanics, 147*(9), 04021056.

Li, T., & Wu, T. (2023). Modeling nonlinear flutter behavior of long-span bridges using knowledge-enhanced long short-term memory network. *Computer-Aided Civil and Infrastructure Engineering, 38*(11), 1504–1519.

Li, T., Wu, T., & Liu, Z. (2020a). Nonlinear unsteady bridge aerodynamics: Reduced-order modeling based on deep LSTM networks. *Journal of Wind Engineering and Industrial Aerodynamics, 198*, Article 104116.

Li, W., Laima, S., Jin, X., Yuan, W., & Li, H. (2020b). A novel long short-term memory neural-network-based self-excited force model of limit cycle oscillations of nonlinear flutter for various aerodynamic configurations. *Nonlinear Dynamics, 100*(3), 2071–2087.

Li, X., & Wu, T. (2026). Statistics knowledge-informed deep learning for simulation of univariate non-Gaussian wind pressure. *Reliability Engineering & System Safety, 269*, Article 112114.

Li, Z., Xiang, X., & Wu, T. (2025b). Long short-term memory-enhanced semi-active control of cable vibrations with a magnetorheological damper. *Frontiers of Structural and Civil Engineering, 19*(2), 163–179.

Linse, D. J., & Stengel, R. F. (1993). Identification of aerodynamic coefficients using computational neural networks. *Journal of Guidance, Control, and Dynamics, 16*(6), 1018–1025.

Litta, A. J., Idicula, S. M., & Francis, C. N. (2012). Artificial neural network model for the prediction of thunderstorms over Kolkata. *International Journal of Computer Applications, 50*(11).

Liu, Z., & Ishihara, T. (2015). Numerical study of turbulent flow fields and the similarity of tornado vortices using large-eddy simulations. *Journal of Wind Engineering and Industrial Aerodynamics, 145*, 42–60.

Manna, S., & Nakai, H. (2019). Effectiveness of word embeddings on classifiers: A case study with tweets. In *2019 IEEE 13th international conference on semantic computing (ICSC)* (pp. 158–161). IEEE.

Mason, M. S., Wood, G. S., & Fletcher, D. F. (2009). Numerical simulation of downburst winds. *Journal of Wind Engineering and Industrial Aerodynamics, 97*(11–12), 523–539.

McCann, D. W. (1992). A neural network short-term forecast of significant thunderstorms. *Weather and Forecasting, 7*(3), 525–534.

Meddage, D. P. P., Mohotti, D., & Wijesooriya, K. (2024). Predicting transient wind loads on tall buildings in three-dimensional spatial coordinates using machine learning. *Journal of Building Engineering, 85*, Article 108725.

Medina, B. L., Carey, L. D., Amiot, C. G., Mecikalski, R. M., Roeder, W. P., McNamara, T. M., & Blakeslee, R. J. (2019). A random forest method to forecast downbursts based on dual-polarization radar signatures. *Remote Sensing, 11*(7), 826.

Memarzadeh, M., & Pozzi, M. (2019). Model-free reinforcement learning with model-based safe exploration: Optimizing adaptive recovery process of infrastructure systems. *Structural Safety, 80*, 46–55.

Meucci, A., Young, I. R., Hemer, M., Kirezci, E., & Ranasinghe, R. (2020). Projected 21st century changes in extreme wind-wave events. *Science Advances, 6*(24), eaaz7295.

Mikhaeil, J., Monfared, Z., & Durstewitz, D. (2022). On the difficulty of learning chaotic dynamics with RNNs. *Advances in Neural Information Processing Systems, 35*, 11297–11312.

Min, A. T. W., Sagarna, R., Gupta, A., Ong, Y. S., & Goh, C. K. (2017). Knowledge transfer through machine learning in aircraft design. *IEEE Computational Intelligence Magazine, 12*(4), 48–60.

Murphy, K. P. (2012). *Machine learning: A probabilistic perspective*. MIT Press.

Mylonas, C., Abdallah, I., & Chatzi, E. N. (2019, May). Deep unsupervised learning for condition monitoring and prediction of high dimensional data with application on windfarm scada data. In *Model validation and uncertainty quantification, volume 3: Proceedings of the 37th IMAC, a conference and exposition on structural dynamics 2019* (pp. 189–196). Springer International Publishing, Cham.

Nishi, A., Kikugawa, H., Matsuda, Y., & Tashiro, D. (1999). Active control of turbulence for an atmospheric boundary layer model in a wind tunnel. *Journal of Wind Engineering and Industrial Aerodynamics, 83*(1–3), 409–419.

Nozhati, S., Ellingwood, B. R., & Chong, E. K. (2020a). Stochastic optimal control methodologies in risk-informed community resilience planning. *Structural Safety, 84*, Article 101920.

Nozhati, S., Sarkale, Y., Chong, E. K., & Ellingwood, B. R. (2020b). Optimal stochastic dynamic scheduling for managing community recovery from natural hazards. *Reliability Engineering & System Safety, 193*, Article 106627.

O'Neal, A., Rodgers, B., Segler, J., Murthy, D., Lakuduva, N., Johnson, M., & Stephens, K. (2018, December). Training an emergency-response image classifier on signal data. In *2018 17th IEEE international conference on machine learning and applications* (pp. 751–756). IEEE.

Oh, B. K., Kim, K. J., Kim, Y., Park, H. S., & Adeli, H. (2017). Evolutionary learning based sustainable strain sensing model for structural health monitoring of high-rise buildings. *Applied Soft Computing, 58*, 576–585.

Pan, S. J., & Yang, Q. (2009). A survey on transfer learning. *IEEE Transactions on Knowledge and Data Engineering, 22*(10), 1345–1359.

Papakonstantinou, K. G., & Shinozuka, M. (2014). Planning structural inspection and maintenance policies via dynamic programming and Markov processes: Part II: POMDP implementation. *Reliability Engineering & System Safety, 130*, 214–224.

Parlak, Z., Şahin, İ, & Parlak, N. (2022). One-way coupled numerical model utilizing Viscoelastic Maxwell model for MR damper. *Journal of Intelligent Material Systems and Structures, 33*(19), 2391–2404.

Perdikaris, P., Raissi, M., Damianou, A., Lawrence, N. D., & Karniadakis, G. E. (2017). Nonlinear information fusion algorithms for data-efficient multi-fidelity modelling. *Proceedings of the Royal Society a: Mathematical, Physical and Engineering Sciences, 473*(2198), 20160751.

Perez-Ramirez, C. A., Amezquita-Sanchez, J. P., Valtierra-Rodriguez, M., Adeli, H., Dominguez-Gonzalez, A., & Romero-Troncoso, R. J. (2019). Recurrent neural network model with Bayesian training and mutual information for response prediction of large buildings. *Engineering Structures, 178*, 603–615.

Peterson, C. J. (2000). Catastrophic wind damage to North American forests and the potential impact of climate change. *Science of the Total Environment, 262*(3), 287–311.

Pielke, R. A., Jr., Gratz, J., Landsea, C. W., Collins, D., Saunders, M. A., & Musulin, R. (2008). Normalized hurricane damage in the United States: 1900–2005. *Natural Hazards Review, 9*(1), 29–42.

Pryor, S. C., & Barthelmie, R. J. (2010). Climate change impacts on wind energy: A review. *Renewable and Sustainable Energy Reviews, 14*(1), 430–437.

Psichogios, D. C., & Ungar, L. H. (1992). A hybrid neural network-first principles approach to process modeling. *AIChE Journal, 38*(10), 1499–1511.

Raissi, M., Perdikaris, P., & Karniadakis, G. E. (2019). Physics-informed neural networks: A deep learning framework for solving forward and inverse problems involving nonlinear partial differential equations. *Journal of Computational Physics, 378*, 686–707.

Rappaport, E. N. (2014). Fatalities in the United States from Atlantic tropical cyclones: New data and interpretation. *Bulletin of the American Meteorological Society, 95*(3), 341–346.

Refan, M., & Hangan, H. (2016). Characterization of tornado-like flow fields in a new model scale wind testing chamber. *Journal of Wind Engineering and Industrial Aerodynamics, 151*, 107–121.

Renn, P. I., & Gharib, M. (2022). Machine learning for flow-informed aerodynamic control in turbulent wind conditions. *Communications Engineering, 1*(1), 45.

Robertson, B. W., Johnson, M., Murthy, D., Smith, W. R., & Stephens, K. K. (2019). Using a combination of human insights and 'deep learning' for real-time disaster communication. *Progress in Disaster Science, 2*, Article 100030.

Russell, S., & Norvig, P. (1995). *Artificial intelligence: A modern approach.* Pearson, Upper Saddle River.

Sanchez-Lengeling, B., & Aspuru-Guzik, A. (2018). Inverse molecular design using machine learning: Generative models for matter engineering. *Science, 361*(6400), 360–365.

Scanlan, R. H., & Tomko, J. J. (1971). Airfoil and bridge deck flutter derivatives. *Journal of the Engineering Mechanics Division, 97*(6), 1717–1737.

Scott, D., Goupil, J., Burton, M., Denoon, R., Larsen, R., Spence, S., & Wu, T. (2023). Advancement in performance-based wind design. In *Workshop Report submitted to National Institute of Standards and Technology (NIST GCR 23-045-upd1)*, Gaithersburg, MD, USA.

Sherman, J., & Morrison, W. J. (1950). Adjustment of an inverse matrix corresponding to a change in one element of a given matrix. *The Annals of Mathematical Statistics, 21*(1), 124–127.

Shields, B. J., Stevens, J., Li, J., Parasram, M., Damani, F., Alvarado, J. I. M., Janey, J. M., Adams, R. P., & Doyle, A. G. (2021). Bayesian reaction optimization as a tool for chemical synthesis. *Nature, 590*(7844), 89–96.

Silver, D., Schrittwieser, J., Simonyan, K., Antonoglou, I., Huang, A., Guez, A., Hubert, T., Baker, L., Lai, M., Bolton, A., & Chen, Y. (2017). Mastering the game of go without human knowledge. *Nature, 550*(7676), 354–359.

Skinner, S. N., & Zare-Behtash, H. (2018). State-of-the-art in aerodynamic shape optimisation methods. *Applied Soft Computing, 62*, 933–962.

Smith, J. T., Masters, F. J., Liu, Z., & Reinhold, T. A. (2012). A simplified approach to simulate prescribed boundary layer flow conditions in a multiple controlled fan wind tunnel. *Journal of Wind Engineering and Industrial Aerodynamics, 109*, 79–88.

Smith, T. M., Elmore, K. L., & Dulin, S. A. (2004). A damaging downburst prediction and detection algorithm for the WSR-88D. *Weather and Forecasting, 19*(2), 240–250.

Snaiki, R., & Wu, T. (2017a). A linear height-resolving wind field model for tropical cyclone boundary layer. *Journal of Wind Engineering and Industrial Aerodynamics, 171*, 248–260.

Snaiki, R., & Wu, T. (2017b). Modeling tropical cyclone boundary layer: Height-resolving pressure and wind fields. *Journal of Wind Engineering and Industrial Aerodynamics, 170*, 18–27.

Snaiki, R., & Wu, T. (2019). Knowledge-enhanced deep learning for simulation of tropical cyclone boundary-layer winds. *Journal of Wind Engineering and Industrial Aerodynamics, 194*, Article 103983.

Snaiki, R., & Wu, T. (2022). Knowledge-enhanced deep learning for simulation of extratropical cyclone wind risk. *Atmosphere, 13*(5), 757.

Solari, G. (2020). Thunderstorm downbursts and wind loading of structures: Progress and prospect. *Frontiers in Built Environment, 6*, 63.

Solari, G., Burlando, M., & Repetto, M. P. (2018 March). THUNDERR: An ERC Project for the "detection, simulation, modelling and loading of thunderstorm outflows to design wind-safer and cost-efficient structures". In *International workshop on wind-related disasters and mitigation*, Sendai, Japan.

Spence, S. M., & Kareem, A. (2014). Performance-based design and optimization of uncertain wind-excited dynamic building systems. *Engineering Structures, 78*, 133–144.

Spencer, B., Jr., Dyke, S. J., Sain, M. K., & Carlson, J. (1997). Phenomenological model for magnetorheological dampers. *Journal of Engineering Mechanics, 123*(3), 230–238.

Stanway, R. S. J. L., Sproston, J. L., & Stevens, N. G. (1987). Non-linear modelling of an electrorheological vibration damper. *Journal of Electrostatics, 20*(2), 167–184.

Stiles, B. W., Danielson, R. E., Poulsen, W. L., Brennan, M. J., Hristova-Veleva, S., Shen, T. P., & Fore, A. G. (2014). Optimized tropical cyclone winds from QuikSCAT: A neural network approach. *IEEE Transactions on Geoscience and Remote Sensing, 52*(11), 7418–7434.

Sutton, R. S., & Barto, A. G. (1998). *Reinforcement learning: An introduction.* MIT Press, Cambridge, MA, USA.

Swischuk, R., Mainini, L., Peherstorfer, B., & Willcox, K. (2019). Projection-based model reduction: Formulations for physics-based machine learning. *Computers & Fluids, 179*, 704–717.

Tang, S., Sapsis, T., & Azizan, N. (2024). Dissipativity-informed learning for chaotic dynamical systems with attractor characterization. In *Machine learning and the physical sciences workshop, NeurIPS*.

Tang, Z., Feng, C., Wu, L., Zuo, D., & James, D. L. (2018). Characteristics of tornado-like vortices simulated in a large-scale ward-type simulator. *Boundary-Layer Meteorology, 166*(2), 327–350.

Turkkan, N., & Srivastava, N. K. (1995). Prediction of wind load distribution for air-supported structures using neural networks. *Canadian Journal of Civil Engineering, 22*(3), 453–461.

Udrescu, S. M., & Tegmark, M. (2020). AI Feynman: A physics-inspired method for symbolic regression. *Science Advances, 6*(16), eaay2631.

Van Kuik, G. A. M., Peinke, J., Nijssen, R., Lekou, D., Mann, J., Sørensen, J. N., Ferreira, C., van Wingerden, J. W., Schlipf, D., Gebraad, P., & Polinder, H. (2016). Long-term research challenges in wind energy–a research agenda by the European Academy of wind energy. *Wind Energy Science, 1*(1), 1–39.

Vapnik, V. N., & Chervonenkis, A. Y. (2015). On the uniform convergence of relative frequencies of events to their probabilities. *Measures of complexity* (pp. 11–30). Springer International Publishing.

Veers, P., Dykes, K., Lantz, E., Barth, S., Bottasso, C. L., Carlson, O., Clifton, A., Green, J., Green, P., Holttinen, H., & Laird, D. (2019). Grand challenges in the science of wind energy. *Science, 366*(6464), eaau2027.

Von Rueden, L., Mayer, S., Beckh, K., Georgiev, B., Giesselbach, S., Heese, R., Kirsch, B., Pfrommer, J., Pick, A., Ramamurthy, R., & Walczak, M. (2021). Informed machine learning—a taxonomy and survey of integrating prior knowledge into learning systems. *IEEE Transactions on Knowledge and Data Engineering, 35*(1), 614–633.

Vyavahare, A. Y., Godbole, P. N., & Nikose, T. (2012). Analysis of tall building for across wind response. *International Journal of Civil & Structural Engineering, 2*(3), 679–986.

Walsh, K. J., McBride, J. L., Klotzbach, P. J., Balachandran, S., Camargo, S. J., Holland, G., Knutson, T. R., Kossin, J. P., Lee, T. C., Sobel, A., & Sugi, M. (2016). Tropical cyclones and climate change. *Wiley Interdisciplinary Reviews: Climate Change, 7*(1), 65–89.

Wang, H., & Wu, T. (2020). Knowledge-enhanced deep learning for wind-induced nonlinear structural dynamic analysis. *Journal of Structural Engineering, 146*(11), 04020235.

Wang, L., Zhang, X., Su, H., & Zhu, J. (2024). A comprehensive survey of continual learning: Theory, method and application. *IEEE Transactions on Pattern Analysis and Machine Intelligence, 46*(8), 5362–5383.

Wei, C.C. (2015). Forecasting surface wind speeds over offshore islands near Taiwan during tropical cyclones: Comparisons of data-driven algorithms and parametric wind representations. *Journal of Geophysical Research: Atmospheres, 120*(5), 1826–1847.

Willard, J., Jia, X., Xu, S., Steinbach, M., & Kumar, V. (2020). Integrating physics-based modeling with machine learning: A survey, *1*(1), 1–34. arXiv preprint arXiv:2003.04919

Willis, D. J., Niezrecki, C., Kuchma, D., Hines, E., Arwade, S. R., Barthelmie, R. J., DiPaola, M., Drane, P. J., Hansen, C. J., Inalpolat, M., & Mack, J. H. (2018). Wind energy research: State-of-the-art and future research directions. *Renewable Energy, 125*, 133–154.

Wolpert, D. H., & Macready, W. G. (2002). No free lunch theorems for optimization. *IEEE Transactions on Evolutionary Computation, 1*(1), 67–82.

Wu, B., Chen, X., Wang, Q., Liao, H., & Dong, J. (2020). Characterization of vibration amplitude of nonlinear bridge flutter from section model test to full bridge estimation. *Journal of Wind Engineering and Industrial Aerodynamics, 197*, Article 104048.

Wu, T., & Kareem, A. (2011). Modeling hysteretic nonlinear behavior of bridge aerodynamics via cellular automata nested neural network. *Journal of Wind Engineering and Industrial Aerodynamics, 99*(4), 378–388.

Wu, T., & Kareem, A. (2013). Aerodynamics and aeroelasticity of cable-supported bridges: Identification of nonlinear features. *Journal of Engineering Mechanics, 139*(12), 1886–1893.

Wu, T., & Kareem, A. (2014). Simulation of nonlinear bridge aerodynamics: A sparse third-order Volterra model. *Journal of Sound and Vibration, 333*(1), 178–188.

Wu, T., & Kareem, A. (2015). A nonlinear analysis framework for bluff-body aerodynamics: A Volterra representation of the solution of Navier-Stokes equations. *Journal of Fluids and Structures, 54*, 479–502.

Wu, T., & Snaiki, R. (2022). Applications of machine learning to wind engineering. *Frontiers in Built Environment, 8*, Article 811460.

Wu, T., He, J., & Li, S. (2023). Active flutter control of long-span bridges via deep reinforcement learning: A proof of concept. *Wind & Structures, 36*(5), 321–331.

Xie, J., Dong, H., & Zhao, X. (2023). Data-driven torque and pitch control of wind turbines via reinforcement learning. *Renewable Energy, 215*, Article 118893.

Xu, Y. L. (2013). *Wind effects on cable-supported bridges*. John Wiley & Sons.

Xu, Y., Lu, X., Cetiner, B., & Taciroglu, E. (2021). Real-time regional seismic damage assessment framework based on long short-term memory neural network. *Computer-Aided Civil and Infrastructure Engineering, 36*(4), 504–521.

Yan, X., Zhu, J., Kuang, M., & Wang, X. (2019). Aerodynamic shape optimization using a novel optimizer based on machine learning techniques. *Aerospace Science and Technology, 86*, 826–835.

Yang, Q., Gao, R., Bai, F., Li, T., & Tamura, Y. (2018). Damage to buildings and structures due to recent devastating wind hazards in East Asia. *Natural Hazards, 92*(3), 1321–1353.

Yang, Q., Luo, Z., Li, K., & Wu, T. (2024). Reproduction of wind speed time series in a two-dimensional numerical multiple-fan wind tunnel using deep reinforcement learning. *Wind & Structures, 39*(4), 271–285.

Yu, X., & Wu, T. (2023). Simulation of unsteady flow around bluff bodies using knowledge-enhanced convolutional neural network. *Journal of Wind Engineering and Industrial Aerodynamics, 236*, Article 105405.

Zhang, A., & Zhang, L. (2004). RBF neural networks for the prediction of building interference effects. *Computers and Structures, 82*(27), 2333–2339.

Zhang, C., Dai, L., Ma, L., Qian, J., & Yang, B. (2017). Objective estimation of tropical cyclone innercore surface wind structure using infrared satellite images. *Journal of Applied Remote Sensing, 11*(4), 046030–046030.

Zhang, M., Xu, F., Wu, T., & Zhang, Z. (2020). Post flutter analysis of bridge decks using aerodynamic-describing functions. *Journal of Bridge Engineering, 25*(8), 04020046.

Zhang, W. S., & Xu, Y. L. (2001). Closed form solution for along wind response of actively controlled tall buildings with LQG controllers. *Journal of Wind Engineering and Industrial Aerodynamics, 89*(9), 785–807.

Zhou, Z., Li, X., & Zare, R. N. (2017). Optimizing chemical reactions with deep reinforcement learning. *ACS Central Science, 3*(12), 1337–1344.

Zhu, X., Jing, X., & Cheng, L. (2012). Magnetorheological fluid dampers: A review on structure design and analysis. *Journal of Intelligent Material Systems and Structures, 23*(8), 839–873.

Zou, Q., & Chen, S. (2019). Enhancing resilience of interdependent traffic-electric power system. *Reliability Engineering & System Safety, 191*, Article 106557.

Zou, Q., & Chen, S. (2021). Resilience-based recovery scheduling of transportation network in mixed traffic environment: A deep-ensemble-assisted active learning approach. *Reliability Engineering & System Safety, 215*, Article 107800.

Zuo, D., & Jones, N. P. (2010). Interpretation of field observations of wind-and rain-wind-induced stay cable vibrations. *Journal of Wind Engineering and Industrial Aerodynamics, 98*(2), 73–87.

MIX
Papier aus verantwortungsvollen Quellen
Paper from responsible sources
FSC® C105338

If you have any concerns about our products,
you can contact us on
ProductSafety@springernature.com

In case Publisher is established outside the EU,
the EU authorized representative is:
Springer Nature Customer Service Center GmbH
Europaplatz 3, 69115 Heidelberg, Germany

Printed by Libri Plureos GmbH
in Hamburg, Germany